PRINCIPLES AND MODERN APPLICATIONS OF MASS TRANSFER OPERATIONS

PRINCIPLES AND MODERN APPLICATIONS OF MASS TRANSFER OPERATIONS

Jaime Benitez

WILEY-INTERSCIENCE

A JOHN WILEY & SONS, INC., PUBLICATION

This text is printed on acid-free paper. ∞

Copyright © 2002 by John Wiley & Sons, Inc., New York. All rights reserved.

Published simultaneously in Canada.

No part of this publication may be reproduced, stored in a retrieval system or transmitted in any form or by any means, electronic, mechanical, photocopying, recording, scanning or otherwise, except as permitted under Section 107 or 108 of the 1976 United States Copyright Act, without either the prior written permission of the Publisher, or authorization through payment of the appropriate per-copy fee to the Copyright Clearance Center, 222 Rosewood Drive, Danvers, MA 01923, (978) 750-8400, fax (978) 750-4744. Requests to the Publisher for permission should be addressed to the Permissions Department, John Wiley & Sons, Inc., 605 Third Avenue, New York, NY 10158-0012, (212) 850-6011, fax (212) 850-6008, E-Mail: PERMREQ @ WILEY.COM.

For ordering and customer service, call 1-800-CALL-WILEY.

Library of Congress Cataloging-in-Publication Data:

Library of Congress Cataloging-in-Publication Data is available.
ISBN 0-471-20344-0

Printed in the United States of America.

10 9 8 7 6 5 4 3 2

Contents

Contents

Contents

A la memoria de mis Viejos
A Elsa Castro, mi Inspiración y Mecenas
A Dan Taylor, mi Maestro

Preface

The importance of the mass-transfer operations in chemical processes is profound. There is scarcely any industrial process that does not require a preliminary purification of raw materials or final separation of products. This is the realm of mass-transfer operations. Frequently, the major part of the cost of a process is that for the separations accomplished in the mass-transfer operations, a good reason for process engineers and designers to master this subject. The mass-transfer operations are largely the responsibility of chemical engineers, but increasingly practitioners of other engineering disciplines are finding them necessary for their work. This is especially true for those engaged in environmental engineering, where separation processes predominate.

My objective in writing this book is to provide a means to teach undergraduate chemical engineering students the basic principles of mass transfer and to apply these principles, aided by modern computational tools, to the design of equipment used in separation processes. The idea for it was born out of my experiences during the last 25 years teaching mass-transfer operations courses at the University of Puerto Rico.

The material treated in the book can be covered in a one-semester course. Chapters are divided into sections with clearly stated objectives at the beginning. Numerous detailed examples follow each brief section of text. Abundant end-of-chapter problems are included, and problem degree of difficulty is clearly labeled for each. Most of the problems are accompanied by their answers. Computer solution is emphasized, both in the examples and in the end-of-chapter problems. The book uses mostly SI units, which virtually eliminates the tedious task of unit conversions and makes it "readable" to the international scientific and technical community.

Following the lead of other authors in the chemical engineering field and related technical disciplines, I decided to incorporate the use of Mathcad into this book. Most readers will probably have a working knowledge of Mathcad. (Even if they don't, my experience is that the basic knowledge needed to begin using Mathcad effectively can be easily taught in a two-hour workshop.) The use of Mathcad simplifies mass-transfer calculations to a point that it allows the instructor and the student to readily try many different combinations of the design variables, a vital experience for the amateur designer.

The Mathcad environment can be used as a sophisticated scientific calculator, can be easily programed to perform a complicated sequence of calculations (for example, to check the design of a sieve-plate column for flooding, pressure drop, entrainment, weeping, and calculating Murphree plate efficiencies), can be used to plot results, and as a word processor to neatly present homework problems. Mathcad can perform calculations using a variety of unit systems, and will give a warning signal when calculations that are not dimensionally consistent are tried. This is a most powerful didactic tool, since dimensional consistency in calculations is one of the most fundamental concepts in chemical engineering education.

The first four chapters of the book present a basic framework of analysis that is applicable to any mass-transfer operation. Chapters 5 to 7 apply this common methodology to the analysis and design of the most popular types of mass-transfer operations. Chapter 5 covers gas absorption and stripping, chapter 6 distillation columns, and chapter 7 liquid extraction. This choice is somewhat arbitrary, and based on my own perception of the relevance of these operations. However, application of the general framework of analysis developed in the first four chapters should allow the reader to master, with relative ease, the peculiarities of any other type of mass-transfer operation.

I wish to acknowledge gratefully the contribution of the University of Puerto Rico at Mayagüez to this project. My students in the course INQU 4002 reviewed the material presented in the book, found quite a few errors, and gave excellent suggestions on ways to improve it. I am grateful to Wilmer Ruiz, who was my first contact with Wiley. Michael Penn and Kristin Cooke, of the Wiley staff in New York, were very important in the successful completion of this project. My special gratitude goes to Teresa, my wife, and my four children who were always around lifting my spirits during the long, arduous hours of work devoted to this volume. They make it all worthwhile!

Jaime Benítez
Mayagüez, Puerto Rico

Nomenclature

LATIN LETTERS

$\mathbf{A}$	first-derivative orthogonal collocation matrix; dimensionless.
A	absorption factor; dimensionless.
A	mass flow rate of species A; kg/s.
A_a	active area of a sieve tray; m^2.
A_d	area taken by the downspout in a sieve tray; m^2.
A_h	area taken by the perforations on a sieve tray; m^2.
A_n	net cross-section area between trays inside a tray column; m^2.
A_t	total cross-section area; m^2.
a	mass-transfer surface area per unit volume; m^{-1}.
a_h	hydraulic, or effective, specific surface area of packing; m^{-1}.
B	mass flow rate of species B; kg/s.
c	total molar concentration; moles/m^3.
c_i	molar concentration of species i; moles/m^3.
C	total number of components in multicomponent distillation.
C_p	specific heat at constant pressure; J/(kg$\times$ K).
C_D	drag coefficient; dimensionless.
$\mathfrak{D}_{ij}$	Maxwell-Stefan diffusivity for pair i-j; m^2/s.
D_{ij}	Fick diffusivity or diffusion coefficient for pair i-j; m^2/s.
d_e	equivalent diameter; m.
$\mathbf{d}_i$	driving force for mass diffusion of species i; m^{-1}.
d_i	inside diameter; m.
d_o	outside diameter; m.
d_o	perforation diameter in a sieve plate; m.
d_p	particle size; m.
d_{vs}	Sauter mean drop diameter defined in equation (7-48); m.
$\mathbf{DM}$	dimensional matrix.
D	tube diameter; m.
D	distillate flow rate; moles/s.
E	fractional entrainment; liquid mass flow rate/gas mass flow rate.
E	extract mass flow rate; kg/s.
E_m	mechanical efficiency of a motor-fan system; dimensionless.
Eo	Eotvos number defined in equation (7-53); dimensionless.
EF	extraction factor defined in equation (7-19); dimensionless.
$\mathbf{E}_{ME}$	Murphree stage efficiency in terms of extract composition; dimensionless.
$\mathbf{E}_{MG}$	Murphree gas-phase tray efficiency; dimensionless.
$\mathbf{E}_{MGE}$	Murphree gas-phase tray efficiency corrected for entrainment.

$\mathbf{E}_O$	overall tray efficiency of a cascade; equilibrium trays/real trays.
$\mathbf{E}_{OG}$	point gas-phase tray efficiency; dimensionless.
f_{12}	proportionality coefficient in equation (1-21).
f	friction factor; dimensionless.
f	fractional approach to flooding velocity; dimensionless.
f_{ext}	fractional extraction; dimensionless.
F	mass-transfer coefficient; moles/(m^2 × s).
F	molar flow rate of the feed to a distillation column; moles/s.
F	mass flow rate of the feed to a liquid extraction process; kg/s.
F_p	packing factor; ft^{-1}.
$FR_{i,D}$	fractional recovery of component i in the distillate; dimensionless.
$FR_{i,W}$	fractional recovery of component i in the residue; dimensionless.
Fr_L	liquid Froude number; dimensionless.
Ga	Galileo number; dimensionless.
G_M	superficial molar velocity; moles/(m^2 × s).
G_{Mx}	superficial liquid-phase molar velocity; moles/(m^2 × s).
G_{My}	superficial gas-phase molar velocity; moles/(m^2 × s).
G_x	superficial liquid-mass velocity; kg/(m^2 × s).
G_y	superficial gas-mass velocity; kg/(m^2 × s).
Gr_D	Grashof number for mass transfer; dimensionless.
Gr_H	Grashof number for heat transfer; dimensionless.
Gz	Graetz number; dimensionless.
g	acceleration due to gravity; 9.8 m/s^2.
g_c	dimensional conversion factor; 1 kg × m/(N × s^2).
H	Henry's law constant; atm, kPa, Pa.
H	molar enthalpy; J/mole.
H	height of mixing vessel; m.
HETS	height equivalent to a theoretical stage in staged liquid extraction columns; m.
HK	heavy-key component in multicomponent distillation.
ΔH_S	heat of solution; J/mole of solution.
H_{tL}	height of a liquid-phase transfer unit; m.
H_{tG}	height of a gas-phase transfer unit; m.
H_{tOG}	overall height of a gas-phase transfer unit; m.
H_{tOL}	overall height of a liquid-phase transfer unit; m.
h	convective heat-transfer coefficient, W/(m^2 × K).
h_d	dry-tray head loss; cm of liquid.
h_l	equivalent head of clear liquid on tray; cm of liquid.
h_L	specific liquid holdup; m^3 holdup/m^3 packed bed.
h_t	total head loss/tray; cm of liquid.
h_w	weir height; m.
h_σ	head loss due to surface tension; cm of liquid.
$h_{2\phi}$	height of two-phase region on a tray; m.

i	number of dimensionless groups needed to describe a situation.
j_D	Chilton-Colburn j-factor for mass transfer; dimensionless.
j_H	Chilton-Colburn j-factor for heat transfer; dimensionless.
$\mathbf{j}_i$	mass diffusion flux of species i with respect to the mass-average velocity; kg/(m$^2 \times$ s).
$\mathbf{J}_i$	molar diffusion flux of species i with respect to the molar-average velocity; moles/(m$^2 \times$ s).
J_0	Bessel function of the first kind and order zero; dimensionless.
J_1	Bessel function of the first kind and order one; dimensionless.
K	distribution coefficient; dimensionless.
K_W	wall factor in Billet-Schultes pressure-drop correlations; dimensionless.
k	thermal conductivity; W/(m $\times$ K).
k_c	convective mass-transfer coefficient for diffusion of A through stagnant B in dilute gas-phase solution with driving force in terms of molar concentrations; m/s.
k'_c	convective mass-transfer coefficient for equimolar counterdiffusion in gas-phase solution with driving force in terms of molar concentrations; m/s.
k_G	convective mass-transfer coefficient for diffusion of A through stagnant B in dilute gas-phase solution with driving force in terms of partial pressure; moles/(m$^2 \times$ s $\times$ Pa).
K_G	overall convective mass-transfer coefficient for diffusion of A through stagnant B in dilute solutions with driving force in terms of partial pressures; moles/(m$^2 \times$ s $\times$ Pa).
k'_G	convective mass-transfer coefficient for equimolar counterdiffusion in gas-phase solution with driving force in terms of partial pressure; moles/(m$^2 \times$ s $\times$ Pa).
k_L	convective mass-transfer coefficient for diffusion of A through stagnant B in dilute liquid-phase solution with driving force in terms of molar concentrations; m/s.
k'_L	convective mass-transfer coefficient for equimolar counterdiffusion in liquid-phase solution with driving force in terms of molar concentrations; m/s.
k_r	reaction rate constant; moles/(m$^2 \times$ s $\times$ mole fraction).
k_x	convective mass-transfer coefficient for diffusion of A through stagnant B in dilute liquid-phase solution with driving force in terms of mole fractions; moles/(m$^2 \times$ s).
K_x	overall convective mass-transfer coefficient for diffusion of A through stagnant B in dilute solutions with driving force in terms of liquid-phase molar fractions; moles/(m$^2 \times$ s).
k'_x	convective mass-transfer coefficient for equimolar counterdiffusion in liquid-phase solution with driving force in terms of mole fractions; moles/(m$^2 \times$ s).

k_y	convective mass-transfer coefficient for diffusion of A through stagnant B in dilute gas-phase solution with driving force in terms of mole fractions; moles/(m^2 × s).
K_y	overall convective mass-transfer coefficient for diffusion of A through stagnant B in dilute solutions with driving force in terms of gas-phase molar fractions; moles/(m^2 × s).
k'_y	convective mass-transfer coefficient for equimolar counterdiffusion in gas-phase solution with driving force in terms of mole fractions; moles/(m^2 × s).
L	characteristic length, m.
L	molar flow rate of the L-phase; moles/s.
L	length of settling vessel; m.
LK	light-key component in multicomponent distillation.
L_S	molar flow rate of the nondiffusing solvent in the L-phase; moles/s.
L'	mass flow rate of the L-phase; kg/s.
L'_S	mass flow rate of the nondiffusing solvent in the L-phase; kg/s.
L_e	entrainment mass flow rate, kg/s.
L_w	weir length; m.
l	characteristic length, m.
l	tray thickness; m.
Le	Lewis number; dimensionless.
M_i	molecular weight of species i.
m	amount of mass; kg.
m	slope of the equilibrium distribution curve; dimensionless.
n	total mass flux with respect to fixed coordinates; kg/(m^2 × s).
n$_i$	mass flux of species i with respect to fixed coordinates; kg/(m^2 × s).
n	number of variables significant to dimensional analysis of a given problem.
n	rate of mass transfer from the dispersed to the continuous phase in liquid extraction; kg/s.
N	total molar flux with respect to fixed coordinates; moles/(m^2 × s).
N$_i$	molar flux of species i with respect to fixed coordinates; moles/(m^2 × s).
N	number of equilibrium stages in a cascade; dimensionless.
N_E	mass of B/(mass of A + mass of C) in the extract liquids.
N_R	number of stages in rectifying section; dimensionless.
N_R	mass of B/(mass of A + mass of C) in the raffinate liquids.
N_S	number of stages in stripping section; dimensionless.
N_{tL}	number of liquid-phase transfer units; dimensionless.
N_{tG}	number of gas-phase transfer units; dimensionless.
N_{tOD}	overall number of dispersed-phase transfer units; dimensionless.
N_{tOG}	overall number of gas-phase transfer units; dimensionless.
N_{tOL}	overall number of liquid-phase transfer units; dimensionless.
Nu	Nusselt number; dimensionless.
n	number of species in a mixture.

O_t	molar oxygen concentration in the air leaving an aeration tank; percent.
O_{eff}	oxygen transfer efficiency; mass of oxygen absorbed by water/total mass of oxygen supplied.
p'	pitch, distance between centers of perforations in a sieve plate; m.
p_i	partial pressure of species i; atm, Pa, kPa, bar.
$p_{B,M}$	logarithmic mean partial pressure of component B; atm, Pa, kPa, bar.
P	total pressure; atm, Pa, kPa, bar.
P	permeate flow (through a membrane); moles/s.
P	impeller power; kW.
P_c	critical pressure, Pa, kPa, bar.
Pe_D	Peclet number for mass transfer.
Pe_H	Peclet number for heat transfer.
P_i	vapor pressure of species i; atm, Pa, kPa, bar.
Po	power number defined in equation (7-37); dimensionless.
Pr	Prandtl number; dimensionless.
Q	volumetric flow rate; m^3/s.
Q	net rate of heating; J/s.
q	parameter defined by equation (6-27); dimensionless.
r	rank of the dimensional matrix, **DM.**
r_A	solute particle radius; m.
R	radius; m.
R	ideal gas constant; Pa $\times$ m^3/(mol $\times$ K).
R	reflux ratio; mole of reflux/moles of distillate.
R	raffinate mass flow rate; kg/s.
R_m	retentate flow (in a membrane); moles/s.
Re	Reynolds number; dimensionless.
R_i	volumetric rate of formation of component i; moles/(m^3 $\times$ s).
S	surface area, cross-sectional area; m^2.
S	stripping factor, reciprocal of absorption factor (A); dimensionless.
S	mass flow rate of the solvent entering a liquid extraction process; kg/s.
Sc	Schmidt number; dimensionless.
Sh	Sherwood number; dimensionless.
St_D	Stanton number for mass transfer; dimensionless.
St_H	Stanton number for heat transfer; dimensionless.
t	tray spacing; m.
t	time; s, hr.
t_{res}	residence time; min.
T	temperature; K.
T_b	normal boiling point temperature; K.
T_c	critical temperature, K
x_i	mole fraction of species i in either liquid or solid phase.
x_i	mass fraction of species i in raffinate (liquid extraction).
$x_{B,M}$	logarithmic mean mole fraction of component B in liquid or solid phase.

x	rectangular coordinate.
x'	mass of C/mass of A in raffinate liquids.
X	mole ratio in phase L; moles of A/mole of A-free L.
X	flow parameter; dimensionless.
X	parameter in Gilliland's correlation, see equation (6-83); dimensionless.
X	mass of C/(mass of A + mass of C) in the raffinate liquids.
X'	mass ratio in phase L; kg of A/kg of A-free L.
y	rectangular coordinate.
y'	mass of C/mass of B in extract liquids.
$y_{B,M}$	logarithmic mean mole fraction of component B in gas phase.
y_i	mole fraction of species i in the gas phase.
y_i	mass fraction of species i in extract (liquid extraction).
Y	mole ratio in phase V; moles of A/mole of A-free V.
Y	pressure-drop parameter defined in equation (4-6); dimensionless.
Y	parameter in Gilliland's correlation, see equation (6-82); dimensionless.
Y	mass of C/(mass of A + mass of C) in the extract liquids.
Y'	mass ratio in phase V; kg of A/kg of A-free V.
u	fluid velocity past a stationary flat plate, parallel to the surface; m/s.
$\mathbf{v}$	mass-average velocity for multicomponent mixture; m/s.
$\mathbf{v}_i$	velocity of species i; m/s.
v_t	terminal velocity of a particle; m/s.
$\mathbf{V}$	molar-average velocity for multicomponent mixture; m/s.
V	volume; m^3.
V	molar flow rate of the V-phase; moles/s.
V_S	molar flow rate of the nondiffusing solvent in the V-phase; moles/s.
V'	mass flow rate of the V-phase; kg/s.
V'_S	mass flow rate of the nondiffusing solvent in the V-phase; kg/s.
V_A	molar volume of a solute as liquid at its normal boiling point; cm^3/mol.
V_B	boilup ratio; moles of boilup/moles of residue.
V_b	molar volume of a substance as liquid at its normal boiling point; cm^3/mol.
V_c	critical volume; cm^3/mol.
w	mass-flow rate; kg/s.
W	work per unit mass; J/kg.
W	molar flow rate of the residue from a distillation column; moles/s.
We	Weber number defined in equation (7-49); dimensionless.
z	rectangular coordinate.
z_i	average mole fraction of component i in a solution or multiphase mixture.
Z	total height; m.
Z_c	compressibility factor at critical conditions; dimensionless.
Z_R	total height of the rectifying section of a packed fractionator; m.
Z_S	total height of the stripping section of a packed fractionator; m.

GREEK LETTERS

α	thermal diffusivity; m^2/s.
α	relative volatility; dimensionless.
α_m	membrane separation factor; dimensionless.
β	volume coefficient of thermal expansion; K^{-1}.
Γ	matrix of thermodynamic factors with elements defined by equation (1-32).
γ_i	activity coefficient of species i in solution.
δ_{ij}	Kronecker delta; 1 if $i = k$, 0 otherwise.
Δ_R	difference in flow rate, equation (7-12); kg/s.
δ	length of the diffusion path; m.
ε	porosity or void fraction; dimensionless.
ε_{AB}	Lennard-Jones parameter; erg.
η	vector of collocation points along the diffusion path; dimensionless.
θ	membrane cut; moles of permeate/moles of feed.
κ	Boltzmann constant; 1.38×10^{-16} erg/K.
κ	constant in equation (4-46), defined in equation (4-47); dimensionless.
λ_i	molar latent heat of vaporization of component i; J/mole.
λ	similar to the stripping factor, S, in equations (4-56) to (4-61).
μ_i	chemical potential of species i; J/mol.
μ_B	solvent viscosity; cP.
ν	momentum diffusivity, or kinematic viscosity; m^2/s.
ξ	reduced inverse viscosity in Lucas method; $(\mu P)^{-1}$.
π	constant; 3.1416
π	Pi groups in dimensional analysis.
ρ	mass density; kg/m^3.
ρ_i	mass density of species i; kg/m^3.
σ_{AB}	Lennard-Jones parameter; Å.
σ	surface tension, dyn/cm, N/m.
τ	shear stress; N/m^2.
Φ_B	association factor of solvent B; dimensionless.
ϕ	packing fraction in hollow-fiber membrane module; dimensionless.
ϕ	root of equation (6-78); dimensionless.
ϕ_e	effective relative froth density; height of clear liquid/froth height.
ϕ_C	fractional holdup of the continuous liquid phase.
ϕ_D	fractional holdup of the dispersed liquid phase.
φ_G	specific gas holdup; m^3 holdup/m^3 total volume.
ω_i	mass fraction of species i.
Ω_D	diffusion collision integral; dimensionless.
Ω	impeller rate of rotation; rpm.
Ψ_A	molar flux fraction of component A; dimensionless.
Ψ_0	dry-packing resistance coefficient in Billet-Schultes pressure-drop correlations; dimensionless.

1

Fundamentals of Mass Transfer

1.1 INTRODUCTION

When a system contains two or more components whose concentrations vary from point to point, there is a natural tendency for mass to be transferred, minimizing the concentration differences within the system and moving it towards equilibrium. The transport of one component from a region of higher concentration to that of a lower concentration is called *mass transfer*.

Many of our daily experiences involve mass-transfer phenomena. The invigorating aroma of a cup of freshly brewed coffee and the sensuous scent of a delicate perfume both reach our nostrils from the source by diffusion through air. A lump of sugar added to the cup of coffee eventually dissolves and then diffuses uniformly throughout the beverage. Laundry hanging under the sun during a breezy day dries fast because the moisture evaporates and diffuses easily into the relatively dry moving air.

Mass transfer plays an important role in many industrial processes. A group of operations for separating the components of mixtures is based on the transfer of material from one homogeneous phase to another. These methods—covered by the term *mass-transfer operations*—include such techniques as distillation, gas absorption, humidification, liquid extraction, adsorption, membrane separations, and others. The driving force for transfer in these operations is a concentration gradient, much as a temperature gradient provides the driving force for heat transfer.

Distillation separates, by partial vaporization, a liquid mixture of miscible and volatile substances into individual componentes or, in some cases, into groups of components. The separation of a mixture of methanol and water into its components; of liquid air into oxygen, nitrogen, and argon; and of crude petroleum into gasoline, kerosene, fuel oil, and lubricating stock are examples of distillation.

In *gas absorption* a soluble vapor is absorbed by means of a liquid in which the solute gas is more or less soluble, from its mixture with an inert gas. The washing of ammonia from a mixture of ammonia and air by means of liquid water is a typical example. The solute is subsequently recovered from the liquid by distillation, and the absorbing liquid can be either discarded or reused. When a solute is transferred from the solvent liquid to the gas phase, the operation is known as *desorption* or *stripping*.

In *humidification* or *dehumidification* (depending upon the direction of transfer) the liquid phase is a pure liquid containing but one component while the gas phase contains two or more substances. Usually the inert or carrier gas is virtually insoluble in the liquid. Removal of water vapor from air by condensation on a cold surface and the condensation of an organic vapor such as carbon tetrachloride out of a stream of nitrogen are examples of dehumidification. In humidification operations the direction of transfer is from the liquid to the gas phase.

The *adsorption* operations exploit the ability of certain solids preferentially to concentrate specific substances from solution onto their surfaces. In this manner, the components of either gaseous or liquid solutions can be separated from each other. A few examples will illustrate the great variety of practical applications of adsorption. It is used to dehumidify air and other gases, to remove objectionable odors and impurities from industrial gases, to recover valuable solvent vapors from dilute mixtures with air and other gases, to remove objectionable taste and odor from drinking water, and many other applications.

Liquid extraction is the separation of the constituents of a liquid solution by contact with another insoluble liquid. If the substances constituting the original solution distribute themselves differently between the two liquid phases, a certain degree of separation will result. The solution which is to be extracted is called the *feed*, and the liquid with which the feed is contacted is called the *solvent*. The solvent-rich product of the operation is called the *extract*, and the residual liquid from which the solute has been removed is called the *raffinate*.

Membrane separations are rapidly increasing in importance. In general, the membranes serve to prevent intermingling of two miscible phases. They also prevent ordinary hydrodynamic flow, and movement of substances through them is by diffusion. Separation of the components of the original solution takes place by selectively controlling their passage from one side of the membrane to the other. An example of a membrane-mediated, liquid–liquid separation process is *dyalisis*. In this process, a colloid is removed from a liquid solution by contacting the solution with a solvent through an intervening membrane which is permeable to the solution, but not to the larger colloidal particles. For example, aqueous beet-sugar solutions containing undesired colloidal material are freed of the latter by contact with water through a semipermeable membrane. Sugar and water diffuse through the membrane, but not the colloid.

Returning to the lump of sugar added to the cup of coffee, it is evident that the time required for the sugar to distribute uniformly depends upon whether the liquid is quiescent or whether it is mechanically agitated by a spoon. In general, the mechanism of mass transfer depends upon the dynamics of the system in which it occurs. Mass can be transferred by random molecular motion in quiescent fluids, or it can be transferred from a surface into a moving fluid, aided by the dynamic characteristics of the flow. These two distinct modes of transport, *molecular mass transfer* and *convective mass transfer*, are analogous to conduction heat transfer and convective heat transfer. Each of these modes of mass transfer will be described and analyzed. The two mechanisms often act simultaneously. Frequently, when this happens, one mechanism can dominate quantitatively so that approximate solutions involving only the dominant mode can be used.

1.2 MOLECULAR MASS TRANSFER

As early as 1815 it was observed qualitatively that whenever a gas mixture contains two or more molecular species, whose relative concentrations vary from point to point, an apparently natural process results which tends to diminish any inequalities in composition. This macroscopic transport of mass, independent of any convection effects within the system, is defined as *molecular diffusion*.

In the specific case of a gaseous mixtures, a logical explanation of this transport phenomenon can be deduced from the kinetic theory of gases. At any temperature above absolute zero, individual molecules are in a state of continual yet random motion. Within dilute gas mixtures, each solute molecule behaves independently of the other solute molecules, since it seldom encounters them. Collisions between the solute and the solvent molecules are continually occurring. As a result of the collisions, the solute molecules move along a zigzag path, sometimes toward a region of higher concentration, sometimes toward a region of lower concentration.

Consider a hypothetical section passing normal to the concentration gradient within an isothermal, isobaric gaseous mixture containing solute and solvent molecules. The two thin, equal elements of volume above and below the section will contain the same number of molecules, as stipulated by Avogadro's law (Welty, et al., 1984).

Although it is not possible to state which way any particular molecule will travel in a given interval of time, a definite number of the molecules in the lower element of volume will cross the hypothetical section from below, and the same number of molecules will leave the upper element and cross the section from above. With the existence of a concentration gradient, there are more solute molecules in one of the elements of volume than in the other; accordingly, an overall net transfer from a region of higher concentration to one of lower concentration will result. The net flow of each molecular species occurs in the direction of a negative concentration gradient.

The laws of mass transfer show the relation between the flux of the diffusing substance and the concentration gradient responsible for this mass transfer. Since diffusion occurs only in mixtures, its evaluation must involve an examination of the effect of each component. For example, it is often desired to know the diffusion rate of a specific component relative to the velocity of the mixture in which it is moving. Since each component may possess a different mobility, the mixture velocity must be evaluated by averaging the velocities of all the components present.

In order to establish a common basis for future discussions, definitions and relations which are often used to explain the role of components within a mixture are considered next.

1.2.1 Concentrations

Your objectives in studying this section are to be able to:

1. Convert a composition given in mass fraction to mole fraction, and the reverse.
2. Transform a material from one measure of concentration to another, including mass/volume and moles/volume.

In a multicomponent mixture, the concentration of particular species can be expressed in many ways. A mass concentration for each species, as well as for the mixture, can be defined. For species A, the *mass concentration*, ρ_A, is defined as the mass of A per unit volume of the mixture. The total mass concentration, or *density*, ρ, is the total mass of the mixture contained in a unit volume; that is:

$$\rho = \sum_{i=1}^{n} \rho_i \tag{1-1}$$

where n is the number of species in the mixture. The *mass fraction*, ω_A, is the mass concentration of species A divided by the total mass density,

$$\omega_A = \frac{\rho_A}{\displaystyle\sum_{i=1}^{n} \rho_i} = \frac{\rho_A}{\rho} \tag{1-2}$$

The sum of the mass fractions, by definition, must be 1;

$$\sum_{i=1}^{n} \omega_i = 1 \tag{1-3}$$

The *molar concentration* of species A, c_A, is defined as the number of moles of A present per unit volume of the mixture. By definition, one mole of any species contains a mass equivalent to its molecular weight; therefore, the mass concentration and the molar concentration are related by

$$c_A = \frac{\rho_A}{M_A} \qquad (1\text{-}4)$$

where M_A is the molecular weight of species A. When dealing with a gas phase under conditions in which the ideal gas law applies, the molar concentration is given by

$$c_A = \frac{p_A}{RT} \qquad (1\text{-}5)$$

where p_A is the partial pressure of the species A in the mixture, T is the absolute temperature, and R is the gas constant. The total molar concentration, c, is the total moles of mixture contained in a unit volume; that is,

$$c = \sum_{i=1}^{n} c_i \qquad (1\text{-}6)$$

For a gaseous mixture that obeys the ideal gas law,

$$c = \frac{P}{RT} \qquad (1\text{-}7)$$

where P is the total pressure. The *mole fraction* for liquid or solid mixtures, x_A, and for gaseous mixtures, y_A, are the molar concentrations of species A divided by the total molar concentration:

$$x_A = \frac{c_A}{c} \qquad \text{(liquids and solids)}$$

$$y_A = \frac{c_A}{c} \qquad \text{(gases)} \qquad (1\text{-}8)$$

For a gaseous mixture that obeys the ideal gas law, the mole fraction, y_A, can be written in terms of pressures,

$$y_A = \frac{c_A}{c} = \frac{p_A}{P} \qquad (1\text{-}9)$$

Equation (1-9) is an algebraic representation of Dalton's law for gas mixtures. The sum of the mole fractions, by definition, must be 1.

$$\sum_{i=1}^{n} y_i = \sum_{i=1}^{n} x_i = 1 \qquad\qquad (1\text{-}10)$$

Example 1.1 Concentration of the Feed to a Gas Absorber

A gas containing 88% (by volume) CH_4, 4% C_2H_6, 5% n-C_3H_8, and 3% n-C_4H_{10} at 300 K and 500 kPa will be scrubbed by contact with a nonvolatile oil in a gas absorber. The objective of the process is to recover in the liquid effluent as much as possible of the heavier hydrocarbons in the feed (see Figure 1.1). Calculate:

(a) Total molar concentration in the gas feed.

(b) Density of the gas feed.

(c) Composition of the gas feed, expressed in terms of mass fractions.

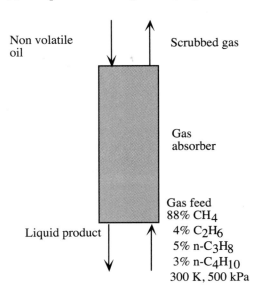

Non volatile oil

Scrubbed gas

Gas absorber

Gas feed
88% CH4
4% C2H6
5% n-C3H8
3% n-C4H10
300 K, 500 kPa

Liquid product

Figure 1.1 Schematic diagram of the gas absorber of Example 1.1.

Solution
(a) Use Equation (1-7) to calculate the total molar concentration:

$$c = \frac{P}{RT} = \frac{500}{8.314 \times 300} = 0.20 \; \frac{kmol}{m^3}$$

(b) Calculate the gas average molecular weight, M_{av}:

Basis:100 kmol of gas mixture			
Component	kmol	Molecular weight	Mass, kg
CH_4	88	16.04	1411.52
C_2H_6	4	30.07	120.28
$n-C_3H_8$	5	44.09	220.45
$n-C_4H_{10}$	3	58.12	174.36
Total	100		1926.61

$$M_{av} = \frac{1926.61}{100} = 19.26 \frac{kg}{kmol}$$

To calculate the mixture mass density:

$$\rho = cM_{av} = 0.20 \times 19.26 = 3.85 \frac{kg}{m^3}$$

(c) Calculate the mass fraction of each component in the gas mixture from the intermediate results of part b

Component	Mass, kg	Mass fraction
CH_4	1411.52	$(1411.52/1926.61) = 0.733$
C_2H_6	120.28	0.062
$n-C_3H_8$	220.45	0.114
$n-C_4H_{10}$	174.36	0.091
Total	1926.61	1.000

Example 1.2 Concentration of a Potassium Nitrate Wash Solution

In the manufacture of potassium nitrate, potassium chloride reacts with a hot aqueous solution of sodium nitrate according to:

$$KCl + NaNO_3 \rightarrow KNO_3 + NaCl$$

The reaction mixture is cooled down to 293 K and pure KNO_3 crystallizes. The resulting slurry contains the KNO_3 crystals and an aqueous solution of both KNO_3 and NaCl. The crystals in the slurry are washed in a multistage process with a saturated KNO_3 solution to free them of NaCl (see Figure 1.2). The equilibrium solubility of KNO_3 in water at 293 K is 24% (by weight); the density of the saturated solution is 1,162 kg/m^3 (Perry and Chilton, 1973). Calculate:

(a) Total molar density of the fresh wash solution.

b) Composition of the fresh wash solution, expressed in terms of molar fractions.

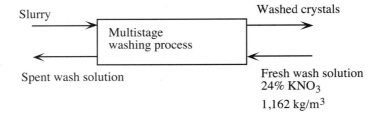

Figure 1.2 Schematic diagram of the washing process in Example 1.2.

Solution

(a) Calculate the average molecular weight, M_{av}, of the wash solution, and then its total molar density.

Basis: 100 kg of fresh wash solution

Component	Mass, kg	Molecular weight	kmol
KNO_3	24	101.10	0.237
H_2O	76	18.02	4.218
Total	100		4.455

Therefore,

$$M_{av} = \frac{100 \text{ kg}}{4.455 \text{ kmol}} = 22.45 \ \frac{\text{kg}}{\text{kmol}}$$

$$c = \frac{\rho}{M_{av}} = \frac{1,162 \text{ kg/m}^3}{22.45 \text{ kg/kmol}} = 51.77 \ \frac{\text{kmol}}{\text{m}^3}$$

(b) Calculate the mole fractions from intermediate results in part a:

Component	kmol	Mole fraction
KNO_3	0.237	$(0.237/4.455) = 0.053$
H_2O	4.218	0.947
Total	4.455	1.000

1.2.2 Velocities and Fluxes

Your objectives in studying this section are to be able to:

1. Define the following terms: mass-average velocity, molar-average velocity, mass (or molar) flux, and diffusion mass (or molar) flux.
2. Write down an expression to calculate the mass (or molar) flux relative to a fixed coordinate system in terms of the diffusion mass (or molar) flux and the bulk motion contibution

---

The basic empirical relation to estimate the rate of molecular difussion, first postulated by Fick (1855) and, accordingly, often referred to as Fick's first law, quantifies the diffusion of component A in an isothermal, isobaric system. According to Fick's law, a species can have a velocity relative to the mass or molar-average velocity (called *diffusion velocity*) only if gradients in the concentration exist. In a multicomponent system the various species will normally move at different velocities; for that reason, an evaluation of a characteristic velocity for the gas mixture requires the averaging of the velocities of each species present.

The *mass-average velocity* for a multicomponent mixture is defined in terms of the mass densities

$$\mathbf{v} = \frac{\sum_{i=1}^{n} \rho_i \mathbf{v}_i}{\sum_{i=1}^{n} \rho_i} = \frac{\sum_{i=1}^{n} \rho_i \mathbf{v}_i}{\rho} = \sum_{i=1}^{n} \omega_i \mathbf{v}_i \qquad (1.11)$$

where $\mathbf{v}_i$ denotes the absolute velocity of species i relative to stationary coordinate axis. The mass-average velocity is the velocity that would be measured by a pitot tube. On the other hand, the *molar-average velocity* for a multicomponent mixture is defined in terms of the molar concentrations of all components by:

$$\mathbf{V} = \frac{\sum_{i=1}^{n} c_i \mathbf{v}_i}{\sum_{i=1}^{n} c_i} = \frac{\sum_{i=1}^{n} c_i \mathbf{v}_i}{c} = \sum_{i=1}^{n} x_i \mathbf{v}_i \qquad (1.12)$$

Diffusion rates are most conveniently described in terms of fluxes. The *mass* (or *molar*) *flux* of a given species is a vector quantity denoting the amount of the particular species, in either mass or molar units, that passes per given unit time through a unit area normal to the vector. The flux may be defined with reference to coordinates that are fixed in space, coordinates which are moving with the mass-average velocity, or coordinates which are moving with the molar-average velocity.

The *mass flux* of species i with respect to coordinates that are fixed in space is defined by

$$\mathbf{n}_i = \rho_i \mathbf{v}_i \tag{1-13}$$

If we sum the component fluxes, we obtain the *total mass flux*

$$\mathbf{n} = \rho \mathbf{v} \tag{1-14}$$

The *molar flux* of species i with respect to coordinates that are fixed in space is given by

$$\mathbf{N}_i = c_i \mathbf{v}_i \tag{1-15}$$

The *total molar flux* is the sum of these quantities

$$\mathbf{N} = c\mathbf{V} \tag{1-16}$$

The *mass diffusion flux* of species i with respect to the mass-average velocity is given by

$$\mathbf{j}_i = \rho_i(\mathbf{v}_i - \mathbf{v}) \quad \text{and} \quad \sum_{i=1}^{n} \mathbf{j}_i = 0 \tag{1-17}$$

The *molar diffusion flux* of species i with respect to the molar-average velocity is given by

$$\mathbf{J}_i = c_i(\mathbf{v}_i - \mathbf{V}) \quad \text{and} \quad \sum_{i=1}^{n} \mathbf{J}_i = 0 \tag{1-18}$$

The mass flux $\mathbf{n}_i$ is related to the mass diffusion flux as

$$\mathbf{n}_i = \mathbf{j}_i + \rho_i \mathbf{v} = \mathbf{j}_i + \omega_i \mathbf{n} \tag{1-19}$$

The molar flux $\mathbf{N}_i$ is related to the molar diffusion flux as

$$\mathbf{N}_i = \mathbf{J}_i + c_i \mathbf{V} = \mathbf{J}_i + y_i \mathbf{N} \tag{1-20}$$

It is important to note that the molar flux, $\mathbf{N}_i$, described by equation (1-20) is a resultant of the two vector quantities:

$$\mathbf{J}_i$$

the molar diffusion flux, $\mathbf{J}_i$, resulting from the concentration gradient; this term is referred to as the *concentration gradient contribution*;

and

$$y_i \mathbf{N} = c_i \mathbf{V}$$

the molar flux resulting as component i is carried in the bulk flow of the fluid; this flux term is designated the *bulk motion contribution*.

Either or both quantities can be a significant part of the total molar flux, $\mathbf{N}_i$. Whenever equation (1-20) is applied to describe molar diffusion, the vector nature of the individual fluxes, $\mathbf{N}_i$, must be considered. The same applies for the mass fluxes in equation (1-19).

1.2.3 The Maxwell-Stefan Relations

Your objectives in studying this section are to be able to:

1. Write down the Maxwell-Stefan equations for a binary system, and for multicomponent systems.
2. Define the concepts Maxwell-Stefan (MS) diffusivity, and thermodynamic factor.
3. Express the driving force for mass transfer in terms of mole fraction gradients for ideal and nonideal systems.

Consider a binary mixture of ideal gases 1 and 2 at constant temperature and pressure. From a momentum balance describing collisions between molecules of species 1 and molecules of species 2, we obtain (Taylor and Krishna, 1993)

$$\nabla p_1 = -f_{12} y_1 y_2 (\mathbf{v}_1 - \mathbf{v}_2) \qquad (1\text{-}21)$$

where f_{12} is an empirical parameter analogous to a friction factor or a drag coefficient. For convenience, we define an inverse drag coefficient $\mathcal{D}_{12} = P/f_{12}$ and rewrite equation (1-21) as

$$\mathbf{d}_1 = \frac{\nabla p_1}{P} = -\frac{y_1 y_2 (\mathbf{v}_1 - \mathbf{v}_2)}{\mathcal{D}_{12}} \qquad (1\text{-}22)$$

where $\mathbf{d}_1 = (1/P)\nabla p_1$ is the driving force for diffusion of species 1 in an ideal gas mixture at constant temperature and pressure.

Equation (1-22) is the Maxwell-Stefan equation for the diffusion of species 1 in a binary ideal gas mixture. The symbol $\mathcal{D}_{12}$ is the Maxwell-Stefan (MS) diffusivity. From a similar analysis for species 2:

$$\mathbf{d}_2 = \frac{\nabla p_2}{P} = -\frac{y_1 y_2 (\mathbf{v}_2 - \mathbf{v}_1)}{\mathcal{D}_{21}} \qquad (1\text{-}23)$$

For all the applications considered in this book the system pressure is constant across the diffusion path. Then, equations (1-22) and (1-23) simplify to

$$\nabla y_1 = -\frac{y_1 y_2 (\mathbf{v}_1 - \mathbf{v}_2)}{\mathcal{D}_{12}}$$

$$\nabla y_2 = -\frac{y_1 y_2 (\mathbf{v}_2 - \mathbf{v}_1)}{\mathcal{D}_{21}} \qquad (1\text{-}24)$$

Example 1.3 Diffusivities in Binary Mixtures

Show that, for a binary mixture, $\mathcal{D}_{12} = \mathcal{D}_{21}$.

Solution
Since for a binary mixture $(y_1 + y_2) = 1.0$,

$$\nabla y_1 = -\nabla y_2$$

Then, from equation (1-24)

$$-\frac{y_1 y_2 (\mathbf{v}_1 - \mathbf{v}_2)}{\mathcal{D}_{12}} = \frac{y_1 y_2 (\mathbf{v}_2 - \mathbf{v}_1)}{\mathcal{D}_{21}}$$

which can be true only if

$$\mathcal{D}_{12} = \mathcal{D}_{21}$$

For multicomponent mixtures equation (1-22) can be generalized to (Taylor and Krishna, 1993)

$$\mathbf{d}_i = -\sum_{j=1}^{n} \frac{y_i y_j (\mathbf{v}_i - \mathbf{v}_j)}{\mathcal{D}_{ij}} \qquad i = 1,2,\cdots,n-1 \qquad (1\text{-}25)$$

Equations (1-25) can be written in terms of the molar fluxes $\mathbf{N}_i = c_i \mathbf{v}_i$ to get

$$\mathbf{d}_i = \sum_{j=1}^{n} \frac{y_i \mathbf{N}_j - y_j \mathbf{N}_i}{c \mathcal{D}_{ij}} \qquad i = 1,2,\cdots,n-1 \qquad (1\text{-}26)$$

or, in terms of the diffusion fluxes, $\mathbf{J}_i$

$$\mathbf{d}_i = \sum_{j=1}^{n} \frac{x_i \mathbf{J}_j - x_j \mathbf{J}_i}{c \mathfrak{D}_{ij}} \qquad i = 1, 2, \cdots, n-1 \tag{1-27}$$

These are the *Maxwell-Stefan* diffusion equations for multicomponent systems. They are named after the Scottish physicist James Clerk Maxwell and the Austrian scientist Josef Stefan who were primarily reponsible for their development around 1870. It is important to point out that only $(n-1)$ of the Maxwell-Stefan equations are independent because the $\mathbf{d}_i$ must sum to zero. Also, for a multicomponent ideal gas mixture a more elaborate analysis than that of Example 1.3 is needed to show that (Taylor and Krishna, 1993)

$$\mathfrak{D}_{ij} = \mathfrak{D}_{ji} \tag{1-28}$$

It is easy to show, for a binary mixture of ideal gases where the driving force for diffusion is the mole fraction gradient, equation (1-27) reduces to

$$\mathbf{J}_1 = -c \mathfrak{D}_{12} \mathbf{d}_1 = -c \mathfrak{D}_{12} \nabla y_1 \tag{1-29}$$

For nonideal fluids the driving force for diffusion must be defined in terms of chemical potential gradients as

$$\mathbf{d}_i \equiv \frac{x_i}{RT} \nabla_{T,P} \mu_i \tag{1-30}$$

The subscripts T,P are to emphasize that the gradient is to be calculated under constant temperature and pressure conditions. We may express equation (1-30) in terms of activity coefficients, γ_i, and mole fraction gradients as

$$\mathbf{d}_i = \sum_{j=1}^{n-1} \Gamma_{ij} \nabla x_j = \Gamma \nabla x_i \tag{1-31}$$

where the *thermodynamic factor matrix* Γ is given by

$$\Gamma_{ij} = \delta_{ij} + x_i \left. \frac{\partial \ln \gamma_i}{\partial x_j} \right|_{T,P,\Sigma} \tag{1-32}$$

where δ_{ij} is the *Kronecker delta* defined as

$$\begin{aligned} \delta_{ij} &= 1 & \text{if } i = j \\ \delta_{ij} &= 0 & \text{if } i \neq j \end{aligned} \tag{1-33}$$

The symbol Σ is used in equation (1-32) to indicate that the differentiation with respect to mole fraction is to be carried out in such a manner that the mole fractions always add up to 1.0. Combining equations (1-26), (1-27), and (1-31) we obtain

$$\Gamma \nabla x_i = \sum_{j=1}^{n} \frac{x_i \mathbf{J}_j - x_j \mathbf{J}_i}{c \mathfrak{D}_{ij}} = \sum_{j=1}^{n} \frac{x_i \mathbf{N}_j - x_j \mathbf{N}_i}{c \mathfrak{D}_{ij}} \qquad i = 1, 2, \cdots, n-1 \tag{1-34}$$

For a multicomponent mixture of ideal gases, equation (1-34) becomes

$$\nabla y_i = \sum_{j=1}^{n} \frac{y_i \mathbf{J}_j - y_j \mathbf{J}_i}{c\mathcal{D}_{ij}} = \sum_{j=1}^{n} \frac{y_i \mathbf{N}_j - y_j \mathbf{N}_i}{c\mathcal{D}_{ij}} \qquad i = 1,2,\cdots,n-1 \tag{1-35}$$

For a binary mixture, equation (1-34) reduces to

$$\mathbf{J}_1 = -c\mathcal{D}_{12}\Gamma \nabla x_1 \tag{1-36}$$

where the *thermodynamic factor* Γ is given by

$$\Gamma = 1 + x_1 \frac{\partial \ln \gamma_1}{\partial x_1} \tag{1-37}$$

The thermodynamic factor is evaluated for liquid mixtures from activity coefficient models. For a *regular solution*, for example,

$$\ln \gamma_1 = A x_2^2 = A(1 - x_1)^2 \tag{1-38}$$

therefore, equation (1-37) yields

$$\Gamma = 1 - 2A x_1 x_2 \tag{1-39}$$

1.2.4 Fick's First Law for Binary Mixtures

Your objectives in studying this section are to be able to:

1. Write down Fick's first law for diffusion in a binary, isothermal, isobaric mixture.
2. Define the Fick diffusivity, and establish the relation between the Fick diffusivity and the Maxwell-Stefan diffusivity for a binary system.

At about the same time that Maxwell and Stefan were developing their ideas of diffusion in multicomponent mixtures, Adolf Fick and others were attempting to uncover the basic diffusion equations through experimental studies involving binary mixtures (Fick, 1855). The result of Fick's work was the "law" that bears his name. The Fick rate equation for a binary mixture in an isothermal, isobaric system is

$$\mathbf{J}_1 = -c D_{12} \nabla x_1 \tag{1-40}$$

where D_{12} is the *Fick diffusivity or diffusion coeficient*. Comparing equations (1-36) and (1-40) we see that, for a binary system, the Fick diffusivity D and the MS diffusivity $\mathcal{D}$ are related by

$$D_{12} = \mathcal{D}_{12}\,\Gamma \qquad (1\text{-}41)$$

For ideal systems, Γ is unity and the diffusivities are identical.

$$D_{12} = \mathcal{D}_{12} \qquad \text{for ideal systems} \qquad (1\text{-}42)$$

   The correlation and prediction of Fick and MS diffusion coefficients is discussed in the section that follows. The Fick diffusivity incorporates two aspects: (1) the significance of an inverse drag coefficient ($\mathcal{D}$), and (2) thermodynamic nonideality (Γ). Consequently, the physical interpretation of the Fick diffusion coefficient is less transparent than for the MS diffusivity.

1.3 THE DIFFUSION COEFFICIENT

   Fick's law proportionality factor, D_{12}, is known as the diffusion coefficient or diffusivity. Its fundamental dimensions, which are obtained from equation (1-40),

$$D_{12} = \frac{-J_1}{c\nabla x_1} = \left(\frac{M}{L^2 t}\right)\left(\frac{1}{M/L^3 \cdot 1/L}\right) = \frac{L^2}{t}$$

are identical to the fundamental dimensions of the other transport properties: kinematic viscosity, ν, and thermal diffusivity, α. The mass diffusivity is usually reported in units of cm²/s; the SI units are m²/s, which is a factor 10^{-4} smaller.

   The diffusion coefficient depends upon the pressure, temperature, and composition of the system. Experimental values for the diffusivities of gases, liquids, and solids are tabulated in Appendix A. As one might expect from consideration of the mobility of the molecules, the diffusivities are generally higher for gases (in the range of 0.5×10^{-5} to 1.0×10^{-5} m²/s) than for liquids (in the range of 10^{-10} to 10^{-9} m²/s) which are higher than the values reported for solids (in the range of 10^{-14} to 10^{-10} m²/s). In the absence of experimental data, semitheoretical expressions have been developed which give approximations, sometimes as valid as experimental values due to the difficulties encountered in their measurement.

1.3.1 Diffusion Coefficients for Binary Ideal Gas Systems

Your objectives in studying this section are to be able to:

1. Estimate diffusion coefficients for binary gas systems using the Wilke-Lee equation with tabulated values of the Lennard-Jones parameters.
2. Estimate diffusion coefficients for binary gas systems using the Wilke-Lee equation with values of the Lennard-Jones parameters estimated from empirical correlations.
3. Use a Mathcad® routine to implement calculation of diffusion coefficients for binary gas systems using the Wilke-Lee equation.

The theory describing diffusion in binary gas mixtures at low to moderate pressures has been well developed. Modern versions of the kinetic theory of gases have attempted to account for the forces of attraction and repulsion between molecules. Hirschfelder, et al. (1949), using the Lennard-Jones potential to evaluate the influence of intermolecular forces, presented an equation for the diffusion coefficient for gas pairs of nonpolar, nonreacting molecules

$$D_{AB} = \frac{0.00266\, T^{3/2}}{PM_{AB}^{1/2}\sigma_{AB}^2\Omega_D}$$
(1-43)

where

$$M_{AB} = 2\left[\frac{1}{M_A} + \frac{1}{M_B}\right]^{-1}$$

D_{AB} = diffusion coefficient, cm²/s
M_A, M_B = molecular weights of A and B
T = temperature, K
P = pressure, bar
σ_{AB} = "collision diameter", a Lennard-Jones parameter, Å
Ω_D = diffusion collision integral, dimensionless

The collision integral, Ω_D, is a function of the temperature and of the intermolecular potential field for one molecule of A and one molecule of B. It is usually tabulated as a function of $T^* = \kappa T/\varepsilon_{AB}$, where κ is the Boltzmann constant (1.38 × 10^-16 erg/K) and ε_{AB}, is the energy of molecular interaction for the binary system A and B—a Lennard-Jones parameter—in erg. A very accurate approximation of Ω_D can be obtained from (Neufield, et al., 1972)

$$\Omega_D = \frac{a}{(T*)^b} + \frac{c}{\exp(dT*)} + \frac{e}{\exp(fT*)} + \frac{g}{\exp(hT*)} \qquad (1\text{-}44)$$

where $T* = \kappa T / \varepsilon_{AB}$, $a = 1.06036$ $b = 0.15610$

$c = 0.19300$ $d = 0.47635$ $e = 1.03587$

$f = 1.52996$ $g = 1.76474$ $h = 3.89411$

For a binary system composed of nonpolar molecular pairs, the Lennard-Jones parameters of the pure components may be combined empirically by the following relations:

$$\sigma_{AB} = \frac{\sigma_A + \sigma_B}{2}, \qquad \varepsilon_{AB} = \sqrt{\varepsilon_A \varepsilon_B} \qquad (1\text{-}45)$$

These relations must be modified for polar-polar and polar-nonpolar molecular pairs; the proposed modifications are discussed by Hirschfelder, et al. (1954).

The Lennard-Jones parameters for the pure components are usually obtained from viscosity data. Appendix B tabulates some of the data available. In the absence of experimental data, the values of the parameters for pure components may be estimated from the following empirical correlations:

$$\sigma = 1.18 V_b^{1/3} \qquad (1\text{-}46)$$

$$\varepsilon_A / \kappa = 1.15 T_b \qquad (1\text{-}47)$$

where V_b is the molar volume of the substance as liquid at its normal boiling point, in cm³/gmole, and T_b is the normal boiling point temperature. Molar volumes at normal boiling point for some commonly encountered compounds are listed in Table 1.1. For other compounds not listed in Table 1.1, if a reliable value of the critical volume (V_c) is available, the Tyn and Calus (1975) method is recommended:

$$V_b = 0.285 V_c^{1.048} \qquad (1\text{-}48)$$

Otherwise, the atomic volume of each element present are added together as per the molecular formula of the compound. Table 1.2 lists the contributions for each of the constituent atoms.

Several proposed methods for estimating D_{AB} in low-pressure binary gas systems retain the general form of equation (1-43), with empirical constants based on experimental data. One of the most widely used methods, shown to be quite general and reliable, was proposed by Wilke and Lee (1955):

$$D_{AB} = \frac{\left[3.03 - \left(0.98 / M_{AB}^{1/2} \right) \right] \left(10^{-3} \right) T^{3/2}}{P M_{AB}^{1/2} \sigma_{AB}^2 \Omega_D} \qquad (1\text{-}49)$$

where all the symbols are as defined under equation (1-43). The use of the Wilke-Lee equation is illustrated in the following examples.

Table 1.1 Molar Volumes at Normal Boiling Point

Compound	Volume cm^3/gmole	Compound	Volume cm^3/gmole
Hydrogen, H_2	14.3	Nitric oxide, NO	23.6
Oxygen, O_2	25.6	Nitrous oxide, N_2O	36.4
Nitrogen, N_2	31.2	Ammonia, NH_3	25.8
Air	29.9	Water, H_2O	18.9
Carbon monoxide, CO	30.7	Hydrogen sulfide, H_2S	32.9
Carbon dioxide, CO_2	34.0	Bromine, Br_2	53.2
Carbonyl sulfide, COS	51.5	Chlorine, Cl_2	48.4
Sulfur dioxide, SO_2	44.8	Iodine, I_2	71.5

Source: Welty, et al., 1984.

Table 1.2 Atomic Volume Contributions of the Elements

Element	Volume cm^3/gmole	Element	Volume cm^3/gmole
Bromine	27.0	Oxygen, except as noted below	7.4
Carbon	14.8		
Chlorine	24.6	Oxygen, in methyl esters	9.1
Hydrogen	3.7		
Iodine	37.0	Oxygen, in methyl ethers	9.9
Nitrogen	15.6		
Nitrogen, in primary amines	10.5	Oxygen, in higher ethers and other esters	11.0
Nitrogen, in secondary amines	12.0	Oxygen, in acids	12.0
		Sulfur	25.6

For three-membered ring, such as ethylene oxide, subtract	6.0
For four-membered ring, such as cyclobutane, subtract	8.5
For five-membered ring, such as furan, subtract	11.5
For pyridine, subtract	15.0
For benzene ring, subtract	15.0
For naphthalene ring, subtract	30.0
For anthracene ring, subtract	47.5

Source: Welty, et al., 1984.

Example 1.4 Calculation of Diffusivity by the Wilke-Lee Equation with Known Values of the Lennard-Jones Parameters

Estimate the diffusivity of carbon disulfide vapor in air at 273 K and 1 bar using the Wilke-Lee equation (1-49). Compare this estimate with the experimental value reported in Appendix A.

Solution

Values of the Lennard-Jones parameters (σ and ϵ/κ) are obtained from Appendix B:

	σ, in Å	ϵ/κ, in K	M, g/mol
CS_2	4.483	467	76
Air	3.620	97	29

Evaluate the various parameters of equation (1-49) as follows:

$$\sigma_{AB} = \frac{\sigma_A + \sigma_B}{2} = \frac{4.483 + 3.620}{2} = 4.052 \text{Å}$$

$$\frac{\epsilon_{AB}}{\kappa} = \sqrt{\frac{\epsilon_A}{\kappa}\frac{\epsilon_B}{\kappa}} = \sqrt{467 \times 97} = 212.8 \text{ K}$$

$$\frac{\kappa T}{\epsilon_{AB}} = \frac{273}{212.8} = 1.283, \qquad \Omega_D = 1.282 \text{ (Equation 1-44)}$$

$$M_{AB} = 2\left[\frac{1}{M_A} + \frac{1}{M_B}\right]^{-1} = \frac{2}{\frac{1}{76} + \frac{1}{29}} = 41.981$$

Substituting these values into the Wilke-Lee equation:

$$D_{AB} = \frac{\left(10^{-3}\right)\left(3.03 - \frac{0.98}{\sqrt{41.981}}\right)(273)^{1.5}}{(1)(4.052)^2(1.282)\sqrt{41.981}} = 0.0952 \ \frac{\text{cm}^2}{\text{s}}$$

$$= 9.52 \times 10^{-6} \ \frac{\text{m}^2}{\text{s}}$$

As evidenced by this example, estimation of binary diffusivities can be quite tedious. Most mass-transfer problems involve the calculation of one or more values of diffusivities. Convenience therefore suggests the use of a computer software package for technical calculations, such as Mathcad, for that purpose. Figure 1.3 shows a Mathcad routine to estimate gas phase binary diffusivities using the Wilke-Lee equation which, for computational purposes, gets the name

$$D_{AB}(T, P, M_A, M_B, \sigma_A, \sigma_B, \varepsilon_{Ak}, \varepsilon_{Bk})$$

When the quantities in parentheses are assigned numerical values, the Mathcad routine yields the value of the diffusivity predicted by the Wilke-Lee equation, in units of cm²/s. Thus, for the evaluation of the diffusivity in this example, we write

$$D_{AB}(273, 1, 76, 29, 4.483, 3.62, 467, 97) = 0.0952 \text{ cm}^2/\text{s}$$

The experimental value is obtained from Appendix A:

$$D_{AB}P = 0.894 \frac{\text{m}^2 \text{Pa}}{\text{s}}$$

$$D_{AB} = 0.894 \frac{\left(\text{m}^2\right)(\text{Pa})(1 \text{ bar})}{(\text{s})(1 \text{ bar})\left(10^5 \text{ Pa}\right)}$$

$$D_{AB} = 8.94 \times 10^{-6} \frac{\text{m}^2}{\text{s}}$$

The error of the estimate, compared to the experimental value, is 6.5%.

Example 1.5 Calculation of Diffusivity by the Wilke-Lee Equation with Estimated Values of the Lennard-Jones Parameters

Estimate the diffusivity of allyl chloride (C_3H_5Cl) in air at 298 K and 1 bar using the Wilke-Lee equation (1-49). The experimental value reported by Lugg (1968) is 0.098 cm²/s.

Solution

Values of the Lennard-Jones parameters for allyl chloride (*A*) are not available in Appendix B. Therefore, they must be estimated from equations (1-46) and (1-47). From Table 1.2, V_b = (3)(14.8) + (5)(3.7) + 24.6 = 87.5 cm³/mol. An alternate method to estimate V_b for allyl chloride is using equation (1-48) with a value of V_c = 234 cm³/mol (Reid, et al., 1987). The two estimates are virtually identical. From equation (1-46), σ_A = 5.24 Å. The normal boiling point temperature for allyl chloride is T_b = 318.3 K (Reid, et al., 1987). From equation (1-47), ε_A/κ = 366 K. The molecular weight of allyl chloride is 76.5. The corresponding values for air are σ_B = 3.62 Å, ε_B/κ = 97 K, and M_B = 29. Substituting these values into the Mathcad routine of Figure 1.3:

$$D_{AB}(298, 1, 76.5, 29, 5.24, 3.62, 366, 97) = 0.0992 \text{ cm}^2/\text{s}$$

The error of this estimate, compared to the experimental value, is 1.2%.

$$\sigma_{AB}(\sigma_A, \sigma_B) := \frac{\sigma_A + \sigma_B}{2}$$

$$\varepsilon_{ABk}(\varepsilon_{Ak}, \varepsilon_{Bk}) := \sqrt{\varepsilon_{Ak} \cdot \varepsilon_{Bk}}$$

$$x(T, \varepsilon_{Ak}, \varepsilon_{Bk}) := \frac{T}{\varepsilon_{ABk}(\varepsilon_{Ak}, \varepsilon_{Bk})}$$

$$a := 1.06036 \quad b := 0.15610 \quad c := 0.19300 \quad d := 0.47635$$

$$e1 := 1.03587 \quad f := 1.52996 \quad g := 1.76474 \quad h := 3.89411$$

$$M_{AB}(M_A, M_B) := 2 \cdot \left(\frac{1}{M_A} + \frac{1}{M_B} \right)^{-1}$$

$$\Omega(T, \varepsilon_{Ak}, \varepsilon_{Bk}) := \frac{a}{x(T, \varepsilon_{Ak}, \varepsilon_{Bk})^b} + \frac{c}{e^{d \cdot x(T, \varepsilon_{Ak}, \varepsilon_{Bk})}} + \frac{e1}{e^{f \cdot x(T, \varepsilon_{Ak}, \varepsilon_{Bk})}} + \frac{g}{e^{h \cdot x(T, \varepsilon_{Ak}, \varepsilon_{Bk})}}$$

$$D_{AB}(T, P, M_A, M_B, \sigma_A, \sigma_B, \varepsilon_{Ak}, \varepsilon_{Bk}) := \frac{10^{-3} \cdot \left(3.03 - \dfrac{0.98}{\sqrt{M_{AB}(M_A, M_B)}} \right) \cdot T^{\frac{3}{2}}}{P \cdot \sigma_{AB}(\sigma_A, \sigma_B)^2 \cdot \Omega(T, \varepsilon_{Ak}, \varepsilon_{Bk}) \cdot \sqrt{M_{AB}(M_A, M_B)}}$$

Figure 1.3 Mathcad routine to estimate gas-phase mass diffusivities using the Wilke-Lee equation.

1.3.2 Diffusion Coefficients for Liquids

Your objectives in studying this section are to be able to:

1. Estimate diffusion coefficients for binary dilute liquid systems using the Wilke and Chang equation.
2. Estimate diffusion coefficients for binary dilute liquid systems using the Hayduk and Minhas equation.
3. Use a Mathcad routine to implement calculation of diffusion coefficients for binary dilute liquid systems using the Hayduk and Minhas equation.

In contrast to the case for gases, where an advanced kinetic theory to explain molecular motion is available, theories of the structure of liquids and their transport characteristics are still inadequate to allow a rigorous treatment. Liquid diffusion coefficients are several orders of magnitude smaller than gas diffusivities, and depend on concentration due to the changes in viscosity with concentration and changes in the degree of ideality of the solution. As the mole fraction of either component in a binary mixture approaches unity, the thermodynamic factor Γ approaches unity and the Fick diffusivity and the MS diffusivity are equal. The diffusion coefficients obtained under these conditions are the infinite dilution diffusion coefficients and are given the symbol $\mathfrak{D}^0$.

The Stokes-Einstein equation is a purely theoretical method of estimating $\mathfrak{D}^0$,

$$\mathfrak{D}^0_{AB} = \frac{\kappa T}{6\pi r_A \mu_B} \tag{1-50}$$

where r_A is the solute particle radius, and μ_B is the solvent viscosity. This equation has been fairly successful in describing diffusion of colloidal particles or large round molecules through a solvent which behaves as a continuum relative to the diffusing species.

Equation (1-50) has provided a useful starting point for a number of semiempirical correlations arranged into the general form

$$\frac{\mathfrak{D}^0_{AB} \mu_B}{\kappa T} = f(V_A) \tag{1-51}$$

in which $f(V_A)$ is a function of the molecular volume of the diffusing solute. Empirical correlations using the general form of equation (1-51) have been developed which attempt to predict the liquid diffusion coefficient in terms of the solute and solvent properties. Wilke and Chang (1955) have proposed the following still widely used correlation for nonelectrolytes in an infinitely dilute solution:

$$\frac{\mathcal{D}_{AB}^{0}\mu_B}{T} = \frac{7.4 \times 10^{-8}(\Phi_B M_B)^{1/2}}{V_A^{0.6}} \tag{1-52}$$

where $\mathcal{D}_{AB}^{0}$ = diffusivity of A in very dilute solution in solvent B, cm²/s

$\qquad M_B$ = molecular weight of solvent B

$\qquad T$ = temperature, K

$\qquad \mu_B$ = viscosity of solvent B, cP

$\qquad V_A$ = solute molar volume at its normal boiling point, cm³/mol

$\qquad\qquad$ = 75.6 cm³/mol for water as *solute*

$\qquad \Phi_B$ = *association factor* of solvent B, dimensionless

$\qquad\qquad$ = 2.26 for water as solvent

$\qquad\qquad$ = 1.9 for methanol as solvent

$\qquad\qquad$ = 1.5 for ethanol as solvent

$\qquad\qquad$ = 1.0 for unassociated solvents, e.g., benzene, ether, heptane

The value of V_A may be the true value or, if necessary, estimated from equation (1-48), or from the data of Table 1.2, except when water is the diffusing solute, as noted above. The association factor for a solvent can be estimated only when diffusivities in that solvent have been experimentally measured. There is also some doubt about the ability of the Wilke-Chang equation to handle solvents of very high viscosity, say 100 P or more.

Hayduk and Minhas (1982) considered many correlations for the infinite dilution binary diffusion coefficient. By regression analysis, they proposed several correlations depending on the type of solute–solvent system.

(a) For *solutes in aqueous* solutions:

$$\mathcal{D}_{AB}^{0} = 1.25 \times 10^{-8}\left(V_A^{-0.19} - 0.292\right)T^{1.52}\mu_B^{\varepsilon}$$

$$\varepsilon = \frac{9.58}{V_A} - 1.12 \tag{1-53}$$

where $\mathcal{D}_{AB}^{0}$ = diffusivity of A in very dilute aqueous solution , cm²/s

$\qquad T$ = temperature, K

$\qquad \mu_B$ = viscosity of water, cP

$\qquad V_A$ = solute molar volume at its normal boiling point, cm³/mol

(b) For *nonaqueous (nonelectrolyte)* solutions:

$$\mathcal{D}_{AB}^0 = 1.55 \times 10^{-8} \frac{V_B^{0.27} \; T^{1.29} \; \sigma_B^{0.125}}{V_A^{0.42} \; \mu_B^{0.92} \; \sigma_A^{0.105}} \qquad (1\text{-}54)$$

where σ is surface tension at the normal boiling point temperature, in dyn/cm. If values of the surface tension are not known, they may be estimated by the Brock and Bird (1955) corresponding states method (limited to nonpolar liquids):

$$\sigma = P_c^{2/3} T_c^{1/3} (0.132\alpha_c - 0.278)(1 - T_{br})^{11/9}$$

$$\alpha_c = 0.9076 \left[1 + \frac{T_{br} \ln(P_c / 1.013)}{1 - T_{br}} \right] \qquad (1\text{-}55)$$

$$T_{br} = T_b / T_c$$

where σ = surface tension at the normal boiling point temperature, in dyn/cm

          P_c = critical pressure, in bars

          T_c = critical temperature, in K

When using the correlation shown in equation (1-54), the authors note several restrictions:

1. The method should not be used for diffusion in viscous solvents. Values of μ_B above about 20 cP would classify the solvent as viscous.

2. If the solute is water, a dimer value of V_A should be used ($V_A = 37.4$ cm³/mol).

3. If the solute is an organic acid and the solvent is other than water, methanol, or butanol, the acid should be considered a dimer with twice the expected value of V_A.

4. For nonpolar solutes diffusing into monohydroxy alcohols, the values of V_B should be multiplied by a factor equal to $8\mu_B$ where μ_B is the solvent viscosity in cP.

Example 1.6 Calculation of Liquid Diffusivity in Aqueous Solution

Estimate the diffusivity of ethanol (C_2H_6O) in a dilute solution in water at 288 K. Compare your estimate with the experimental value reported in Appendix A.

Solution

(a) Use the Wilke-Chang correlation, equation (1-52). Calculate the molar volume of ethanol using equation (1-48) with $V_c = 167.1$ cm³/mol (Reid et al., 1987):

$$V_A = 0.285(167.1)^{1.048} = 60.9 \text{ cm}^3 / \text{mol}$$

The viscosity of liquid water at 288 K is $\mu_B = 1.153$ cP (Reid et al., 1987). Substituting in equation (1-52):

$$\mathcal{D}_{AB}^0 = \frac{\left(7.4 \times 10^{-8}\right)(288)\left[(2.26)(18)\right]^{1/2}}{(1.153)(60.9)^{0.6}} = 1.002 \times 10^{-5} \ \frac{cm^2}{s}$$

The experimental value reported in Appendix A is 1.0×10^{-5} cm²/s. Therefore, the error of the estimate is only 0.2%. (This is not typical!)

(b) Using the Hayduk and Minhas correlation for aqueous solutions:

$$\varepsilon = \frac{9.58}{60.9} - 1.12 = -0.963$$

$$\mathcal{D}_{AB}^0 = 1.25 \times 10^{-8} \left[60.9^{-0.19} - 0.292\right](288)^{1.52}(1.153)^{-0.963}$$

$$\mathcal{D}_{AB}^0 = 0.991 \times 10^{-5} \ cm^2 \ / \ s$$

The error of the estimate in this case is – 0.9%.

Example 1.7 Calculation of Liquid Diffusivity in Dilute Nonaqueous Solution

Estimate the diffusivity of acetic acid ($C_2H_4O_2$) in a dilute solution in acetone (C_3H_6O) at 313 K. Compare your estimate with the experimental value reported in Appendix A. The following data are available (Reid, et al., 1987):

Parameter	Acetic acid	Acetone
T_b , K	390.4	329.2
T_c , K	594.8	508.0
P_c , bars	57.9	47.0
V_c , cm³/mol	171	209
μ, cP	---	0.264
M	60	58

Solution

a) Use the Wilke-Chang correlation, equation (1-52). Calculate the molar volume of acetic acid using equation (1-48) with $V_c = 171$ cm³/mol: $V_A = 62.4$ cm³/mol. The association factor for acetone is $\Phi = 1.0$. From equation (1-52):

$$\mathscr{D}_{AB}^{0} = \frac{7.4 \times 10^{-8} [(1)(58)]^{1/2} (313)}{(0.264)(62.4)^{0.6}} = 5.595 \times 10^{-5} \text{ cm}^2 / \text{s}$$

From Appendix A, the experimental value is 4.04×10^{-5} cm²/s. Therefore, the error of the estimate is 38.5%.

(b) Use the Hayduk and Minhas correlation for nonaqueous solutions. According to restriction 3 mentioned above, the molar volume of the acetic acid to be used in equation (1-54) should be $V_A = 2 \times 62.4 = 124.8$ cm³/mol. The molar volume of acetone is calculated from equation (1-48): $V_B = 77.0$ cm³/mol. Estimates of the surface tension at the normal boiling point of each component are calculated from equation (1-55) as follows.

For acetone (B):

$$T_{br} = \frac{329.2}{508} = 0.648$$

$$\alpha_c = 7.319$$

$$\sigma_B = 20.0 \text{ dyn / cm}$$

For acetic acid (A):

$$T_{br} = \frac{390.4}{594.8} = 0.656$$

$$\alpha_c = 7.91$$

$$\sigma_A = 26.2 \text{ dyn / cm}$$

Substituting numerical values in equation (1-54):

$$\mathscr{D}_{AB}^{0} = \frac{1.55 \times 10^{-8} (77)^{0.27} (313)^{1.29} (20)^{0.125}}{(124.8)^{0.42} (0.264)^{0.92} (26.2)^{0.105}} = 3.840 \times 10^{-5} \text{ cm}^2 / \text{s}$$

From Appendix A, the experimental value is 4.04×10^{-5} cm²/s. Therefore, the error of the estimate is – 5.0%.

Figure 1.4 shows a Mathcad routine to estimate liquid-phase binary diffusivities for nonaqueous solutions using the Hayduk and Minhas equation which, for computational purposes, gets the name

$$D_{L0} (T_{bA}, T_{cA}, P_{cA}, V_A, T_{bB}, T_{cB}, P_{cB}, V_B, T, \mu_B)$$

When the quantities in parentheses are assigned numerical values, the Mathcad routine yields the value of the diffusivity predicted by the Hayduk and Minhas equation, in units of cm²/s.

$$T_{brA}(T_{bA}, T_{cA}) := \frac{T_{bA}}{T_{cA}} \qquad T_{brB}(T_{bB}, T_{cB}) := \frac{T_{bB}}{T_{cB}}$$

$$\alpha_{cA}(T_{bA}, T_{cA}, P_{cA}) := 0.9076 \cdot \left[1 + \frac{T_{brA}(T_{bA}, T_{cA}) \cdot \ln\left[\dfrac{P_{cA}}{1.013}\right]}{1 - T_{brA}(T_{bA}, T_{cA})}\right]$$

$$\alpha_{cB}(T_{bB}, T_{cB}, P_{cB}) := 0.9076 \cdot \left[1 + \frac{T_{brB}(T_{bB}, T_{cB}) \cdot \ln\left[\dfrac{P_{cB}}{1.013}\right]}{1 - T_{brB}(T_{bB}, T_{cB})}\right]$$

$$\sigma_A(T_{bA}, T_{cA}, P_{cA}) := P_{cA}^{\frac{2}{3}} \cdot T_{cA}^{\frac{1}{3}} \cdot \left[0.132 \cdot \alpha_{cA}(T_{bA}, T_{cA}, P_{cA}) - 0.278\right] \cdot \left[1 - T_{brA}(T_{bA}, T_{cA})\right]^{\frac{11}{9}}$$

$$\sigma_B(T_{bB}, T_{cB}, P_{cB}) := P_{cB}^{\frac{2}{3}} \cdot T_{cB}^{\frac{1}{3}} \cdot \left[0.132 \cdot \alpha_{cB}(T_{bB}, T_{cB}, P_{cB}) - 0.278\right] \cdot \left[1 - T_{brB}(T_{bB}, T_{cB})\right]^{\frac{11}{9}}$$

$$D_{L0}(T_{bA}, T_{cA}, P_{cA}, V_A, T_{bB}, T_{cB}, P_{cB}, V_B, T, \mu_B) := \frac{1.55 \cdot 10^{-8} \cdot V_B^{0.27} \cdot T^{1.29} \cdot \sigma_B(T_{bB}, T_{cB}, P_{cB})^{0.125}}{V_A^{0.42} \cdot \mu_B^{0.92} \cdot \sigma_A(T_{bA}, T_{cA}, P_{cA})^{0.105}}$$

Figure 1.4 Mathcad® routine to estimate liquid-phase mass diffusivities in dilute nonaqueous solutions using the Hayduk and Minhas equation.

Most methods for predicting $\mathcal{D}$ in concentrated liquid solutions attempt to combine the infinite dilution diffusion coefficients $(\mathcal{D}_{12})^0$ and $(\mathcal{D}_{21})^0$ in a single function of composition. The Vignes formula is recommended by Reid et al. (1987):

$$\mathcal{D}_{12} = \left(\mathcal{D}_{12}^0\right)^{x_2}\left(\mathcal{D}_{21}^0\right)^{x_1}$$

$$D_{12} = \mathcal{D}_{12}\,\Gamma \tag{1-56}$$

Example 1.8 Diffusion Coefficients in the System Acetone-Benzene (Taylor and Krishna, 1993)

Estimate the MS and Fick diffusion coefficients for an acetone(1)-benzene(2) mixture of composition $x_1 = 0.7808$ at 298 K. The infinite dilution diffusivities are:

$$\mathcal{D}_{12}^0 = 2.75 \times 10^{-9}\ \text{m}^2\ /\ \text{s}$$

$$\mathcal{D}_{21}^0 = 4.15 \times 10^{-9}\ \text{m}^2\ /\ \text{s}$$

From the NRTL equation, for this system at the given temperature and concentration, the thermodynamic correction factor $\Gamma = 0.871$. The experimental value of D_{12} at this concentration is 3.35×10^{-9} m²/s.

Solution
Substituting in equation (1-56):

$$\mathcal{D}_{12} = \left(2.75 \times 10^{-9}\right)^{0.2192} \times \left(4.15 \times 10^{-9}\right)^{0.7808}$$

$$= 3.792 \times 10^{-9}\ \text{m}^2\ /\ \text{s}$$

$$D_{12} = 3.792 \times 10^{-9} \times 0.871$$

$$= 3.30 \times 10^{-9}\ \text{m}^2\ /\ \text{s}$$

The predicted value of the Fick diffusivity is in excellent agreement with the experimental result.

1.3.3 Effective Diffusivities in Multicomponent Mixtures

Your objectives in studying this section are to be able to:

1. Estimate the effective diffusivity of a component in a multicomponent mixture of gases from its binary diffusion coefficients with each of the other constituents of the mixture.

2. Estimate the effective diffusivity of a dilute solute in a homogeneous solution of mixed solvents from its infinite dilution binary diffusion coefficients into each of the individual solvents.

The Maxwell-Stefan equations for diffusion in multicomponent systems can become very complicated, and they are sometimes handled by using an *effective diffusivity or pseudobinary approach*. The effective diffusivity is defined by assuming that the rate of diffusion of component i depends only on its own composition gradient; that is,

$$\mathbf{J}_i = -cD_{i,eff}\nabla x_i \qquad\qquad i = 1,2,\cdots,n$$

$$(1\text{-}57)$$

where $D_{i,eff}$ is some characteristic diffusion coefficient of species i in the mixture which must be synthesized from the binary MS diffusivities.

In gases, the binary MS diffusivities are normally assumed independent of composition. With this approximation, the effective diffusivity of component i in a multicomponent gas mixture can be derived from the Stefan-Maxwell equation to give (Treybal, 1980):

$$D_{i,eff} = \frac{\mathbf{N}_i - y_i \sum\limits_{j=1}^{n} \mathbf{N}_j}{\sum\limits_{\substack{j=1 \\ j\neq i}}^{n} \frac{1}{\mathcal{D}_{ij}}\left(y_j \mathbf{N}_i - y_i \mathbf{N}_j\right)}$$

$$(1\text{-}58)$$

In order to use this equation, the relation between the values of the $\mathbf{N}$'s must be known. This is usually fixed by other considerations, such as reaction stoichiometry or energy balances. For example, if ammonia is being cracked on a solid catalyst according to

$$NH_3 \rightarrow \tfrac{1}{2}N_2 + \tfrac{3}{2}H_2$$

under circumstances such that NH_3 (1) diffuses to the catalyst surface, and N_2 (2) and H_2 (3) diffuse back, the reaction stoichiometry fixes the relationships

$$N_2 = -\frac{1}{2}N_1$$

$$N_3 = -\frac{3}{2}N_1$$

Equation (1-58) suggests that the effective diffusivity may vary considerably along the diffusion path, but a linear variation with distance can usually be assumed (Bird, et al., 1960). Therefore, if $D_{i,eff}$ changes significantly from one end of the diffusion path to the other, the arithmetic average of the two values should be used.

A common situation in multicomponent diffusion is when all the $\mathbf{N}$'s except $\mathbf{N}_1$ are zero; i.e., all but component 1 is stagnant. Equation (1-58) then becomes

$$D_{1,eff} = \frac{1-y_1}{\sum\limits_{i=2}^{n} \dfrac{y_i}{\mathcal{D}_{1i}}} = \frac{1}{\sum\limits_{i=2}^{n} \dfrac{y_i'}{\mathcal{D}_{1i}}} \tag{1-59}$$

where y_i' is the mole fraction of component i on a component 1-free base, that is

$$y_i' = \frac{y_i}{1-y_1} \qquad i = 2,3,\cdots,n \tag{1-60}$$

Example 1.9 Calculation of Effective Diffusivity in a Multicomponent Gas Mixture

Ammonia is being cracked on a solid catalyst according to the reaction

$$NH_3 \rightarrow \tfrac{1}{2}N_2 + \tfrac{3}{2}H_2$$

At one place in the apparatus where the pressure is 1 bar and the temperature 300 K, the analysis of the gas is 40% NH_3 (1), 20% N_2 (2), and 40% H_2 (3) by volume. Estimate the effective diffusivity of ammonia in the gaseous mixture.

Solution

Calculate the binary diffusivities of ammonia in nitrogen ($\mathcal{D}_{12}$) and ammonia in hydrogen ($\mathcal{D}_{13}$) using the Mathcad routine developed in Example 1.4 to implement the Wilke-Lee equation. Values of the Lennard-Jones parameters (σ and ε/κ) are obtained from Appendix B. The data needed are:

	σ, in Å	ε/κ, in K	M, g/mol
NH_3	2.900	558.3	17
N_2	3.798	71.4	28
H_2	2.827	59.7	2

Therefore,

$$\mathcal{D}_{12} = D_{AB}(300, 1, 17, 28, 2.9, 3.798, 558.3, 71.4) = 0.237 \text{ cm}^2/\text{s}$$

$$\mathcal{D}_{13} = D_{AB}(300, 1, 17, 2, 2.9, 2.827, 558.3, 59.7) = 0.728 \text{ cm}^2/\text{s}$$

Figure 1.5 shows the flux of ammonia (1) toward the catalyst surface where it is consumed by chemical reaction, and the fluxes of nitrogen (2) and hydrogen (3) produced by the reaction migrating away from the same surface.

Gas mixture

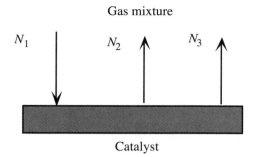

Figure 1.5 Catalytic cracking of ammonia (1) into N_2 (2) and H_2 (3).

Catalyst

As was shown before, the stoichiometry of the reaction fixes the relationship between the individual fluxes, which, in this particular case, becomes $N_2 = -1/2N_2$, $N_3 = -3/2N_2$; $N_1 + N_2 + N_3 = -N_1$. Substituting in equation (1-58):

$$D_{1,eff} = \frac{N_1(1+y_1)}{\dfrac{y_2 N_1 + 1/2 y_1 N_1}{\mathcal{D}_{12}} + \dfrac{y_3 N_1 + 3/2 y_1 N_1}{\mathcal{D}_{13}}}$$

$$= \frac{1+y_1}{\dfrac{y_2 + 1/2 y_1}{\mathcal{D}_{12}} + \dfrac{y_3 + 3/2 y_1}{\mathcal{D}_{13}}}$$

$$= \frac{1.4}{\dfrac{0.2 + 0.5 \times 0.4}{0.237} + \dfrac{0.4 + 1.5 \times 0.4}{0.728}}$$

$$= 0.457 \text{ cm}^2 / \text{s}$$

Example 1.10 Calculation of Effective Diffusivity in a Multicomponent Stagnant Gas Mixture

Ammonia is being absorbed from a stagnant mixture of nitrogen and hydrogen by contact with a 2 N sulfuric acid solution. At one place in the apparatus where the pressure is 1 bar and the temperature 300 K, the analysis of the gas is 40% NH_3 (1), 20% N_2 (2), and 40% H_2 (3) by volume. Estimate the effective diffusivity of ammonia in the gaseous mixture.

Solution

The binary diffusivities are the same as for the previous example. However, in this case, only the ammonia diffuses toward the liquid interphase since neither nitrogen nor hydrogen are soluble in the sulfuric acid solution. Therefore, equation (1-59) applies. Calculate the mole fractions of nitrogen (2) and hydrogen (3) on an ammonia (1)-free base from equation (1-60):

$$y_2' = \frac{y_2}{1-y_1} = \frac{0.2}{1-0.4} = 0.333$$

$$y_3' = \frac{y_3}{1-y_1} = \frac{0.4}{1-0.4} = 0.667$$

Substituting in equation (1-59):

$$D_{1,eff} = \frac{1}{\dfrac{y_2'}{\mathcal{D}_{12}} + \dfrac{y_3'}{\mathcal{D}_{13}}}$$

$$= \frac{1}{\dfrac{0.333}{0.237} + \dfrac{0.667}{0.728}}$$

$$= 0.431 \text{ cm}^2/\text{s}$$

For multicomponent liquid mixtures, it is usually very difficult to obtain numerical values of the diffusion coefficients relating fluxes to concentration gradients. One important case of multicomponent diffusion for which simplified empirical correlations have been proposed is when a dilute solute diffuses through a homogeneous solution of mixed solvents. Perkins and Geankoplis (1969) evaluated several methods and suggested

$$D_{1,eff}^0 \mu_M^{0.8} = \sum_{j=2}^{n} x_j \mathcal{D}_{1j}^0 \mu_j^{0.8} \tag{1-61}$$

where

$D_{1,eff}^0$ = effective diffusivity for a dilute solute A into the mixture, cm^2/s

$\mathcal{D}_{Aj}^0$ = infinite dilution binary diffusivity of solute A into solvent j, cm^2/s

x_j = mole fraction of component j

μ_M = mixture viscosity, cP

μ_j = pure component viscosity, cP

When tested with data for eight ternary systems, errors were normally less than 20%, except for cases where CO_2 was the solute. For CO_2 as a dilute solute diffusing into mixed solvents, Takahashi et al. (1982) recommend

$$D^0_{CO_2,eff}\left(\frac{\mu_M}{V_M}\right)^{1/3} = \sum_{\substack{j=1 \\ j\neq CO_2}}^{n} x_j \mathcal{D}^0_{CO_2 j}\left(\frac{\mu_j}{V_j}\right)^{1/3} \qquad (1.62)$$

where

V_M = molar volume of the mixture at T, cm³/mol

V_j = molar volume of the pure component j at T, cm³/mol

Tests with a number of ternary systems involving CO_2 led to deviations from experimental values usually less than 4 percent. (Reid, et al., 1987).

Example 1.11 Calculation of Effective Diffusivity of a Dilute Solute in a Homogeneous Mixture of Solvents

Estimate the diffusion coefficient of acetic acid at very low concentrations diffusing into a mixed solvent containing 40.0 mass percent ethyl alcohol in water at a temperature of 298 K.

Solution

Let 1 = acetic acid, 2 = water, and 3 = ethyl alcohol. Estimate the infinite dilution diffusion coefficient of acetic acid in water at 298 K using equation (1-53). The required data are (Reid, et al., 1987): (1) at 298 K, μ_B = 0.894 cP; (2) V_{cA} = 171.0 cm³/mol. From equation (1-48), V_A = 62.4 cm³/mol. Substituting in equation (1-53), $\mathcal{D}^0_{12}$= 1.32×10^{-5} cm²/s . Estimate the infinite dilution diffusion coefficient of acetic acid in ethanol at 298 K. The following data are available (Reid, et al., 1987):

Parameter	Acetic acid	Ethanol
T_b , K	390.4	351.4
T_c , K	594.8	513.9
P_c , bars	57.9	61.4
V_c , cm³/mol	171	167
μ @ 298 K, cP	---	1.043

Use the Hayduk and Minhas correlation for nonaqueous solutions. According to restriction 3 mentioned above, the molar volume of the acetic acid to be used in equation (1-54) should be V_A = 2 × 62.4 = 124.8 cm³/mol. The molar volume of ethanol is calculated from equation (1-48): V_B = 60.9 cm³/mol. Using the Mathcad routine developed in Example 1.7:

$$\mathfrak{D}_{13}^{0} = D_{L0}(390.4, 594.8, 57.9, 124.8, 351.4, 513.9, 61.4, 60.9, 298, 1.043)$$

$$= 1.0 \times 10^{-5} \ cm^2/s$$

The viscosity of a 40% by mass aqueous ethanol solution at 298 K is $\mu_M = 2.35$ cP (Perkins and Geankoplis, 1969). Before substituting in equation (1-61), the solution composition must be changed from mass to molar fractions following a procedure similar to that illustrated in Example 1.2. Accordingly, a 40% by mass aqueous ethanol solution converts to 20.7 molar percent. Substituting in equation (1-61):

$$D_{1,eff}^{0} = \frac{10^{-5}}{(2.35)^{0.8}} \left[(0.207)(1.0)(1.043)^{0.8} + (1 - 0.207)(1.32)(0.894)^{0.8} \right]$$

$$= 0.591 \times 10^{-5} \ cm^2/s$$

The experimental value reported by Perkins and Geankoplis (1969) is 0.571×10^{-5} cm^2/s. Therefore, the error of the estimate is 3.5%.

1.4 STEADY-STATE MOLECULAR DIFFUSION IN FLUIDS

1.4.1 Molar Flux and the Equation of Continuity

Your objectives in studying this section are to be able to:

1. State the equation of continuity for a system of variable composition.
2. Simplify the equation of continuity for steady-state diffusion in only one direction in rectangular coordinates, and without chemical reaction.

The *equation of continuity for component A* in a mixture describes the change of concentration of A with respect to time at a fixed point in space resulting from mass transfer of A and chemical reactions producing A. In vector symbolism, the resulting equation is (Bird et al., 1960):

$$\nabla \cdot \mathbf{N_A} + \frac{\partial c_A}{\partial t} - R_A = 0 \tag{1-63}$$

The first term in equation (1-63) is called "the divergence" of the molar flux vector and it represents the net rate of molar efflux per unit volume. In rectangular coordinates, this term can be expanded as:

$$\nabla \cdot \mathbf{N_A} = \frac{\partial N_{A,x}}{\partial x} + \frac{\partial N_{A,y}}{\partial y} + \frac{\partial N_{A,z}}{\partial z} \qquad (1\text{-}64)$$

The second term in equation (1-63) represents the molar rate of accumulation of A per unit volume. The term R_A represents the volumetric rate of formation of A by chemical reaction (moles of A formed/volume-time). An expression similar to equation (1-63) can be written for each component of the mixture:

$$\nabla \cdot \mathbf{N}_i + \frac{\partial c_i}{\partial t} - R_i = 0 \qquad i = 1,2,\cdots,n \qquad (1\text{-}65)$$

Consider, now, steady-state diffusion (no accumulation) without chemical reaction. In that case, equation (1-65) reduces to

$$\nabla \cdot \mathbf{N}_i = 0 \qquad i = 1,2,\cdots,n \qquad (1\text{-}66)$$

In rectangular coordinates, with diffusion only in the z direction, equation (1-66) simplifies to

$$\frac{dN_{i,z}}{dz} = 0 \qquad i = 1,2,\cdots,n \qquad (1\text{-}67)$$

This relation stipulates a constant molar flux of each component in the mixture, and a constant total molar flux. Therefore, for steady-state, one-dimensional diffusion without chemical reaction:

$$N_i = N_{i,z} = \text{constant}_i \qquad i = 1,2,\cdots,n$$

$$\sum_{i=1}^{n} N_i = \text{constant} \qquad (1\text{-}68)$$

1.4.2 Steady-State Molecular Diffusion in Gases

Your objectives in studying this section are to be able to:

1. Calculate molar fluxes during steady-state molecular, one-dimensional diffusion in gases.
2. Calculate molar fluxes for steady-state gaseous diffusion of A through stagnant B, and for equimolar counterdiffusion.

Consider steady-state diffusion only in the z direction without chemical reaction in a binary gaseous mixture. For the case of one-dimensional diffusion, equation (1-20) becomes:

$$N_A = -cD_{AB}\frac{dy_A}{dz} + y_A(N_A + N_B) \qquad (1-69)$$

Separating the variables in equation (1-69):

$$\frac{-dy_A}{N_A - y_A(N_A + N_B)} = \frac{dz}{cD_{AB}} \qquad (1-70)$$

For a gaseous mixture, if the temperature and pressure are constant, c and D_{AB} are constant, independent of position and composition. According to equation (1-68), all the molar fluxes are also constant. Equation (1-70) may, then, be integrated between the two boundary conditions:

$$\text{at } z = z_1 \qquad y_A = y_{A_1}$$

$$\text{at } z = z_2 \qquad y_A = y_{A_2}$$

where 1 indicates the beginning of the diffusion path (y_A high) and 2 the end of the diffusion path (y_A low). Letting $z_2 - z_1 = z$, we get

$$\frac{1}{(N_A + N_B)}\ln\left[\frac{N_A - y_{A_2}(N_A + N_B)}{N_A - y_{A_1}(N_A + N_B)}\right] = \frac{z}{cD_{AB}} \qquad (1-71)$$

Multiplying both sides of equation (1-71) by N_A and rearranging:

$$N_A = \frac{N_A}{(N_A + N_B)}\frac{cD_{AB}}{z}\ln\left[\frac{\frac{N_A}{(N_A + N_B)} - y_{A_2}}{\frac{N_A}{(N_A + N_B)} - y_{A_1}}\right] \qquad (1-72)$$

Define the *molar flux fraction*, Ψ_A:

$$\Psi_A = \frac{N_A}{(N_A + N_B)} \qquad (1-73)$$

In this expression, N_A gives the positive direction of z. Any flux in the opposite direction will carry a negative sign. Notice that the molar flux fraction Ψ_A depends exclusively on the *relation* between N_A and the other fluxes, a relation that is known *a priori* based on other considerations, as was shown before. Therefore, equation (1-72) becomes

$$N_A = \Psi_A \frac{cD_{AB}}{z}\ln\left[\frac{\Psi_A - y_{A_2}}{\Psi_A - y_{A_1}}\right] \qquad (1-74)$$

Equation (1-74) is one of the most fundamental relations in the analysis of mass transfer phenomena. It was derived by integration assuming that all the molar fluxes were constant, independent of position. Integration under conditions where the fluxes are not constant is also possible. Consider, for example, steady-state radial diffusion from the surface of a solid sphere into a fluid. Equation (1-70) can be applied, but the fluxes are now a function of position owing to the geometry. Most practical problems which deal with such matters, however, are concerned with diffusion under turbulent conditions, and the transfer coefficients which are then used are based upon a flux expressed in terms of some arbitrarily chosen area, such as the surface of the sphere. These matters are discussed in detail in Chapter 2.

Most practical problems of mass transfer in the gas phase are at conditions such that the mixture behaves as an ideal gas. Then, combining equations (1-7) and (1-74):

$$N_A = \Psi_A \frac{D_{AB}P}{RTz} \ln\left[\frac{\Psi_A - y_{A_2}}{\Psi_A - y_{A_1}}\right] \qquad (1\text{-}75)$$

Example 1.12 Production of Nickel Carbonyl: Steady-State, One-Dimensional Binary Flux Calculation

Nickel carbonyl (A) is produced by passing carbon monoxide (B) at 323 K and 1 atm over a nickel slab. The following reaction takes place at the solid surface:

$$Ni(s) + 4CO(g) \rightarrow Ni(CO)_4(g)$$

The reaction is very rapid, so that the partial pressure of CO at the metal surface is essentially zero. The gases diffuse through a film 0.625-mm thick. At steady-state, estimate the rate of production of nickel carbonyl, in mole/m^2 of solid surface-s. The composition of the bulk gas phase is 50 mol% CO. The binary gas diffusivity under these conditions is $D_{AB} = 20.0$ mm^2/s.

Solution
Figure 1.6 is a schematic diagram of the diffusion process. The stoichiometry of the reaction determines the relation between the fluxes, namely $N_B = -4N_A$, $N_A + N_B = -3N_A$. Calculate the molar flux fraction, Ψ_A. From equation (1-73):

$$\Psi_A = \frac{N_A}{N_A + N_B} = \frac{N_A}{-3N_A} = -\frac{1}{3}$$

Combining this result with equation (1-75)

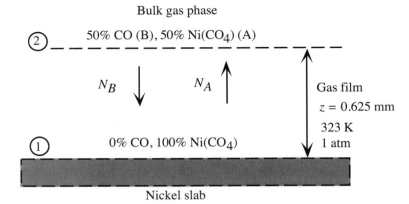

Bulk gas phase

50% CO (B), 50% Ni(CO₄) (A)

0% CO, 100% Ni(CO₄)

Nickel slab

Figure 1.6 Schematic diagram for Example 1.12.

$$N_A = -\frac{D_{AB}P}{3RTz}\ln\left[\frac{-1/3-y_{A_2}}{-1/3-y_{A_1}}\right]$$

$$= \frac{D_{AB}P}{3RTz}\ln\left[\frac{1/3+y_{A_1}}{1/3+y_{A_2}}\right]$$

Substituting numerical values:

$$N_A = \frac{\left(2.0\times10^{-5}\,\mathrm{m^2/s}\right)\left(1.013\times10^5\,\mathrm{Pa}\right)}{3\times\left(8.314\,\dfrac{\mathrm{m^3-Pa}}{\mathrm{mol\text{-}K}}\right)\left(323\,\mathrm{K}\right)\left(0.625\times10^{-3}\,\mathrm{m}\right)}\ln\left(\frac{1/3+1.0}{1/3+0.5}\right)$$

$$= 0.189\,\mathrm{mol/m^2\text{-}s}$$

Example 1.13 Steady-State Diffusion of A Through Stagnant B

Consider steady-state, one-dimensional diffusion of A through nondiffusing B. This might occur, for example, if ammonia (A) is absorbed from air (B) into water. In the gas phase, since air does not dissolve appreciably in water, and neglecting the evaporation of water, only the ammonia diffuses. Thus, $N_B = 0$, $N_A =$ constant, and

$$\Psi_A = \frac{N_A}{N_A + N_B} = 1$$

Equation (1-75) then becomes

$$N_A = \frac{D_{AB}P}{RTz} \ln\left[\frac{1-y_{A_2}}{1-y_{A_1}}\right] \qquad (1\text{-}76)$$

Substituting equation (1-9) into equation (1-76), the driving force for mass transfer is expressed in terms of partial pressures:

$$N_A = \frac{D_{AB}P}{RTz} \ln\left[\frac{P-p_{A_2}}{P-p_{A_1}}\right] \qquad (1\text{-}77)$$

Notice that in equation (1-77), as in all the relations derived so far to calculate the molecular diffusion flux, the driving force is of a logarithmic nature. It will be shown in Chapter 2 that, for diffusion under turbulent conditions, the flux is usually calculated as the product between an empirical mass-transfer coefficient and a linear driving force. For comparison purposes, we will now manipulate equation (1-77) to rewrite it in terms of a linear driving force. Since

$$P - p_{A_2} = p_{B_2} \qquad P - p_{A_1} = p_{B_1}$$

$$p_{B_2} - p_{B_1} = p_{A_1} - p_{A_2}$$

then

$$N_A = \frac{D_{AB}P}{RTz}\left[\frac{p_{A_1} - p_{A_2}}{p_{B_2} - p_{B_1}}\right]\ln\left[\frac{p_{B_2}}{p_{B_1}}\right] \qquad (1\text{-}78)$$

Define the *logarithmic mean partial pressure* of B, $p_{B,M}$:

$$p_{B,M} = \frac{p_{B_2} - p_{B_1}}{\ln\left(\dfrac{p_{B_2}}{p_{B_1}}\right)} \qquad (1\text{-}79)$$

then

$$N_A = \frac{D_{AB}P}{RTp_{B,M}z}\left(p_{A_1} - p_{A_2}\right) \qquad (1\text{-}80)$$

Figure 1.7 shows schematically the concentration profile for diffusion of A through stagnant B. Substance A diffuses by virtue of its concentration gradient, $-dy_A/dz$. Substance B is also diffusing relative to the average molar velocity at a flux J_B which depends upon $-dy_B/dz$, but like a fish which swims upstream at the same velocity as the water flows downstream, $N_B = 0$ relative to a fixed place in space.

Example 1.14 Steady-State Equimolar Counterdiffusion

This is a situation frequently encountered in distillation operations. In this case, $N_B = -N_A = $ const., $\Sigma N_i = 0$. Equations (1-73) and (1-74) become indeterminate. We

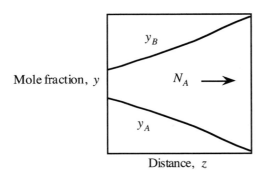

Mole fraction, y

Distance, z

Figure 1.7 Diffusion of A through stagnant B.

can go back to equation (1-69) which, for gases, becomes

$$N_A = -\frac{D_{AB}P}{RT}\frac{dy_A}{dz} \qquad (1\text{-}81)$$

Integrating, with constant N_A

$$N_A = \frac{D_{AB}P}{RTz}\left(y_{A_1} - y_{A_2}\right)$$

$$= \frac{D_{AB}}{RTz}\left(p_{A_1} - p_{A_2}\right) \qquad (1\text{-}82)$$

Notice that equation (1-82), developed for diffusion of A through stagnant B, becomes identical to equation (1-80) for dilute solutions of A in B where $p_{B,M}$ is approximately equal to P.

Example 1.15 Multicomponent, Steady-State, Gaseous Diffusion in a Stefan Tube (Taylor and Krishna, 1993).

The Stefan tube, depicted schematically in Figure 1.8, is a simple device sometimes used for measuring diffusion coefficients in binary vapor mixtures. In the bottom of the tube is a pool of quiescent liquid. The vapor that evaporates from this pool diffuses to the top of the tube. A stream of gas across the top of the tube keeps the mole fraction of diffusing vapor there to esentially nothing. the mole fraction of the vapor at the vapor–liquid interface is its equilibrium value. Carty and Schrodt (1975) evaporated a binary liquid mixture of acetone (1) and methanol (2) in a Stefan tube. Air (3) was used as the carrier gas. In one of their experiments the composition of the vapor at the liquid interface was $y_1 = 0.319$, $y_2 = 0.528$, and $y_3 = 0.153$. The pressure and temperature in the gas phase were 99.4 kPa and 328.5 K, respectively. The length of the diffusion path was 0.24 m. The MS diffusion coefficients of the three binary pairs are:

$$\mathcal{D}_{12} = 8.48 \text{ mm}^2 / \text{s}$$

$$\mathcal{D}_{13} = 13.72 \text{ mm}^2 / \text{s}$$

$$\mathcal{D}_{23} = 19.91 \text{ mm}^2 / \text{s}$$

Calculate the molar fluxes of acetone and methanol, and the composition profiles predicted by the Maxwell-Stefan equations. Compare your results with the experimental data.

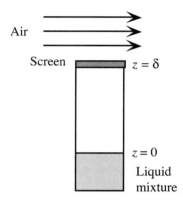

Figure 1.8 Schematic diagram of a Stefan tube.

Solution

For a multicomponent mixture of ideal gases such as this, the Maxwell-Stefan equations (1-35) must be solved simultaneously for the fluxes and the composition profiles. At constant temperature and pressure the total molar density c and the binary diffusion coefficients are constant. Furthermore, diffusion in the Stefan tube takes place in only one direction, up the tube. Therefore, the continuity equation simplifies to equation (1-68). Air (3) diffuses down the tube as the evaporating mixture diffuses up, but because air does not dissolve in the liquid its flux N_3 is zero (i.e., the diffusion flux J_3 of the gas down the tube is exactly balanced by the diffusion-induced convective flux x_3N up the tube).

With the above assumptions the Maxwell-Stefan relations given by equation (1-35) reduce to a system of two first-order differential equations

$$\frac{dy_1}{dz} = \frac{y_1 N_2 - y_2 N_1}{c\mathcal{D}_{12}} + \frac{y_1 N_3 - (1 - y_1 - y_2)N_1}{c\mathcal{D}_{13}}$$

$$\frac{dy_2}{dz} = \frac{y_2 N_1 - y_1 N_2}{c\mathcal{D}_{12}} + \frac{y_2 N_3 - (1 - y_1 - y_2)N_2}{c\mathcal{D}_{23}}$$

(1-83)

subject to the boundary conditions

$$z = 0, \qquad y_i = y_{i0} \qquad z = \delta, \qquad y_i = y_{i\delta} \qquad i = 1,2$$

Define:

$$\eta = \frac{z}{\delta} \qquad F_{12} = \frac{c\mathcal{D}_{12}}{\delta} \qquad F_{13} = \frac{c\mathcal{D}_{13}}{\delta} \qquad F_{23} = \frac{c\mathcal{D}_{23}}{\delta}$$

The Maxwell-Stefan relations then become

$$\frac{dy_1}{d\eta} = \frac{y_1 N_2 - y_2 N_1}{F_{12}} + \frac{y_1 N_3 - (1 - y_1 - y_2)N_1}{F_{13}}$$

$$\frac{dy_2}{d\eta} = \frac{y_2 N_1 - y_1 N_2}{F_{12}} + \frac{y_2 N_3 - (1 - y_1 - y_2)N_2}{F_{23}} \tag{1-84}$$

An analytical solution of these equations for Stefan diffusion ($N_3 = 0$) in a ternary mixture is available (Carty and Schrodt, 1975). However—with the objective of presenting a general method of solution that can be used for any number of components, and any relation between the fluxes—we solved these equations numerically using the orthogonal collocation method (Rice and Do, 1995). This method approximates the differential equations with a system of simultaneous algebraic equations, easily solved using the "solve block" feature of Mathcad.

The first step is to choose the number of collocation points as $NC + 2$, where NC is the number of interior nodes and the 2 represents the two boundary points $\eta = 0, \eta = 1$. (We have found that solving the Maxwell-Stefan equations with $NC = 3$ the results are accurate enough for most engineering calculations.) Next, equations (1-84) are written in terms of the first-derivative collocation matrix $\mathbf{A}$ at each of the NC interior nodes and at the boundary point $\eta = 1$:

$$\sum_{j=1}^{NC+2} A_{i,j} y_{1,j} = \frac{y_{1,i} N_2 - y_{2,i} N_1}{F_{12}}$$

$$+ \frac{y_{1,i} N_3 - (1 - y_{1,i} - y_{2,i})N_1}{F_{13}} \qquad i = 2, \cdots, NC + 2$$

$$\sum_{j=1}^{NC+2} A_{i,j} y_{2,j} = \frac{y_{2,i} N_1 - y_{1,i} N_2}{F_{12}}$$

$$+ \frac{y_{2,i} N_3 - (1 - y_{1,i} - y_{2,i})N_2}{F_{23}} \qquad i = 2, \cdots, NC + 2 \tag{1-85}$$

For $NC = 3$, the vector of collocation points η, and the first-derivative collocation matrix $\mathbf{A}$ are (Rice and Do, 1995):

$$\eta = \begin{bmatrix} 0.000 \\ 0.173 \\ 0.500 \\ 0.827 \\ 1.000 \end{bmatrix} \qquad \mathbf{A} = \begin{bmatrix} -10.0 & 13.514 & -5.352 & 2.827 & -1.0 \\ -2.472 & -0.016 & 3.500 & -1.524 & 0.517 \\ 0.747 & -2.672 & 0.0 & 2.672 & -0.747 \\ -0.517 & 1.529 & -3.500 & 0.016 & 2.472 \\ 1.0 & -2.827 & 5.352 & -13.514 & 10.0 \end{bmatrix}$$

These values are fixed, regardless of the problem being solved, and depend exclusively on the value of NC chosen.

For a ternary system and three interior collocation points, equations (1-85) constitute a system of $2 \times (NC + 1) = 8$ simultaneous algebraic equations. The eight unknown values are the fluxes N_1 and N_2, and the mole fractions of acetone (1) and methanol (2) at each of the three interior nodes (y_{12}, y_{22}, y_{13}, y_{23}, y_{14}, and y_{24}). Appendix C-1 is a Mathcad program to solve this system of equations.

The total molar concentration is estimated from the ideal gas law as $c = 36.4$ mole/m^3. Then,

$$F_{12} = 1.286 \times 10^{-3} \text{ mole / m}^2\text{–s}$$

$$F_{13} = 2.081 \times 10^{-3} \text{ mole / m}^2\text{–s}$$

$$F_{23} = 3.019 \times 10^{-3} \text{ mole / m}^2\text{–s}$$

The "solve block" feature of Mathcad requires initial estimates of all the unknown values. We shall take the following values:

$$N_1 = N_2 = F_{12} = 1.286 \times 10^{-3} \text{ mole / m}^2\text{–s}$$

$$N_3 = 0.0 \text{ mole / m}^2\text{–s}$$

$$y_{11} = 0.319 \quad y_{12} = 0.30 \quad y_{13} = 0.20 \quad y_{14} = 0.00 \quad y_{15} = 0.00$$

$$y_{21} = 0.528 \quad y_{22} = 0.40 \quad y_{23} = 0.25 \quad y_{24} = 0.00 \quad y_{25} = 0.00$$

There is a total of 13 variables in the formulation of this problem: 3 fluxes and 10 mole fractions. The problem specifies the value of 5 of them: one flux (N_3) and 4 mole fractions ($y_{11}, y_{21}, y_{15}, y_{25}$). The unknown variables are identified in the "Find" statement of the solve block, in this case:

$$\text{Find } (N1, N2, y1_2, y1_3, y1_4, y2_2, y2_3, y2_4)$$

The program quickly converges to a solution that can be compared to the experimental results obtained by Carty and Schrodt (1975). The following table compares the predicted and observed fluxes.

	Predicted	Experimental Carty and Schrodt (1975)	Error, %
N_1 mole/m²-s	1.791×10^{-3}	1.779×10^{-3}	$+0.67$
N_2 mole/m²-s	3.095×10^{-3}	3.121×10^{-3}	-0.83

Figure 1.9 shows the concentration profiles predicted by the numerical solution of the Stefan-Maxwell equations, in excellent agreement with the experimental values observed by Carty and Schrodt (1975).

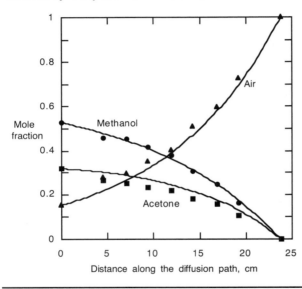

Figure 1.9 Concentration profiles in a Stefan tube. Lines represent calculated profiles; points represent the experimental data of Carty and Schrodt (1975).

Example 1.16 Multicomponent, Steady-State, Gaseous Diffusion with Chemical Reaction.

Consider the vapor phase dehydrogenation of ethanol (1) to produce acetaldehyde (2) and hydrogen (3):

$$C_2H_6O(g) \rightarrow C_2H_4O(g) + H_2(g)$$

carried out at a temperature of 548 K and 101.3 kPa (Taylor and Krishna, 1993). The reaction takes place at a catalyst surface with a reaction rate that is first order in the ethanol mole fraction, with a rate constant $k_r = 10$ mol/(m²-s-mole fraction). Reactant and products diffuse through a gas film 1–mm thick. The bulk gas–phase composition is 60 mol% ethanol (1), 20% acetaldehyde (2), and 20% hydrogen (3). Estimate the overall rate of reaction at steady-state. The Maxwell-Stefan diffusion coefficients are

$$\mathscr{D}_{12}= 72 \text{ mm}^2 / \text{s}$$

$$\mathscr{D}_{13}= 230 \text{ mm}^2 / \text{s}$$

$$\mathscr{D}_{23}= 230 \text{ mm}^2 / \text{s}$$

Solution
The flux of ethanol is determined by the rate of chemical reaction at the catalyst surface. Furthermore, for every mole of ethanol that diffuses to the catalyst surface, one mole of acetaldehyde and one mole of hydrogen diffuse in the opposite direction. That is,

$$N_1 = k_r y_{1\delta}$$
$$N_2 = N_3 = -N_1 \tag{1-86}$$

Equations (1-85) also apply to this problem; however, there are three additional unknown quantities: N_3, y_{15}, and y_{25}. The mole fractions at the interface cannot be specified arbitrarily, but are fixed by the reaction stoichiometry. Thus, the interface mole fractions must be determined simultaneously with the molar fluxes. Therefore, there are 11 unknown quantities: 3 molar fluxes, and 8 molar fractions. Three additional relations, equations (1-86), must be incorporated into the "solve block":

$$N1 = k_r \cdot y_{15}$$
$$N2 = -N1$$
$$N3 = -N1$$

The "solve block" feature of Mathcad requires initial estimates of all the unknown values. We shall take the following values:

$$N_1 = F_{12} = \frac{c\mathscr{D}_{12}}{\delta} = 1.601 \text{ mole} \cdot \text{m}^{-2} \cdot \text{s}^{-1}$$

$$N_3 = N_2 = -N_1 = -1.601 \text{ mole} \cdot \text{m}^{-2} \cdot \text{s}^{-1}$$

$$y_{11} = 0.60 \quad y_{12} = 0.50 \quad y_{13} = 0.40 \quad y_{14} = 0.30 \quad y_{15} = 0.20$$
$$y_{21} = 0.20 \quad y_{22} = 0.30 \quad y_{23} = 0.40 \quad y_{24} = 0.50 \quad y_{25} = 0.60$$

The unknown variables are identified in the "Find" statement of the solve block, in this case:

$$\text{Find } (N1, N2, N3, y1_2, y1_3, y1_4, y1_5, y2_2, y2_3, y2_4, y2_5)$$

The converged solution predicts an overall rate of reaction $N_1 = 0.888$ mole of ethanol/m² surface area-s, and an interface composition of 8.8 mole% ethanol, 58.3% acetaldehyde, and 32.9% hydrogen. More details of the solution, including the complete concentration profiles, can be found in Appendix C-2.

1.4.3 Steady-State Molecular Diffusion in Liquids

Your objectives in studying this section are to be able to:

1. Calculate molar fluxes during steady-state molecular, one-dimensional diffusion in liquids.
2. Calculate molar fluxes for steady-state diffusion of A through stagnant B, and for equimolar counterdiffusion, both in the liquid phase.

The integration of equation (1-70) to put it in the form of equation (1-74) requires the assumption that D_{AB} and c are constant. This is satisfactory for gas mixtures but not in the case of liquids, where both may vary considerably with concentration. Nevertheless, it is customary to use equation (1-74) to predict molecular diffusion in liquids with an average c and the best average value of D_{AB} available. Equation (1-74) is conveniently written for dilute solutions as

$$N_A = \Psi_A \frac{D_{AB}^0}{z} \left(\frac{\rho}{M} \right)_{av} \ln \left[\frac{\Psi_A - x_{A_2}}{\Psi_A - x_{A_1}} \right] \tag{1-87}$$

where ρ and M are the solution density and molecular weight respectively. As for gases, the value of Ψ_A must be established by the prevailing circumstances. For the most commonly occurring cases, we have, as for gases:

1. Steady-state diffusion of A through stagnant B
   For this case, $N_A = \text{const}$, $N_B = 0$, $\Psi_A = 1$, then equation (1-87) becomes

$$N_A = \frac{D_{AB}^0}{z x_{BM}} \left(\frac{\rho}{M} \right)_{av} \left(x_{A_1} - x_{A_2} \right) \tag{1-88}$$

   where

$$x_{BM} = \frac{x_{B_2} - x_{B_1}}{\ln\left(x_{B_2} / x_{B_1}\right)} \tag{1-89}$$

2. Steady-state equimolar counterdiffusion
   For this case, $N_A = -N_B$ and

$$N_A = \frac{D_{AB}^0}{z} \left(\frac{\rho}{M}\right)_{av} \left(x_{A_1} - x_{A_2}\right) \tag{1-90}$$

Example 1.17 Steady-State Molecular Diffusion in Liquids

A crystal of chalcanthite ($CuSO_4 \cdot 5H_2O$) dissolves in a large tank of pure water at 273 K. Estimate the rate at which the crystal dissolves by calculating the flux of $CuSO_4$ from the crystal surface to the bulk solution. Assume that molecular diffusion occurs through a liquid film uniformly 0.01 mm thick surrounding the crystal. At the inner side of the film—adjacent to the crystal surface—the solution is saturated with $CuSO_4$, while at the outer side of the film the solution is virtually pure water. The solubility of chalcanthite in water at 273 K is 24.3 g of crystal/100 g of water and the density of the corresponding saturated solution is 1,140 kg/m^3 (Perry and Chilton, 1973). The diffusivity of $CuSO_4$ in dilute aqueous solution at 273 K can be estimated as 3.6×10^{-10} m^2/s (see Problem 1.20). The density of pure liquid water at 273 K is 999.8 kg/m^3.

Solution
The crystal dissolution process may be represented by the equation

$$CuSO_4 \cdot 5H_2O_{(s)} \rightarrow CuSO_{4(aq)} + 5H_2O_{(l)}$$

Accordingly, for each mole of chalcanthite (molecular weight 249.71) that dissolves, one mole of $CuSO_4$ (molecular weight 159.63) and 5 moles of hydration water diffuse through the liquid film from the surface of the crystal to the bulk of the liquid phase. This is shown schematically in Figure 1.10 (substance A is $CuSO_4$ and substance B is water). Notice that both fluxes are in the same direction; therefore they are both positive and $N_B = 5N_A$. From equation (1-73), $\Psi_A = 1/6 = 0.167$.

Next, we calculate the mole fraction of component A at the inner side of the film (x_{A1}), a saturated solution of chalcanthite in water at 273 K. The solubility of the salt under these conditions is 24.3 g/100 g H_2O:

Basis: 100 g of H_2O (24.3 g of $CuSO_4 \cdot 5H_2O$)

The mass of $CuSO_4$ in 24.3 g of the crystal is (24.3)(159.63)/(249.71) = 15.53 g. The mass of hydration water in the crystal is 24.3 − 15.53 = 8.77 g. Then, the total mass of water is 100 + 8.77 = 108.77 g. Therefore,

$$x_{A_1} = \frac{\dfrac{15.53}{159.63}}{\dfrac{15.53}{159.63} + \dfrac{108.77}{18}} = 0.0158$$

The other end of the film is virtually pure water, therefore,

$$x_{A_2} \cong 0.0$$

Next, we calculate the film average molar density. At point 1, the average molecular weight is $M_1 = (0.0158)(159.63) + (1 - 0.0158)(18) = 20.24$ kg/kmol. The corresponding molar density is $\rho_1/M_1 = 1{,}140/20.24 = 56.32$ kmol/m^3. At point 2, the molar density is $\rho_2/M_2 = 999.8/18 = 55.54$ kmol/m^3. Then

$$\left(\frac{\rho}{M}\right)_{av} = \frac{56.32 + 55.54}{2} = 55.93 \ \text{kmol/m}^3$$

Substituting in equation (1-87):

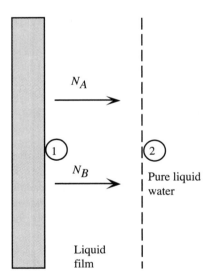

Crystal
surface

Figure 1.10 Schematic diagram for the crystal dissolution process of Example 1.17.

$$N_A = \frac{0.167 \times 3.6 \times 10^{-10} \times 55.93 \times \ln\left[\dfrac{0.167 - 0}{0.167 - 0.0158}\right]}{0.01 \times 10^{-3}}$$

$$= 3.342 \times 10^{-5} \frac{kmol}{m^2\text{-}s}$$

The rate of dissolution of the crystal is $(3.342 \times 10^{-5})(249.71)(3,600) = 30$ kg/m^2 of crystal surface area-hr.

1.5 ANALOGIES AMONG MOLECULAR MOMENTUM, HEAT, AND MASS TRANSFER

Your objectives in studying this section are to be able to:

1. Identify the analogies among molecular momentum, heat, and mass transfer for the simple case of fluid flow past a flat plate.
2. Define and interpret the following dimensionless numbers: Schmidt, Prandtl, and Lewis.

In the flow of a fluid past a phase boundary, there will be a velocity gradient within the fluid which results in a transfer of momentum through it. In some cases, there is also a transfer of heat by virtue of a temperature gradient. The processes of momentum, heat, and mass transfer under these conditions are intimately related, and it is useful to consider at this point the analogies among them.

Consider the velocity profile for the case of a fluid flowing past a stationary flat plate. Since the velocity at the solid surface is zero, there must necessarily be a sublayer adjacent to the surface where the flow is predominantly laminar. Within this region, the shearing stress τ required to maintain the velocity gradient is given by

$$\tau = -\mu \frac{du}{dz} \qquad (1\text{-}91)$$

where u is the fluid velocity parallel to the surface, and z is measured as increasing toward the surface. This can be written as

$$\tau = -\frac{\mu}{\rho} \frac{d(u\rho)}{dz} = -v \frac{d(u\rho)}{dz} \qquad (1\text{-}92)$$

where v is the kinematic viscosity, μ/ρ, also known as the momentum diffusivity.

The kinematic viscosity has the same dimensions as the mass diffusivity, length^2/time, while the quantity $u\rho$ can be interpreted as a volumetric momentum concentration. The shearing stress τ may also be interpreted as a viscous momentum flux towards the solid surface. Equation (1-92) is, therefore, a rate equation analogous to Fick's first law, equation (1-40). The Schmidt number is defined as the dimensionless ratio of the momentum and mass diffusivities:

$$Sc = \frac{\mu}{\rho D_{AB}} \tag{1-93}$$

When a temperature gradient exists between the fluid and the plate, the rate of heat transfer in the laminar region is

$$q = -k\frac{dT}{dz} \tag{1-94}$$

where k is the thermal conductivity of the fluid. This can also be written as

$$q = -\frac{k}{C_p\rho}\frac{d(TC_p\rho)}{dz} = -\alpha\frac{d(TC_p\rho)}{dz} \tag{1-95}$$

where C_p is the specific heat at constant pressure. The quantity $TC_p\rho$ may be looked upon as a volumetric thermal concentration, and $\alpha = k/C_p\rho$ is the thermal diffusivity, which, like the momentum and mass diffusivities, has dimensions of length^2/time. Equation (1-95) is therefore a rate equation analogous to the corresponding equations for momentum and mass transfer.

The Prandtl number is defined as the dimensionless ratio of the momentum and thermal diffusivities:

$$Pr = \frac{\nu}{\alpha} = \frac{C_p\mu}{k} \tag{1-96}$$

A third dimensionless group, the Lewis number, is formed by dividing the thermal by the mass diffusivity:

$$Le = \frac{\alpha}{D_{AB}} = \frac{Sc}{Pr} \tag{1-97}$$

The Lewis number plays an important role in problems of simultaneous heat and mass transfer, such as humidification operations.

An elementary consideration of the processes of momenum, heat, and mass transfer leads to the conclusion that in certain simplified situations there is a direct analogy between them. In general, however, modification of the simple analogy is necessary when mass and momentum transfer occur simultaneously. Nevertheless, even the limited analogies which exist are put to important practical use, as will be seen in Chapter 2.

PROBLEMS

The problems at the end of each chapter have been grouped into four classes (designated by a superscript after the problem number).

Class a: Illustrates direct numerical application of the formulas in the text.
Class b: Requires elementary analysis of physical situations, based on the subject material in the chapter.
Class c: Requires somewhat more mature analysis.
Class d: Requires computer solution.

1.1^a. Concentration of a gas mixture.

A mixture of noble gases (helium, argon, krypton, and xenon) is at a total pressure of 200 kPa and a temperature of 400 K. If the mixture has equal mole fractions of each of the gases, determine:

(a) The composition of the mixture in terms of mass fractions.

(b) The average molecular weight of the mixture.

(c) The total molar concentration.

(d) The mass density.

Answer: 3.9 kg/m^3

1.2^a. Concentration of a liquid solution fed to a distillation column.

A solution of carbon tetrachloride and carbon disulfide containing 50% by weight of each is to be continuously distilled at the rate of 4,000 kg/h.

(a) Determine the concentration of the mixture in terms of mole fractions.

(b) Determine the average molecular weight of the mixture.

(c) Calculate the feed rate in kmol/h.

Answer: 39.27 kmol/h

1.3ª. Concentration of liquified natural gas.

A sample of liquified natural gas, LNG, from Alaska has the following molar composition: 93.5% CH_4, 4.6% C_2H_6, 1.2% C_3H_8, and 0.7% CO_2. Calculate:

(a) Average molecular weight of the LNG mixture.

(b) Weight fraction of CH_4 in the mixture.

Answer: 87.11%

(c) The LNG is heated to 300 K and 140 kPa, and vaporizes completely. Estimate the density of the gas mixture under these conditions.

Answer: 0.967 kg/m^3

1.4ᵇ. Concentration of a flue gas.

A flue gas consists of carbon dioxide, oxygen, water vapor, and nitrogen. The *molar* fractions of CO_2 and O_2 in a sample of the gas are 12% and 6%, respectively. The *weight* fraction of H_2O in the gas is 6.17%. Estimate the density of this gas at 500 K and 110 kPa.

Answer: 0.772 kg/m^3

1.5ᵇ. Material balances around an ammonia gas absorber.

A gas stream flows at the rate of 10.0 m^3/s at 300 K and 102 kPa. It consists of an equimolar mixture of ammonia and air. The gas enters through the bottom of a packed-bed gas absorber, where it flows countercurrent to a stream of pure liquid water that absorbs 90% of all of the ammonia, and virtually no air. The absorber vessel is cylindrical with an internal diameter of 2.5 m.

(a) Neglecting the evaporation of water, calculate the ammonia mol fraction in the gas leaving the absorber.

(b) Calculate the outlet gas mass velocity (defined as mass flow rate per unit empty tube cross-sectional area).

Answer: 1.28 kg/m²-s

1.6ᵇ. Velocities and fluxes in a gas mixture.

A gas mixture at a total pressure of 150 kPa and 295 K contains 20%

H_2, 40% O_2, and 40% H_2O by volume. The absolute velocities of each species are –10 m/s, –2 m/s, and 12 m/s, respectively, all in the direction of the z-axis.

(a) Determine the mass average velocity, $\mathbf{v}$, and the molar average velocity, $\mathbf{V}$, for the mixture.

(b) Evaluate the four fluxes: $j_{O_2}, n_{O_2}, J_{O_2}, N_{O_2}$.

Answer: $j_{O_2} = -3.74$ kg/m²-s

1.7^b. Properties of air saturated with water vapor.

Air, stored in a 30-m³ container at 340 K and 101.3 kPa, is saturated with water vapor. Determine the following properties of the gas mixture:

(a) Mole fraction of water vapor.

(b) Average molecular weight of the mixture.

(c) Total mass contained in the tank.

(d) Mass of water vapor in the tank

Answer: 5.22 kg

1.8^c. Water balance around an industrial cooling tower.

The cooling water flow rate to the condensers of a big coal-fired power plant is 8,970 kg/s. The water enters the condensers at 29 °C and leaves at 45 °C. From the condensers, the water flows to a cooling tower where it is cooled down back to 29 °C by countercurrent contact with air (see Figure 1.11). The air enters the cooling tower at the rate of 6,500 kg/s of dry air, at a dry-bulb temperature of 30 °C and a humidity of 0.016 kg of water/kg of dry air. It leaves the cooling tower saturated with water vapor at 38 °C.

(a) Calculate the water losses by evaporation in the cooling tower.

(b) To account for water losses in the cooling tower, part of the effluent from a nearby municipal wastewater treatment plant will be used as makeup water. This makeup water contains 500 mg/L of dissolved solids. To avoid fouling of the condenser heat-transfer surfaces, the circulating water is to contain no more than 2,000 mg/L of dissolved solids. Therefore, a small amount of the circulating water must be deliberately discarded (blowdown). Windage losses from the tower are estimated at 0.2% of the recirculation rate. Estimate the makeup-water requirement.

Answer: 236 kg/s

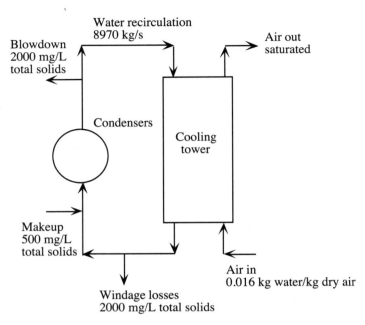

Figure 1.11 Flowsheet for Problem 1.8.

1.9[b.] Water balance around a soap dryer.

It is desired to dry 10 kg/min of soap continuously from 17% moisture by weight to 4% moisture in a countercurrent stream of hot air. The air enters the dryer at the rate of 30.0 m³/min at 350 K, 101.3 kPa, and initial water-vapor partial pressure of 1.6 kPa. The dryer operates at constant temperature and pressure.

(a) Calculate the moisture content of the entering air, in kg of water/kg of dry air.

(b) Calculate the flow rate of dry air, in kg/min.

(c) Calculate the water-vapor partial pressure and relative humidity in the air leaving the dryer.

Answer: 19.8% relative humidity

1.10[b]. Activated carbon adsorption; material balances.

A waste gas contains 0.3% toluene in air, and occupies a volume of 2,500 m³ at 298 K and 101.3 kPa. In an effort to reduce the toluene content of this gas, it is exposed to 100 kg of activated carbon, initially free of toluene. The system is allowed to reach equilibrium at constant temperature and pressure. Assuming that the air does not adsorb on the carbon, calculate the equilibrium concentration of toluene in the gaseous phase, and the amount of toluene adsorbed by the carbon. The adsorption equilibrium for this system is given by the Freundlich isotherm (USEPA, 1987):

$$W = 0.208(p*)^{0.11}$$

where W is the carbon equilibrium adsorptivity, in kg of toluene/kg of carbon, and $p*$ is the equilibrium toluene partial pressure, in Pa, and must be between 0.7 and 345 Pa.

1.11[b]. Activated carbon adsorption; material balances.

It is desired to adsorb 99.5% of the toluene originally present in the waste gas of Problem 1.10. Estimate how much activated carbon should be used if the system is allowed to reach equilibrium at constant temperature and pressure.

Answer: 129 kg

1.12[a, d]. Estimation of gas diffusivity by the Wilke-Lee equation.

Larson (1964) measured the diffusivity of chloroform in air at 298 K and 1 atm and reported its value as 0.093 cm²/s. Estimate the diffusion coefficient by the Wilke-Lee equation and compare it with the experimental value.

1.13[a, d]. Estimation of gas diffusivity by the Wilke-Lee equation.

(a) Estimate the diffusivity of naphthalene ($C_{10}H_8$) in air at 303 K and 1 bar. Compare it with the experimental value of 0.087 cm²/s reported in Appendix A. The normal boiling point of naphthalene is 491.1 K, and its critical volume is 413 cm³/mol.

Answer: Error = –20.1%

(b) Estimate the diffusivity of pyridine (C_5H_5N) in hydrogen at 318 K and 1 atm. Compare it with the experimental value of 0.437 cm²/s reported in Appendix A. The normal boiling point of pyridine is 388.4 K, and its critical volume is 254 cm^3/mol.

Answer: Error = −9.9%

(c) Estimate the diffusivity of aniline (C_6H_7N) in air at 273 K and 1 atm. Compare it with the experimental value of 0.061 cm^2/s (Guilliland, 1934). The normal boiling point of aniline is 457.6 K, and its critical volume is 274 cm^3/mol.

Answer: Error = 15.1%

1.14^d. Diffusivity of polar gases

If one or both components of a binary gas mixture are polar, a modified Lennard-Jones relation is often used. Brokaw (1969) has suggested an alternative method for this case. Equation (1-49) is still used, but the collision integral is now given by

$$\Omega_D = \Omega_D \left[\text{Eq. (1-44)} \right] + \frac{0.19\delta_{AB}^2}{T^*} \qquad (1\text{-}98)$$

where

$$\delta = \frac{1.94 \times 10^3 \mu_p^2}{V_b T_b} \qquad (1\text{-}99)$$

μ_p = dipole moment, debyes [1 debye = 3.162 × 10^{-25} (J-m^3)$^{1/2}$]

$$\frac{\varepsilon}{k} = 1.18\left(1 + 1.3\delta^2\right)T_b \qquad (1\text{-}100)$$

$$\sigma = \left(\frac{1.585 V_b}{1 + 1.3\delta^2}\right)^{1/3} \qquad (1\text{-}101)$$

$$\delta_{AB} = \left(\delta_A \delta_B\right)^{1/2}, \quad \sigma_{AB} = \left(\sigma_A \sigma_B\right)^{1/2}, \quad \frac{\varepsilon_{AB}}{k} = \left(\frac{\varepsilon_A}{k}\frac{\varepsilon_B}{k}\right)^{1/2} \qquad (1\text{-}102)$$

(a) Modify the Mathcad routine of Figure 1.3 to implement Brokaw's method. Use the function name

$$D_{ABp}(T, P, M_A, M_B, \mu_A, V_A, T_{bA}, \mu_B, V_B, T_{bB})$$

(b) Estimate the diffusion coefficient for a mixture of methyl chloride and sulfur dioxide at 1 bar and 323 K, and compare it to the experimental value of 0.078 cm^2/s. The data required to use Brokaw's relation are shown below (Reid, et al., 1987):

Parameter	Methyl chloride	Sulfur dioxide
T_b, K	249.1	263.2
V_b, cm^3/mol	50.6	43.8
μ_p, debyes	1.9	1.6
M	50.5	64.06

Answer: Error = 7.8%

1.15[d]. Diffusivity of polar gases

Evaluate the diffusion coefficient of hydrogen chloride in water at 373 K and 1 bar. The data required to use Brokaw's relation (see Problem 1.14) are shown below (Reid, et al., 1987):

Parameter	Hydrogen chloride	Water
T_b, K	188.1	373.2
V_b, cm^3/mol	30.6	18.9
μ_p, debyes	1.1	1.8
M	36.5	18

Answer: 0.266 cm^2/s

1.16[d]. Diffusivity of polar gases

Evaluate the diffusion coefficient of hydrogen sulfide in sulfur dioxide at 298 K and 1.5 bar. The data required to use Brokaw's relation (see Problem 1.14) are shown below (Reid, et al., 1987):

Parameter	Hydrogen sulfide	Sulfur dioxide
T_b, K	189.6	263.2
V_b, cm^3/mol	35.03	43.8
μ_p, debyes	0.9	1.6
M	34.08	64.06

Answer: 0.065 cm^2/s

1.17^{a,d}. Effective diffusivity in a multicomponent stagnant gas mixture.

Calculate the effective diffusivity of nitrogen through a stagnant gas mixture at 373 K and 1.5 bar. The mixture composition is:

O_2	15 mole %
CO	30%
CO_2	35%
N_2	20%

Answer: 0.192 cm^2/s

1.18^{a,d}. Mercury removal from flue gases by sorbent injection.

Mercury is considered for possible regulation in the electric power industry under Title III of the 1990 Clean Air Act Amendments. One promising approach for removing mercury from fossil-fired flue gas involves the direct injection of activated carbon into the gas. Meserole, et al. (1999) describe a theoretical model for estimating mercury removal by the sorbent injection process. An important parameter of the model is the effective diffusivity of mercuric chloride vapor traces in the flue gas. If the flue gas is at 1.013 bar and 408 K, and its composition (on a mercuric chloride-free basis) is 6% O_2, 12% CO_2, 7% H_2O, and 75% N_2, estimate the effective diffusivity of mercuric chloride in the flue gas. Assume that only the HgCl$_2$ is adsorbed by the activated carbon. Meserole et al. reported an effective diffusivity value of 0.22 cm^2/s.

Answer: 0.156 cm^2/s

1.19^a. Wilke-Chang method for liquid diffusivity.

Estimate the liquid diffusivity of carbon tetrachloride in dilute solution into ethanol at 298 K. Compare to the experimental value reported by Reid, et al. (1987) as 1.5×10^{-5} cm^2/s. The critical volume of carbon tetrachloride is 275.9 cm^3/mol. The viscosity of liquid ethanol at 298 K is 1.08 cP.

Answer: Error = –29.8%

1.20^b. Diffusion in electrolyte solutions.

When a salt dissociates in solution, ions rather than molecules diffuse.

In the absence of an electric potential, the diffusion of a single salt may be treated as molecular diffusion. For dilute solutions of a *single salt*, the diffusion coefficient is given by the Nernst-Haskell equation (Harned and Owen, 1950):

$$\mathcal{D}_{AB}^0 = \frac{RT\left[\left(1/n_+\right)+\left(1/n_-\right)\right]}{F^2\left[\left(1/\lambda_+^0\right)+\left(1/\lambda_-^0\right)\right]} \tag{1-103}$$

where $\mathcal{D}_{AB}^0$ = diffusivity at infinite dilute solution, based on molecular
             concentration, cm^2/s

R = gas constant, 8.314 J/mol-K

T = temperature, K

λ_+^0, λ_-^0 = limiting (zero concentration) ionic conductances,
              $(A/cm^2)(V/cm)(g\text{-equiv}/cm^3)$

n_+, n_- = valences of cation and anion, respectively

F   = Faraday's constant = 96,500 C/g-equiv

Values of limiting ionic conductances for some ionic species at 298 K are included in Table 1.3. If conductance values at other temperatures are needed, an approximate correction factor is $T/(334\mu_w)$, where μ_w is the viscosity of water at T, in cP.

(a) Estimate the diffusion coefficient at 298 K for a very dilute solution of HCl in water.

Answer: 3.33×10^{-5} cm^2/s

(b) Estimate the diffusion coefficient at 273 K for a very dilute solution of $CuSO_4$ in water. The viscosity of liquid water at 273 K is 1.79 cP.

Answer: 3.59×10^{-6} cm^2/s

1.21[a]. Oxygen diffusion in water: Hayduk and Minhas correlation.

Estimate the diffusion coefficient of oxygen in liquid water at 298 K. Use the Hayduk and Minhas correlation for solutes in aqueous solutions. At this temperature, the viscosity of water is 0.9 cP. The critical volume of oxygen is 73.4 cm^3/mol. The experimental value of this diffusivity was reported as 2.1×10^{-5} cm^2/s (Cussler, 1997).

Answer: Error = $- 8.1\%$

Table 1.3 Limiting Ionic Conductances in Water at 298

Anion	λ_-^0	Cation	λ_+^0
OH^-	197.6	H^+	349.8
Cl^-	76.3	Li^+	38.7
Br^-	78.3	Na^+	50.1
NO_3^-	71.4	NH_4^+	73.4
HCO_3^-	44.5	Mg^{2+}	53.1
$CH_3CO_2^-$	40.9	Ca^{2+}	59.5
SO_4^{2-}	80.0	Cu^{2+}	54.0
$C_2O_4^{2-}$	74.2	Zn^{2+}	53.0

Units: $(A/cm^2)(V/cm)(g\text{-equiv}/cm^3)$.
Source: Harned and Owen, 1950.

1.22[a, d]. Liquid diffusivity: Hayduk and Minhas correlation.

Estimate the diffusivity of carbon tetrachloride in a dilute solution in n-hexane at 298 K using the Hayduk and Minhas correlation for nonaqueous solutions. Compare the estimate to the reported value of 3.7×10^{-5} cm^2/s. The following data are available (Reid, et al., 1987):

Parameter	Carbon tetrachloride	n-Hexane
T_b, K	349.9	341.9
T_c, K	556.4	507.5
P_c, bars	45.6	30.1
V_c, cm^3/mol	275.9	370
μ, cP	—	0.3
M	153.82	86.2

Answer: Error = 8.1%

1.23[b]. Estimating molar volumes from liquid diffusion data.

The diffusivity of allyl alcohol (C_3H_6O) in dilute aqueous solution at 288 K is 0.9×10^{-5} cm^2/s (Reid, et al., 1987). Based on this result, and the Hayduk and Minhas correlation for aqueous solutions, estimate the molar volume of allyl alcohol at its normal boiling point. Compare to the result obtained using Table 1.2. The viscosity of water at 288 K is 1.15 cP.

1.24[b, d]. Concentration dependence of binary liquid diffusivities.

(a) Estimate the diffusivity of ethanol in water at 298 K when the mol fraction of ethanol in solution is 40%. Under these conditions (Hammond and Stokes, 1953):

$$\left(1 + \frac{\partial \ln \gamma_A}{\partial \ln x_A} \right)_{T,P} = 0.355$$

The experimental value reported by Hammond and Stokes (1953) is 0.42×10^{-5} cm^2/s. The critical volume of ethanol is 167.1 cm^3/mole; the viscosity of water at 298 K is 0.91 cP.

Answer: Error = 8.3%

(b) Estimate the diffusivity of acetone in water at 298 K when the mol fraction of acetone in solution is 35%. For this system at 298 K, the activity coefficient for acetone is given by Wilson equation (Smith, J. M., et al., 1996):

$$\ln \gamma_A = -\ln(x_A + x_B \Lambda_{12}) + x_B \left(\frac{\Lambda_{12}}{x_A + x_B \Lambda_{12}} - \frac{\Lambda_{21}}{x_B + x_A \Lambda_{21}} \right)$$

$$\Lambda_{12} = 0.149 , \quad \Lambda_{21} = 0.355 \quad @ \ 298 \ K$$

The critical volume for acetone is 209 cm^3/mol.

Answer: 0.273×10^{-5} cm^2/s

1.25[b, d]. Steady-state, one-dimensional, gas-phase flux calculation.

A flat plate of solid carbon is being burned in the presence of pure oxygen according to the reaction

$$2C_{(s)} + \tfrac{3}{2}O_{2(g)} \rightarrow CO_{(g)} + CO_{2(g)}$$

Molecular diffusion of gaseous reactant and products takes place through a gas film adjacent to the carbon surface; the thickness of this film is 1.0 mm. On the outside of the film, the gas concentration is 40% CO, 20% O_2, and 40% CO_2. The reaction at the surface may be assumed to be instantaneous, therefore, next to the carbon surface, there is virtually no oxygen.The temperature of the gas film is 600 K, and the pressure is 1 bar. Estimate the rate of combustion of the carbon, in kg/m^2-min, and the interface concentration.

Answer: 0.243 kg/m^2-min

1.26[b]. Steady-state, one-dimensional, liquid-phase flux calculation.

A crystal of Glauber's salt ($Na_2SO_4 \cdot 10H_2O$) dissolves in a large tank of pure water at 288 K. Estimate the rate at which the crystal dissolves by calculating the flux of Na_2SO_4 from the crystal surface to the bulk solution. Assume that molecular diffusion occurs through a liquid film 0.085 mm thick surrounding the crystal. At the inner side of the film—adjacent to the crystal surface—the solution is saturated with Na_2SO_4, while at the outer side of the film the solution is virtually pure water. The solubility of Glauber's salt in water at 288 K is 36 g of crystal/100 g of water and the density of the corresponding saturated solution is 1,240 kg/m^3 (Perry and Chilton, 1973). The diffusivity of Na_2SO_4 in dilute aqueous solution at 288 K can be estimated as suggested in Problem 1.20. The density of pure liquid water at 288 K is 999.8 kg/m^3; the viscosity is 1.153 cP.

Answer: 12.9 kg of crystal/m^2-hr

1.27[c, d]. Molecular diffusion through a gas–liquid interface.

Ammonia, NH_3, is being selectively removed from an air-NH_3 mixture by absorption into water. In this steady-state process, ammonia is transferred by molecular diffusion through a stagnant gas layer 5 mm thick and then through a stagnant water layer 0.1 mm thick. The concentration of ammonia at the outer boundary of the gas layer is 3.42 mol percent and the concentration at the lower boundary of the water layer is essentially zero.

The temperature of the system is 288 K and the total pressure is 1 atm. The diffusivity of ammonia in air under these conditions is 0.215 cm^2/s and in liquid water is 1.77×10^{-5} cm^2/s. Neglecting water evaporation, determine the rate of diffusion of ammonia, in kg/m^2-hr. Assume that the gas and liquid are in equilibrium at the interface.

The equilibrium data for ammonia over very dilute aqueous solution at 288 K and 1 atm can be represented by (Welty, et al., 1984)

$$y_A = 0.121 + 0.013 \ln x_A$$

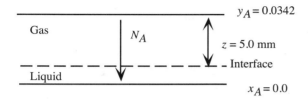

Hint: At steady-state, the ammonia rate of diffusion through the gas layer must equal the rate of diffusion through the liquid layer.

Answer: 0.053 kg/m²-hr

1.28[c]. Steady-state molecular diffusion in gases.

A mixture of ethanol and water vapor is being rectified in an adiabatic distillation column. The alcohol is vaporized and transferred from the liquid to the vapor phase. Water vapor condenses—enough to supply the latent heat of vaporization needed by the alcohol being evaporated—and is transferred from the vapor to the liquid phase. Both components diffuse through a gas film 0.1 mm thick. The temperature is 368 K and the pressure is 1 atm. The mole fraction of ethanol is 0.8 on one side of the film and 0.2 on the other side of the film. Calculate the rate of diffusion of ethanol and of water, in kg/m²-s. The latent heat of vaporization of the alcohol and water at 368 K can be estimated by the Pitzer acentric factor correlation (Reid, et al., 1987)

$$\frac{\Delta H_v}{RT_c} = 7.08(1 - T_r)^{0.354} + 10.95\omega(1 - T_r)^{0.456}$$

$$0.5 \le T_r = T / T_c \le 1.0$$

where ω is the acentric factor.

Answer: 0.17 kg ethanol/m²-s

1.29[a, d]. Analogy between molecular heat and mass transfer.

It has been observed that for the system air-water vapor at near-ambient conditions, Le = 1.0 (Treybal, 1980). This observation, called the *Lewis rela-*

tion, has profound implications in humidification operations, as will be seen later. Based on the Lewis relation, estimate the diffusivity of water vapor in air at 300 K and 1 atm. Compare your result with the value predicted by the Wilke-Lee equation. For air at 300 K and 1 atm: $C_p = 1.01$ kJ/kg-K, $k = 0.0262$ W/m-K, $\mu = 1.846 \times 10^{-5}$ kg/m-s, and $\rho = 1.18$ kg/m^3.

1.30[b, d]. Steady-state molecular diffusion in gases.

Water evaporating from a pond at 298 K does so by molecular diffusion across an air film 1.5 mm thick. If the relative humidity of the air at the outer edge of the film is 20%, and the total pressure is 1 bar, estimate the drop in the water level per day, assuming that conditions in the film remain constant. The vapor pressure of water as a function of temperature can be accurately estimated from the Wagner equation (Reid, et al., 1987)

$$\ln\left(\frac{P_W}{P_c}\right) = \frac{-7.7645\tau + 1.4584\tau^{1.5} - 2.7758\tau^3 - 1.2330\tau^6}{T_r}$$

where $\tau = 1 - T_r$

$$T_r = \frac{T}{T_c}$$

P_W = water vapor pressure

Answer: 2.81 cm

1.31[b, d]. Steady-state molecular diffusion in a ternary gas system.

Calculate the fluxes and concentration profiles for the ternary system hydrogen (1), nitrogen (2), and carbon dioxide (3) under the following conditions. The temperature is 308 K and the pressure is 1 atm. The diffusion path length is 86 mm. At one end of the diffusion path the concentration is 20 mole% H_2, 40% N_2, 40% CO_2; at the other end, the concentration is 50% H_2, 20% N_2, 30% CO_2. The total molar flux is zero, $N = 0$. The MS diffusion coefficients are $\mathcal{D}_{12} = 83.8$ mm^2/s, $\mathcal{D}_{13} = 68.0$ mm^2/s, $\mathcal{D}_{23} = 16.8$ mm^2/s.

Answer: $N_1 = -0.0104$ mole/m^2-s

REFERENCES

Bird, R. B., W. E. Stewart, and E. N. Lightfoot, *Transport Phenomena*, Wiley New York (1960).

Brock, J. R. and R. B. Bird, *AIChE J.*, **1**, 174 (1955).

Brokaw, E., *Ind. Eng. Chem. Process Design Develop.*, **8**:240, (1969)

Carty, R., and T. Schrodt, *Ind. Eng. Chem., Fundam.*, **14**, 276 (1975).

Cussler E. L., *Diffusion,* 2nd ed., Cambridge University Press, Cambridge, UK, 1997.

de Groot, S. R., *Thermodynamics of Irreversible Processes*, North-Holland, Amsterdam (1951).

Fick, A., *Ann. Physik.*, **94**, 59 (1855).

Guilliland, E. R., *Ind. Eng. Chem.*, **26**:681 (1934).

Hammond, B. R., and R. H. Stokes, *Trans. Faraday Soc.*, **49**, 890 (1953).

Harned, H. S., and B. B. Owen, "The Physical Chemistry of Electrolytic Solutions," *ACS Monogr.* 95 (1950).

Hayduk, W., and B. S. Minhas, *Can. J. Chem. Eng.*, **60**, 295 (1982).

Hirschfelder, J. O., R. B. Bird, and E. L. Spotz, *Chem. Revs.*, **44**, 205–231 (1949).

Hirschfelder, J. O., C. F. Curtiss, and R. B. Bird, *Molecular Theory of Gases and Liquids*, Wiley, New York (1954).

Larson, E. M., MS thesis, Oregon State University (1964).

Lugg, G. A., *Anal. Chem.*, **40**, 1072 (1968).

Meserole, et al., *J. Air & Waste Manage. Assoc.*, **49**:694 (1999).

Neufield, P. D., A. R. Janzen, and R. A. Aziz, *J. Chem. Phys.*, **57**, 1100 (1972).

Perkins, L. R., and C. J. Geankoplis, *Chem. Eng. Sci.*, **24**, 1035 (1969).

Perry, R. H., and C. H. Chilton (eds.), *Chemical Engineers' Handbook*, 5th ed., McGraw-Hill, New York (1973).

Reid, R. C., J. M. Prausnitz, and B. E. Poling, *The Properties of Gases and Liquids*, 4th ed., McGraw-Hill, Boston (1987).

Rice, R. G., and D. D. Do, *Applied Mathematics and Modeling for Chemical Engineers,* Wiley, New York (1995).

Smith, J. M., et al., *Introduction to Chemical Engineering Thermodynamics,* 5th ed, McGraw-Hill, New York (1996).

Taylor, R. and R. Krishna, *Multicomponent Mass Transfer,* Wiley, New York (1993).

Takahashi, M., Y. Kobayashi, and H. Takeuchi, *J. Chem. Eng. Data*, **27**, 328 (1982).

Treybal, R. E., *Mass-Transfer Operations*, 3rd ed., McGraw-Hill, New York, (1980).

Tyn, M. T., and W. F. Calus, *Processing*, **21** (4), 16 (1975).

USEPA, *EAB Control Cost Manual*, 3rd. ed., USEPA., Research Triangle Park, NC (1987).

Welty, J. R., C. E. Wicks, and R. E. Wilson, *Fundamentals of Momentum, Heat, and Mass Transfer*, 3rd ed., Wiley, New York (1984).

Wilke, C. R., and P. Chang, *AIChE J.*, **1**, 264 (1955).

Wilke, C. R., and C. Y. Lee, *Ind. Eng. Chem.*, **47**, 1253 (1955).

2

Convective
Mass Transfer

2.1 INTRODUCTION

Mass transfer by convection involves the transport of material between a boundary surface and a moving fluid, or between two relatively immiscible moving fluids. This mode of transfer depends both on the transport properties and on the dynamic characteristics of the flowing fluid. When a fluid flows past a solid surface under conditions such that turbulence generally prevails, there is a region close to the surface where the flow is predominantly laminar, and the fluid particles immediately adjacent to the solid boundary are at rest. Molecular diffusion is responsible for mass transfer through the stagnant and laminar flowing fluid layers. The controlling resistance to convective mass transfer is often the result of this "film" of fluid. However, under most convective conditions, this film is extremely thin and its thickness is virtually impossible to measure or predict theoretically. Therefore, the mass transfer rates cannot be calculated based on the concepts of molecular diffusion alone.

With increasing distance from the surface, the character of the flow gradually changes, becoming increasingly turbulent, until in the outermost region of the fluid fully turbulent conditions prevail. In the turbulent region, particles of fluid no longer flow in the orderly manner found in the laminar sublayer. Instead, relatively large portions of the fluid, called *eddies*, move rapidly from one position to another with an appreciable component of their velocity in the direction perpendicular to the surface past which the fluid is flowing. These eddies contribute considerably to the mass-transfer process. Because the eddy motion is rapid, mass transfer in the turbulent region is much more rapid than it would be under laminar flow conditions. This situation cannot be modeled in terms of Fick's law. Instead, it is explained in terms of a *mass-transfer coefficient*, an approximate engineering idea that simplifies the analysis of a very complex problem.

66

2.2 MASS-TRANSFER COEFFICIENTS

Your objectives in studying this section are to be able to:

1. Explain the concept of a mass-transfer coefficient for turbulent diffusion by analogy with molecular diffusion.
2. Define and use special mass-transfer coefficients for diffusion of A through stagnant B, and for equimolar counterdiffusion.

The mechanism of the flow process involving the movement of the eddies in the turbulent region is not thoroughly understood. On the other hand, the mechanism of molecular diffusion, at least for gases, is fairly well known, since it can be described in terms of a kinetic theory to give results which agree well with experience. It is therefore natural to attempt to describe the rate of mass transfer through the various regions from the surface to the turbulent zone in the same manner found useful for molecular diffusion. Thus, the cD_{AB}/z of equation (1-74), which is characteristic of molecular diffusion, is replaced by F, a mass-transfer coefficient (Treybal, 1980). Then,

$$N_A = \Psi_A F \ln\left[\frac{\Psi_A - c_{A2}/c}{\Psi_A - c_{A1}/c}\right] \qquad (2\text{-}1)$$

where c_A/c is the mole-fraction concentration, x_A for liquids, y_A for gases. As in molecular diffusion, Ψ_A is ordinarily established by nondiffusional considerations.

Since the surface through which the transfer takes place may not be plane, so that the diffusion path in the fluid may be of variable cross section, N_A is defined as the flux at the phase interface or boundary where substance A enters or leaves the phase for which F has been defined. When c_{A1} is at the beginning of the transfer path and c_{A2} is at the end, N_A is positive. In any case, one of these concentrations will be at the phase boundary. The manner of defining the concentration of A in the fluid will influence the value of F, and it must be clearly established when the mass-transfer coefficient is defined for a particular situation.

The two situations noted in Chapter 1, equimolar counterdiffusion and diffusion of A through stagnant B, occur so frequently that special mass-transfer coefficients are usually defined for them. These are defined by equations of the form

Flux = (coefficient)(concentration difference)

Since concentration may be defined in a number of ways and standards have not been established, there is a variety of coefficients that can be defined for each situation.

2.2.1 Diffusion of A Through Stagnant B ($N_B = 0$, $\Psi_A = 1$)

In this case, equation (2-1) becomes:

$$N_A = F_G \ln\left[\frac{1-y_{A2}}{1-y_{A1}}\right] \quad \text{for gases} \tag{2-2}$$

$$N_A = F_L \ln\left[\frac{1-x_{A2}}{1-x_{A1}}\right] \quad \text{for liquids} \tag{2-3}$$

Equations (2-2) and (2-3) can be manipulated algebraically in a manner similar to that applied to equations (1-77) and (1-78) to yield

$$N_A = \frac{F_G}{y_{B,M}}(y_{A1} - y_{A2}) = k_y(y_{A1} - y_{A2}) \tag{2-4}$$

$$N_A = \frac{F_L}{x_{B,M}}(x_{A1} - x_{A2}) = k_x(x_{A1} - x_{A2}) \tag{2-5}$$

where

$$y_{B,M} = \frac{y_{B2} - y_{B1}}{\ln\left(\frac{y_{B2}}{y_{B1}}\right)} \qquad x_{B,M} = \frac{x_{B2} - x_{B1}}{\ln\left(\frac{x_{B2}}{x_{B1}}\right)}$$

$$k_y = \frac{F_G}{y_{B,M}} \qquad k_x = \frac{F_L}{x_{B,M}} \tag{2-6}$$

Since concentrations may be defined in a number of equivalent ways, other definitions of mass-transfer coefficients for this case ($N_B = 0$; dilute solutions) are frequently used, such as

$$N_A = k_G(p_{A1} - p_{A2}) = k_c(c_{A1} - c_{A2}) \tag{2-7}$$

$$N_A = k_L(c_{A1} - c_{A2}) \tag{2-8}$$

where it is easily shown that

$$F_G = k_G p_{B,M} = k_c \frac{p_{B,M}}{RT} = k_y \frac{p_{B,M}}{P} \tag{2-9}$$

$$F_L = k_L x_{B,M} c = k_x x_{B,M} \tag{2-10}$$

Notice that in the case of diffusion of A through stagnant B there is a bulk-motion contribution to the total flux that results in a logarithmic form of the concentration driving force. For that reason, the newly defined k-type mass-transfer coefficients are not constant, but usually depend on the concentration driving force. For very dilute solutions, the bulk-motion contibution is negligible and the driving force becomes approximately linear. In mathematical terms, this follows from the fact that for very dilute solutions $y_{B,M} \cong x_{B,M} \cong 1.0$.

> Therefore, for diffusion of A through stagnant B, equations (2-4), (2-5), (2-7), and (2-8) in terms of the k-type mass-transfer coefficients are recommended *only for very dilute solutions*, otherwise equations (2-2) and (2-3) should be used.

Example 2.1 Mass-Transfer Coefficients in a Blood Oxygenator (Cussler, 1997)

Blood oxygenators are used to replace the human lungs during open-heart surgery. To improve oxygenator design, you are studying mass transfer of oxygen into water at 310 K in one specific blood oxygenator. From published correlations of mass-transfer coefficients, you expect that the coefficient based on the oxygen concentration difference in the water is 3.3×10^{-5} m/s. Calculate the corresponding mass-transfer coefficient based on the mole fraction of oxygen in the liquid.

Solution

Neglecting water evaporation, in the liquid phase this is a case of diffusion of A (oxygen) through stagnant B (water). Since the solubility of oxygen in water at 310 K is extremely low, we are dealing with dilute solutions. Therefore, the use of k-type coefficients is appropriate. From the information given (coefficient based on the oxygen concentration difference in the water), $k_L = 3.3 \times 10^{-5}$ m/s. You are asked to calculate the coefficient based on the mole fraction of oxygen in the liquid, or k_x. From equation (2.10)

$$k_x = k_L c \qquad\qquad (2\text{-}11)$$

Since we are dealing with very dilute solutions

$$c = \frac{\rho}{M_{av}} \cong \frac{\rho_B}{M_B} = \frac{993}{18} = 55.2 \text{ kmol/m}^3$$

Therefore,

$$k_x = 3.3 \times 10^{-5} \times 55.2 = 1.82 \times 10^{-3} \frac{\text{kmole}}{\text{m}^2\text{-s}}$$

Example 2.2 Mass-Transfer Coefficient in a Gas Absorber

A gas absorber is used to remove ammonia from air by scrubbing the gas mixture with water at 300 K and 1 atm. At a certain point in the absorber, the ammonia mole fraction in the bulk of the gas phase is 0.80, while the corresponding interfacial ammonia gas-phase concentration is 0.732. The ammonia flux at that point is measured as 4.3×10^{-4} kmol/m^2-s. Neglecting the evaporation of water, calculate the mass-transfer coefficient in the gas phase at that point in the equipment.

Solution

Neglecting water evaporation and the solubility of air in water, in the gas phase this is a case of diffusion of A (ammonia) through stagnant B (air). However, because of the high ammonia concentration, k-type coefficients should not be used; instead, use equation (2-2) to solve for F_G. In this case, since the ammonia flux is from the bulk of the gas phase to the interface, $y_{A1} = 0.8, y_{A2} = 0.732$.

$$F_G = \frac{4.3 \times 10^{-4}}{\ln[(1-0.732)/(1-0.8)]} = 1.47 \times 10^{-4} \, \text{kmol/m}^2\text{-s}$$

2.2.2 Equimolar Counterdiffusion ($N_B = -N_A$, Ψ_A = undefined)

In this case, there is no bulk-motion contribution to the flux, and the flux is related linearly to the concentration difference driving force. Special k'-type mass-transfer coefficients are defined specifically for equimolar counterdiffusion as follows:

$$N_A = k'_G(p_{A1} - p_{A2}) = k'_c(c_{A1} - c_{A2}) = k'_y(y_{A1} - y_{A2})$$
$$= F_G(y_{A1} - y_{A2}) \quad (2\text{-}12)$$

$$N_A = k'_L(c_{A1} - c_{A2}) = k'_x(x_{A1} - x_{A2})$$
$$= F_L(x_{A1} - x_{A2}) \quad (2\text{-}13)$$

Equations (2-12) and (2-13) are always valid for equimolar counterdiffusion, regardless of whether the solutions are dilute or concentrated. It is easily shown that the various mass-transfer coefficients defined for this particular case are related through:

$$F_G = k'_G P = k'_c \frac{P}{RT} = k'_y \quad (2\text{-}14)$$

$$F_L = k'_L c = k'_x \quad (2\text{-}15)$$

Example 2.3 Mass-Transfer Coefficient in a Packed-Bed Distillation Column

A packed-bed distillation column is used to adiabatically separate a mixture of methanol and water at a total pressure of 1 atm. Methanol—the more volatile of the two components—diffuses from the liquid phase toward the vapor phase, while water diffuses in the opposite direction. Assuming that the molar latent heat of vaporization of both components is similar, this process is usually modeled as one of equimolar counterdiffusion. At a point in the column, the mass-transfer coefficient is estimated as 1.62×10^{-5} kmol/m^2-s-kPa. The gas-phase methanol mole fraction at the interface is 0.707, while at the bulk of the gas it is 0.656. Estimate the methanol flux at that point.

Solution

Equimolar counterdiffusion can be assumed in this case (as will be shown in a later chapter, this is the basis of the *McCabe-Thiele method* of analysis of distillation columns). Methanol diffuses from the interface towards the bulk of the gas phase, therefore, $y_{A1} = 0.707$ and $y_{A2} = 0.656$. Since they are not limited to dilute solutions, k'-type mass-transfer coefficients may be used to estimate the methanol flux. From the units given, it can be inferred that the coefficient given in the problem statement is k'_G. From equations (2-14) and (2-12),

$$k'_y = k'_G P = \left(1.62 \times 10^{-5}\right)(101.3) = 1.64 \times 10^{-3} \text{ kmol / m}^2-\text{s}$$

$$N_A = 1.64 \times 10^{-3}(0.707 - 0.656) = 8.36 \times 10^{-5} \text{ kmol / m}^2\text{-s}$$

2.3 DIMENSIONAL ANALYSIS

Your objectives in studying this section are to be able to:

1. Explain the concept and importance of dimensional analysis in correlating experimental data on convective mass-transfer coefficients.
2. Use the Buckingham method to determine the dimensionless groups significant to a given mass-transfer problem.

Most practically useful mass-transfer situations involve turbulent flow, and for these it is generally not possible to compute mass-transfer coefficients from theoretical considerations. Instead, we must rely principally on experimental data. The data are limited in scope, however, with respect to circumstances and situations as well as to range of fluid properties. Therefore, it is important to be able to extend their applicability to conditions not covered experimentally and to draw upon knowledge of other transport processes (of heat, particularly) for help. A very useful procedure to this end is *dimensional analysis*.

In dimensional analysis, the significant variables in a given situation are grouped into dimensionless parameters which are less numerous than the original variables. Such a procedure is very helpful in experimental work in which the very number of significant variables presents an imposing task of correlation. By combining the variables into a smaller number of dimensionless parameters, the work of experimental data reduction is considerably reduced.

2.3.1 The Buckingham Method

Dimensional analysis predicts the various dimensionless parameters which are helpful in correlating experimental data. Certain dimensions must be established as fundamental, with all others expressible in terms of these. One of these fundamental dimensions is length, symbolized L. Thus, area and volume may dimensionally be expressed as L^2 and L^3, respectively. A second fundamental dimension is time, symbolized t. Velocity and acceleration may de expressed as L/t and L/t^2, respectively. Another fundamental dimension is mass, symbolized M. The mole is included in M. An example of a quantity whose dimensional expression involves mass is the density (mass or molar) which would be expressed as M/L^3.

If the differential equation describing a given situation is known, then dimensional homogeneity requires that each term in the equation have the same units. The ratio of one term in the equation to another must then, of necessity, be dimensionless. With knowledge of the physical meaning of the various terms in the equation we are then able to give some physical interpretation to the dimensionless parameters thus formed. A more general situation in which dimensional analysis may be profitably employed is one in which there is no governing differential equation which clearly applies. In such cases, the *Buckingham method* is used.

The initial step in applying the Buckingham method requires the listing of the variables significant to a given problem. It is then necessary to determine the number of dimensionless parameters into which the variables may be combined. This number may be determined using the *Buckingham pi theorem*, which states (Buckingham, 1914):

> The number of dimensionless groups used to describe a situation, i, involving n variables is equal to $n - r$, where r is the rank of the dimensional matrix of the variables.

Thus,

$$i = n - r \qquad (2\text{-}16)$$

The dimensional matrix is simply the matrix formed by tabulating the exponents of the fundamental dimensions M, L, and t, which appear in each of the variables involved. The rank of a matrix is the number of rows in the largest nonzero determinant which can be formed from it. An example of the evaluation of r and i, as well as the application of the Buckingham method, follows.

Example 2.4 Mass Transfer Into a Dilute Stream Flowing Under Forced Convection in a Circular Conduit ($N_B = 0$).

Consider the transfer of mass from the walls of a circular conduit to a dilute stream flowing through the conduit. The transfer of A through stagnant B is a result of the concentration driving force, $c_{A1} - c_{A2}$. Use the Buckingham method to determine the dimensionless groups formed from the variables significant to this problem.

Solution
The first step is to construct a table of the significant variables in the problem and their dimensions. For this case, the important variables, their symbols, and their dimensional representations are listed below:

Variable	Symbol	Dimensions
tube diameter	D	L
fluid density	ρ	M/L^3
fluid viscosity	μ	M/Lt
fluid velocity	v	L/t
mass diffusivity	D_{AB}	L^2/t
mass-transfer coefficient	k_c	L/t

The above variables include terms descriptive of the system geometry, flow, fluid properties, and the quantity which is of primary interest, k_c. To determine the number of dimensionless parameters to be formed, we must know the rank, r, of the dimensional matrix. The matrix is formed from the following tabulation:

	k_c	v	ρ	μ	D_{AB}	D
M	0	0	1	1	0	0
L	1	1	-3	-1	2	1
t	-1	-1	0	-1	-1	0

The numbers in the table represent the exponent of M, L, and t in the dimensional expression of each of the six variables involved. For example, the dimensional expression of μ is M/Lt, hence the exponents 1, -1, and -1 are tabulated versus

M, L, and t, respectively, the dimensions with which they are associated. The dimensional matrix, **DM**, is then the array of numbers

$$\mathbf{DM} = \begin{pmatrix} 0 & 0 & 1 & 1 & 0 & 0 \\ 1 & 1 & -3 & -1 & 2 & 1 \\ -1 & -1 & 0 & -1 & -1 & 0 \end{pmatrix}$$

The rank of the matrix is easily obtained using the rank (**A**) function of Mathcad. Therefore, $r = $ rank (**DM**) $= 3$. From equation (2-16), $i = 6 - 3 = 3$, which means that there will be three dimensionless groups.

The three dimensionless parameters will be symbolized π_1, π_2, and π_3 and may be formed in several different ways. Initially, a *core group* of r variables must be chosen which will appear in each of the pi groups and, among them, contain all of the fundamental dimensions. One way to choose a core is to exclude from it those variables whose effect one wishes to isolate. In the present problem, it would be desirable to have the mass-transfer coefficient in only one dimensionless group, hence it will not be in the core. Let us arbitrarily exclude the fluid velocity and viscosity from the core. The core group now consists of D_{AB}, D, and ρ, which include M, L, and t among them.

We now know that all π_1, π_2, and π_3 contain D_{AB}, D, and ρ, that one of them includes k_c, one includes μ, and the other includes v; and that all must be dimensionless. For each to be dimensionless, the variables must be raised to certain exponents. Therefore,

$$\pi_1 = D_{AB}^a \rho^b D^c k_c$$
$$\pi_2 = D_{AB}^d \rho^e D^f v$$
$$\pi_3 = D_{AB}^g \rho^h D^i \mu$$

Writing π_1 in dimensional form,

$$M^0 L^0 t^0 = 1 = \left(\frac{L^2}{t}\right)^a \left(\frac{M}{L^3}\right)^b (L)^c \left(\frac{L}{t}\right)$$

Equating the exponents of the fundamental dimensions on both sides of the equation, we have for

$$
\begin{aligned}
L: &\quad 0 = 2a - 3b + c + 1 \\
t: &\quad 0 = -a - 1 \\
M: &\quad 0 = b
\end{aligned}
$$

The solution of these equations for the three unknown exponents yields $a = -1$, $b = 0$, $c = 1$, thus

$$\pi_1 = \frac{k_c D}{D_{AB}} = \text{Sh}$$

where Sh represents the *Sherwood number*, the mass-transfer analog to the Nusselt number of heat transfer. The other two pi groups are determined in the same manner, yielding

$$\pi_2 = \frac{vD}{D_{AB}} = \text{Pe}_D$$

where Pe_D represents the *Peclet number for mass transfer*, analogous to Pe_H —the Peclet number for heat transfer—and

$$\pi_3 = \frac{\mu}{\rho D_{AB}} = \text{Sc}$$

where Sc represents the Schmidt number, already defined in Chapter 1. Dividing π_2 by π_3, we obtain

$$\frac{\pi_2}{\pi_3} = \frac{\dfrac{Dv}{D_{AB}}}{\dfrac{\mu}{\rho D_{AB}}} = \frac{Dv\rho}{\mu} = \text{Re}$$

the Reynolds number. The result of the dimensional analysis of forced-convection mass transfer in a circular conduit indicates that a correlating relation could be of the form

$$\text{Sh} = f(\text{Re, Sc}) \tag{2-17}$$

which is analogous to the heat-transfer correlation

$$\text{Nu} = f(\text{Re, Pr}) \tag{2-18}$$

Example 2.5 Mass Transfer with Natural Convection

Natural convection currents will develop if there exists a significant variation in density within a liquid or gas phase. The density variations may be due to temperature differences or to relatively large concentration differences. Consider natural convection involving mass transfer from a vertical plane wall to an adjacent fluid. Use the Buckingham method to determine the dimensionless groups formed from the variables significant to this problem.

Solution

The important variables in this case, their symbols, and dimensional representations are listed below.

Variable	Symbol	Dimensions
characteristic length	L	L
fluid density	ρ	M/L^3
fluid viscosity	μ	M/Lt
buoyant force	$g\Delta\rho_A$	M/L^2t^2
mass diffusivity	D_{AB}	L^2/t
mass-transfer coefficient	F	M/L^2t

By the Buckingham theorem, there will be three dimensionless groups. With D_{AB}, L, and ρ as the core variables, the three pi groups to be formed are

$$\pi_1 = D_{AB}^a \, L^b \, \rho^c F$$

$$\pi_2 = D_{AB}^d \, L^e \, \rho^f \mu$$

$$\pi_3 = D_{AB}^g \, L^h \, \rho^i g\Delta\rho_A$$

Solving for the three pi groups, we obtain

$$\pi_1 = \frac{FL}{D_{AB}\rho} = \text{Sh}$$

$$\pi_2 = \frac{\mu}{D_{AB}\rho} = \text{Sc}$$

and

$$\pi_3 = \frac{g\Delta\rho_A L^3}{D_{AB}^2\rho}$$

Dividing π_3 by the square of π_2, we obtain

$$\frac{\pi_3}{\pi_2^2} = \frac{\dfrac{g\Delta\rho_A L^3}{D_{AB}^2\rho}}{\left(\dfrac{\mu}{\rho D_{AB}}\right)^2} = \frac{g\rho\Delta\rho_A L^3}{\mu^2} = \text{Gr}_D$$

the Grashof number for mass transfer, analogous to Gr_H —the Grashof number in natural-convection heat transfer. The result of the dimensional analysis of natural-convection mass transfer indicates that a correlating relation could be of the form

$$Sh = f(Gr_D, Sc) \qquad (2\text{-}19)$$

which is analogous to the heat-transfer correlation

$$Nu = f(Gr_H, Pr) \qquad (2\text{-}20)$$

For both forced and natural convection, relations have been obtained by dimensional analysis which suggest that a correlation of experimental data may be in terms of three variables instead of the original six. This reduction in variables has aided investigators who have developed correlations for estimating convective mass-transfer coefficients in a variety of situations.

2.4 MASS- AND HEAT-TRANSFER ANALOGIES

Your objectives in studying this section are to be able to:

1. Define the dimensionless groups of mass transfer corresponding to those for heat transfer.
2. Convert correlations of data on heat transfer to correlations on mass transfer.
3. Use the Chilton-Colburn analogy in problems where the heat-transfer data are correlated exclusively in terms of the Reynolds number.

There are many more experimental data available for heat transfer than for mass transfer. In the previous dimensional analyses of mass-transfer situations, we have recognized the similarities between the dimensionless groups for heat and mass transfer suggested by application of the Buckingham method. There are also similarities in the differential equations that govern convective mass and heat transfer and in the boundary conditions when the transport gradients are expressed in terms of dimensionless variables. In this section, we will consider several analogies among transfer phenomena which have been proposed because of the similarity in their mechanisms. The analogies are useful in understanding the underlying transport phenomena and as a satisfactory means of predicting behavior of systems for which limited quantitative data exist.

For analogous circumstances, temperature and concentration profiles in dimensionless form and heat- and mass-transfer coefficients in the form of dimensionless groups, repectively, are given by the same functions (Treybal, 1980).

> Therefore, to convert equations or correlations of data on heat transfer to correlations on mass transfer, the dimensionless groups of heat transfer are replaced by the corresponding groups of mass transfer.

Table 2.1 lists the commonly appearing dimensionless groups. The limitations to the above rule are:

(a) The flow conditions and geometry must be the same.

(b) Most heat-transfer data are based on situations involving no mass transfer. Use of the analogy would then produce mass-transfer coefficients corresponding to no net mass transfer, in turn corresponding most closely to k'_G , k'_c , or k'_y ($= F$). Sherwood numbers are commonly written in terms of any of the coefficients, but when derived by replacement of Nusselt numbers for use where the net mass transfer is not zero, they should be taken as Sh = Fl/cD_{AB} , and the F used with equation (2-1).

Example 2.6 Mass Transfer to Fluid Flow Normal to a Cylinder

For flow of a fluid at right angle to a circular cylinder, the average heat-transfer coefficient—averaged around the periphery of the cylinder—is given by (Eckert and Drake, 1959)

$$Nu_{av} = 0.43 + 0.532\,Re^{0.5}\,Pr^{0.31}$$
$$1 \le Re \le 4,000$$

where Nu_{av} and Re are computed using the cylinder diameter as the characteristic length, and the fluid properties are evaluated at the average conditions of the stagnant fluid film surrounding the solid (average temperature and average concentration). Estimate the rate of sublimation of a cylinder of UF_6 (molecular weight = 352), 1.0 cm diameter and 10 cm long exposed to an airstream that flows normal to the cylinder axis at a velocity of 1.0 m/s. The surface temperature of the solid is 303 K, at which temperature the vapor pressure of UF_6 is 27 kPa (Perry and Chilton, 1973). The bulk air is at 1 atm and 325 K.

Solution
Replacing Nu_{av} with Sh_{av} and Pr with Sc in the given heat-transfer correlation, the analogous expression for the mass-transfer coefficient is

Table 2.1 Dimensionless Groups for Mass and Heat Transfer

Mass transfer	Heat transfer
Reynolds number $$\mathrm{Re} = \frac{lv\rho}{\mu}$$	Reynolds number $$\mathrm{Re} = \frac{lv\rho}{\mu}$$
Schmidt number $$\mathrm{Sc} = \frac{\mu}{\rho D_{AB}}$$	Prandtl number $$\mathrm{Pr} = \frac{C_p\mu}{k}$$
Sherwood number $$\mathrm{Sh} = \frac{Fl}{cD_{AB}},\ \frac{k_G P_{B,M} RTl}{PD_{AB}},$$ $$\frac{k_c P_{B,M} l}{PD_{AB}},\ \frac{k_y' l}{D_{AB}},\ \frac{k_y RTl}{PD_{AB}},\ \mathrm{etc.}$$	Nusselt number $$\mathrm{Nu} = \frac{hl}{k}$$
Grashof number $$\mathrm{Gr}_D = \frac{gl^3\rho\Delta\rho}{\mu^2}$$	Grashof number $$\mathrm{Gr}_H = \frac{gl^3\rho^2\beta\Delta T}{\mu^2}$$
Peclet number $$\mathrm{Pe}_D = \mathrm{Re}\,\mathrm{Sc} = \frac{lv}{D_{AB}}$$	Peclet number $$\mathrm{Pe}_H = \mathrm{Re}\,\mathrm{Pr} = \frac{C_p lv\rho}{k}$$
Stanton number $$\mathrm{St}_D = \frac{\mathrm{Sh}}{\mathrm{Re}\,\mathrm{Sc}} = \frac{\mathrm{Sh}}{\mathrm{Pe}_D} = \frac{F}{cv}$$	Stanton number $$\mathrm{St}_H = \frac{\mathrm{Nu}}{\mathrm{Re}\,\mathrm{Pr}} = \frac{\mathrm{Nu}}{\mathrm{Pe}_H} = \frac{h}{C_p v\rho}$$
Chilton-Colburn j-factor $$j_D = \mathrm{St}_D \mathrm{Sc}^{2/3}$$	Chilton-Colburn j-factor $$j_H = \mathrm{St}_H \mathrm{Pr}^{2/3}$$

l = characteristic length β = volume coefficient of expansion
Taken from Treybal (1980).

$$\mathrm{Sh}_{av} = 0.43 + 0.532\,\mathrm{Re}^{0.5}\,\mathrm{Sc}^{0.31}$$
$$1 \le \mathrm{Re} \le 4{,}000$$

Along the mass-transfer path [cylinder surface (point 1) to bulk air (point 2) in Figure 2.1] the average temperature is $T_{av} = (303 + 325)/2 = 314$ K. At point 1, the gas is saturated with UF_6 vapor, while at point 2 the gas is virtually free of UF_6. Then, the average partial pressure of UF_6 is $(27 + 0)/2 = 13.5$ kPa corresponding to $y_A = 13.5/101.3 = 0.133$ mole fraction UF_6.

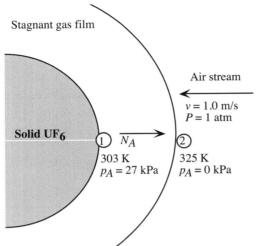

Stagnant gas film

Air stream

$v = 1.0$ m/s
$P = 1$ atm

Solid UF_6

① $\overline{N_A}$ ②

303 K   325 K
$p_A = 27$ kPa $p_A = 0$ kPa

Figure 2.1 Schematic diagram for Example 2.6.

Use the ideal gas law to estimate the average density of the gas film. The average molecular weight is $M_{av} = 352 (0.133) + 29 (1 - 0.133) = 72$. The average density is

$$\rho = \frac{PM_{av}}{RT_{av}} = \frac{101.3 \times 72}{8.314 \times 314} = 2.8 \text{ kg/m}^3$$

To calculate the average viscosity of the mixture, μ, use the corresponding states method of Lucas (1980) for nonpolar gases:

$$\mu_{rM} = \xi_M \mu = f(T_{rM}) \tag{2-21}$$

where:

$$\xi_M = 0.176 \left[\frac{T_{cM}}{M_{av}^3 P_{cM}^4} \right]^{1/6} \tag{2-22}$$

ξ_M = reduced inverse viscosity of the mixture, $(\mu P)^{-1}$

T_{cM} = critical temperature of the mixture, K

P_{cM} = critical pressure of the mixture, bars

$T_{rM} = T/T_{cM}$

$$f(T_{rM}) = 0.807T_{rM}^{0.618} - 0.357\exp(-0.449T_{rM})$$
$$+ 0.340\exp(-4.058T_{rM}) + 0.018 \qquad (2\text{-}23)$$

where the mixture properties are defined as follows

$$T_{cM} = \sum_i y_i T_{ci} \qquad (2\text{-}24)$$

$$P_{cM} = RT_{cM} \frac{\sum_i y_i Z_{ci}}{\sum_i y_i V_{ci}} \qquad (2\text{-}25)$$

$$M_{av} = \sum_i y_i M_i \qquad (2\text{-}26)$$

Polar and quantum effects (in H_2, D_2, and He) can be taken into account multiplying equation (2-23) by factors given by Lucas (1980). The data needed to apply the Lucas method are tabulated next (Reid, et al., 1987).

Parameter	O_2	N_2	UF_6
y_i	$(0.21)(0.867) = 0.182$	$0.79)(0.867) = 0.685$	0.133
T_c, K	154.6	126.2	505.8
P_c, bar	50.4	33.9	46.6
M, g/mol	32	28	352
V_c, cm^3/mol	73.4	89.8	250.0
Z_c,	0.288	0.290	0.277

Substituting in equations (2-24) and (2-25), T_{cM} = 181.9 K, P_{cM} = 40.3 bars, T_{rM} = 314/181.9 = 1.726 . From equation (2-21), $\xi_M = 4.207 \times 10^{-3}$ (μP)$^{-1}$; from equation (2-23), $f(1.726) = 0.962$; from equation (2-22), $\mu = 0.962/4.207 \times 10^{-3} = 229$ μP = 2.29×10^{-5} kg/m-s. Calculate the Reynolds number, based—as specified—on the cylinder diameter, d.

$$Re = \frac{dv\rho}{\mu} = \frac{0.01 \times 1.0 \times 2.8}{2.29 \times 10^{-5}} = 1,223$$

Notice that this value of Re is within the range specified for the original correlation.

The next step is to calculate the diffusivity of UF_6 vapors in air at 314 K and 1 atm. The Lennard-Jones parameters for these gases are obtained from Appendix B. The Mathcad routine developed in Example 1.4 gives

$$D_{AB}(314, 1.013, 352, 29, 5.967, 3.711, 236.8, 78.6) = 0.0904 \text{ cm}^2/\text{s}$$

Calculate the Schmidt number:

$$Sc = \frac{2.29 \times 10^{-5}}{2.8 \times 0.904 \times 10^{-5}} = 0.905$$

Substituting in the mass-transfer coefficient correlation,

$$Sh_{av} = 0.43 + 0.532(1,223)^{0.5}(0.905)^{0.31} = 18.4$$

To decide what type of mass-transfer coefficient is needed in this problem, we must explore the details of migration of the two species through the stagnant gas film surrrounding the solid surface. Evidently, UF_6 vapors (A) are generated at the interface by sublimation of the solid, and these vapors migrate through the film toward the bulk of the gas phase. On the other hand, there is no source or sink of air at the interface since air does not react with UF_6, nor does the solid adsorb air. Therefore, air does not migrate through the stagnant gas film surrounding the solid surface, and $N_B = 0$. Notice that there is a net movement in the *bulk of the gas phase* toward and around the solid, but not so in the stagnant air film! Therefore, this is a case of diffusion of A through stagnant B ($\Psi_A = 1$). However, the solution is not dilute since the average A content along the diffusion path is 0.133 mole fraction. For that reason, we must use the F form of the mass-transfer coefficient, instead of the k-type. From Table 2.1,

$$Sh_{av} = \frac{F_{av}d}{cD_{AB}} = 18.4$$

From equation (1-7)

$$c = \frac{P}{RT_{av}} = \frac{101.3}{8.314 \times 314} = 0.039 \text{ kmol/m}^3$$

Then,

$$F_{av} = \frac{18.4 \times 0.039 \times 0.904 \times 10^{-5}}{0.01} = 6.52 \times 10^{-4} \text{ kmol/m}^2\text{-s}$$

The UF_6 flux is from equation (2-2),

$$N_{Aav} = 6.52 \times 10^{-4} \ln\left[\frac{1-0}{1-\frac{27}{101.3}}\right] = 2.021 \times 10^{-4} \text{ kmol/m}^2\text{-s}$$

This average flux is based on the total surface area of the cylinder, S, given by

$$S = \frac{2\pi d^2}{4} + \pi dL = \pi d\left[\frac{d}{2} + L\right] = 3.3 \times 10^{-3} \text{ m}^2$$

where L is the cylinder length. Therefore, the mass rate of sublimation of the solid, w_A, is given by

$$w_A = N_{Aav} S M_A = 2.35 \times 10^{-4} \text{ kg / s}$$

The calculated rate of sublimation is an instantaneous value. Mass transfer will rapidly reduce the cylinder diameter, so that Re and hence F_{av} will change with time. Furthermore, the surface will not remain in the form of a circular cylinder owing to the variation of the local F about the perimeter, so that the correlation for Nu_{av} and Sh_{av} will no longer apply.

Example 2.7 The Chilton-Colburn Analogy

Chilton and Colburn (1934), based on experimental data, defined the *j-factor for mass transfer,* and established the analogy with heat transfer

$$j_D = St_D Sc^{2/3} = \psi(Re) = j_H = St_H Pr^{2/3} \qquad (2\text{-}27)$$

This is an extremely powerful analogy to relate heat and mass transfer data when the effects of Pr and Sc are not given explicitly, as will be shown next.

We wish to estimate the rate at which benzene will evaporate from a wetted surface of unusual shape when nitrogen at 1 bar and 300 K blows over the surface at a velocity of 10 m/s. No mass-transfer data are available for this situation, but heat-transfer measurements indicate that for CO_2 at 300 K and 1 bar, the heat-transfer coefficient h between the gas and the surface is given empirically by

$$h = 20 \left(G_y \right)^{0.5}$$

where h is in units of W/m²-K, and $G_y = \rho v$ is the superficial gas mass velocity, in kg/m²-s. Estimate the required mass-transfer coefficient.

Solution
The experimental heat-transfer correlation given does not include the effect of changing the Prandtl number. The j_H factor will satisfactorily describe the effect of Prandtl on the heat-transfer coefficient. Combining the given correlation with the definitions of j_H and St_H from Table 2.1,

$$j_H = \frac{h}{C_p \rho v} Pr^{2/3} = \frac{h}{C_p G_y} Pr^{2/3} = \psi(Re)$$

$$h = \frac{C_p G_y}{Pr^{2/3}} \psi(Re) = 20 \left(G_y \right)^{0.5} \qquad \text{for carbon dioxide}$$

Since $Re = \rho v l/\mu = G_y l/\mu$, where l is a characteristic length, the function $\psi(Re)$ must be compatible with $20G_y^{0.5}$. Therefore, let $\psi(Re) = bRe^n$, where b and n are constants to be evaluated. Then,

$$h = \frac{C_p G_y}{Pr^{2/3}} b \left(\frac{lG_y}{\mu} \right)^n = 20 \left(G_y \right)^{0.5}$$

$$\frac{bC_p}{Pr^{2/3}} \left(\frac{l}{\mu} \right)^n (G_y)^{n+1} = 20 \left(G_y \right)^{0.5}$$

Comparing both sides of the equation,

$$n + 1 = 0.5 \qquad\qquad n = -0.5$$

$$\frac{bC_p}{Pr^{2/3}} \left(\frac{l}{\mu} \right)^{-0.5} = 20$$

$$b = 20 \left(\frac{l}{\mu} \right)^{0.5} \frac{Pr^{2/3}}{C_p}$$

To evaluate b, we need data on the properties of CO_2 at 300 K and 1 bar. The values are: $\mu = 150$ µP, $Pr = 0.77$, and $C_p = 853$ J/kg-K (Welty, et al., 1984). Substituting these values in the previous equation,

$$b = 20 \left(\frac{l}{1.5 \times 10^{-5}} \right)^{0.5} \frac{(0.77)^{2/3}}{853} = 5.086 l^{0.5}$$

Therefore,

$$j_D = j_H = \psi(Re) = 5.086 l^{0.5} \, Re^{-0.5}$$

or

$$F = \frac{5.086 l^{0.5} cv}{Re^{0.5} \, Sc^{2/3}} = \frac{5.086 (\rho v \mu)^{0.5}}{M_{av} Sc^{2/3}}$$

The physical-property data in the previous equation will correspond to the average film conditions. The inner edge of the film is a saturated mixture of benzene vapors and nitrogen at 300 K and 1 bar, while the outer edge is virtually pure nitrogen at the same temperature and pressure. The vapor pressure of benzene, P_A, at 300 K is from Antoine equation (Himmelblau, 1989)

$$\ln P_A = 15.9008 - \frac{2788.51}{300 - 52.36}$$

$$P_A = 104 \text{ mm Hg} = 13.9 \text{ kPa}$$

Therefore, the average partial pressure of benzene in the film is 13.9/2 = 6.95 kPa, corresponding to 6.95/100 = 0.07 mole fraction. The data needed to estimate the physical properties are tabulated next (Reid, et al., 1987).

Parameter	C_6H_6	N_2
y_i	0.07	0.93
T_c, K	562.2	126.2
P_c, bar	48.9	33.9
M, g/mol	78.1	28
V_c, cm^3/mol	259	89.8
σ, Å	5.349	3.798
ε/k, K	412.3	71.4
Z_c,	0.271	0.290

The estimated average properties of the film at 300 K and 1 bar are:

M_{av} = 31.4 kg/kmol
ρ = 1.26 kg/m^3 (ideal gas law)
μ = 161 μP (method of Lucas)
D_{AB} = 0.0986 cm^2/s (Mathcad routine of Example 1.4)
Sc = 1.3

Substituting in the equation for F derived above,

$$F = \frac{5.086\left(1.26 \times 10 \times 1.61 \times 10^{-5}\right)^{0.5}}{31.4 \times (1.3)^{2/3}} = 1.94 \times 10^{-3} \text{ kmol / m}^2\text{- s}$$

2.5 CONVECTIVE MASS-TRANSFER CORRELATIONS

Your objectives in studying this section are to be able to:

1. Estimate convective mass-transfer coefficients for the following situations: (a) flow paralell to a flat surface, (b) flow past a single sphere, (c) flow normal to a single cylinder, (d) turbulent flow in circular pipes, (e) flow through packed and fluidized beds, and (f) flow through the shell side of a hollow-fiber membrane module.
2. Use the corresponding coefficients to solve typical mass-transfer problems.

Thus far, we have considered mass-transfer correlations developed from analogies with heat transfer. In this section, we shall present a few of the correlations developed directly from experimental mass-transfer data in the literature. Others more appropriate to particular types of mass-transfer equipment will be introduced as needed.

Experimental data are usually obtained by blowing gases over various shapes wet with evaporating liquids, or causing liquids to flow past solids which dissolve. Average, rather than local, mass-transfer coefficients are usually obtained. In most cases, the data are reported in terms of the k-type coefficients applicable to the binary systems used with $N_B = 0$, without details concerning the actual concentrations of solute during the experiments. Fortunately, the experimental concentrations are usually fairly low, so that if necessary, conversion of the data to the corresponding F is possible by taking $p_{B,M}/P$, $x_{B,M}$, etc., equal to unity.

2.5.1 Mass-Transfer Coefficients for Flat Plates

Several investigators have measured the evaporation from a free liquid surface—or the sublimation from a flat solid surface—of length L into a controlled air stream under both laminar and turbulent conditions. These data are correlated by (Welty, et al., 1984)

$$\mathrm{Sh}_L = 0.664 \, \mathrm{Re}_L^{0.5} \, \mathrm{Sc}^{1/3} \quad \text{(laminar)} \quad \mathrm{Re}_L < 3 \times 10^5 \quad (2\text{-}28)$$

$$\mathrm{Sh}_L = 0.036 \, \mathrm{Re}_L^{0.8} \, \mathrm{Sc}^{1/3} \quad \text{(turbulent)} \quad \mathrm{Re}_L \geq 3 \times 10^5 \quad (2\text{-}29)$$

where the characteristic length in Re_L and Sh_L is L. These equations may be used if the Schmidt number is in the range $0.6 < \mathrm{Sc} < 2500$. Equations (2-28) and (2-29) give the average mass-transfer coefficient along the surface, and can be expressed in terms of the j_D-factor

$$j_D = 0.664 \, \mathrm{Re}_L^{-1/2} \quad \text{(laminar)} \quad \mathrm{Re}_L < 3 \times 10^5 \quad (2\text{-}30)$$

$$j_D = 0.036 \, \mathrm{Re}_L^{-0.2} \quad \text{(turbulent)} \quad \mathrm{Re}_L \geq 3 \times 10^5 \quad (2\text{-}31)$$

Example 2.8 Benzene Evaporation Along a Vertical Flat Plate

Liquid benzene, C_6H_6, flows in a thin film down the outside surface of a vertical plate, 1.5 m wide and 3 m long. The liquid temperature is 300 K. Benzene-free nitrogen at 300 K and 1 bar pressure flows across the width of the plate parallel to the surface at a speed of 5 m/s. Calculate the rate at which the liquid should be supplied at the top of the plate so that evaporation will just prevent it from reaching the bottom of the plate. The density of liquid benzene at 300 K is 0.88 g/cm^3 (Himmelblau, 1989).

Solution

The film conditions, and average properties, are identical to those in Example 2.7, only the geometry is different. To determine whether the flow is laminar or turbulent, calculate Re_L. Notice that, since the flow is across the width of the plate, L actually refers to the width, 1.5 m:

$$Re_L = \frac{\rho v L}{\mu} = \frac{1.2 \times 5 \times 1.5}{1.61 \times 10^{-5}} = 5.59 \times 10^5 \quad \text{(turbulent)}$$

The Schmidt number calculated in Example 2.7, Sc = 1.3, is within the limits of applicability of equations (2-28) and (2-29). Substituting in (2-29)

$$Sh_L = 0.036 \left(5.59 \times 10^5\right)^{0.8} (1.3)^{1/3} = 1,557$$

Nitrogen (component B) does not react with benzene (component A), neither dissolves in the liquid, therefore, $N_B = 0$ and $\Psi_A = 1$. Since, at least at the inner edge of the film, the benzene concentration is relatively high (0.139 mole fraction), the F-form of the mass-transfer coefficient should be used:

$$F = \frac{Sh_L c D_{AB}}{L} = \frac{Sh_L \rho D_{AB}}{M_{av} L} = \frac{1,557 \times 1.26 \times 0.986 \times 10^{-5}}{31.4 \times 1.5}$$

$$= 4.11 \times 10^{-4} \text{ kmol / m}^2\text{-s}$$

$$N_A = F \ln\left[\frac{1 - y_{A_2}}{1 - y_{A_1}}\right] = 4.11 \times 10^{-4} \ln\left[\frac{1 - 0}{1 - 0.139}\right]$$

$$= 6.15 \times 10^{-5} \text{ kmol / m}^2\text{-s}$$

Calculate the total mass rate of evaporation over the surface of the plate,

$$w_A = N_A S M_A = 6.15 \times 10^{-5} \times 3 \times 1.5 \times 78.1 \times 60 \times 1,000$$
$$= 1,296 \text{ g / min}$$

Therefore, liquid benzene should be supplied at the top of the plate at the rate of 1,296/0.88 = 1,470 mL/min, or about 1.5 L/min, so that evaporation will just prevent it from reaching the bottom of the plate.

2.5.2 Mass-Transfer Coefficients for a Single Sphere

Investigators have studied the mass transfer from single spheres and have correlated the Sherwood number by direct addition of terms representing transfer by purely molecular diffusion and transfer by forced convection in the form

$$Sh = Sh_0 + C\,Re^m\,Sc^{1/3} \tag{2-32}$$

where C and m are correlating constants. For very low Reynolds number, when there are no natural convection effects, the Sherwood number approaches a theoretical value of 2.0 (Bird, et al., 1960). Accordingly, the generalized correlation becomes

$$Sh = 2.0 + C\,Re^m\,Sc^{1/3} \tag{2-33}$$

For transfer into liquid streams, the equation of Brian and Hales (1969)

$$Sh = \left(4 + 1.21 Pe_D^{2/3}\right)^{1/2} \tag{2-34}$$

correlates data that are obtained for $Pe_D < 10{,}000$. For $Pe_D > 10{,}000$, Levich (1962) recommended

$$Sh = 1.01\,Pe_D^{1/3} \tag{2-35}$$

The equation by Froessling (1939) and Evnochides and Thodos (1959)

$$Sh = 2.0 + 0.552\,Re^{1/2}\,Sc^{1/3} \tag{2-36}$$

correlates the data for transfer into gases at Reynolds numbers ranging from 2 to 12,000, and Schmidt numbers ranging from 0.6 to 2.7.

Equations (2-34), (2-35), and (2-36) can be used only when the effects of natural convection are negligible, that is, when

$$Re \geq 0.4 Gr_D^{1/2} Sc^{-1/6} \tag{2-37}$$

The following equations by Steinberger and Treybal (1960) are recommended when the transfer occurs in the presence of natural convection

$$Sh = Sh_{nc} + 0.347(Re\,Sc^{1/2})^{0.62}$$

$$1 \le Re \le 3 \times 10^4 \qquad (2\text{-}38)$$

$$0.6 \le Sc \le 3,200$$

where

$$Sh_{nc} = 2.0 + 0.569(Gr_D Sc)^{0.25} \qquad Gr_D Sc < 10^8$$

$$Sh_{nc} = 2.0 + 0.0254(Gr_D Sc)^{1/3} Sc^{0.244} \qquad Gr_D Sc > 10^8 \qquad (2\text{-}39)$$

Example 2.9 Evaporation of a Drop of Water Falling in Air

Estimate the distance a spherical drop of water, originally 1.0 mm in diameter, must fall in quiet, dry air at 323 K and 1 atm in order to reduce its volume by 50%. Assume that the velocity of the drop is its terminal velocity evaluated at its mean diameter during the process, and that the water temperature remains at 293 K.

Solution
The arithmetic mean diameter during the process is evaluated by

$$d_p = \frac{d_{p1} + d_{p2}}{2} = \frac{d_{p1} + (1/2)^{1/3} d_{p1}}{2}$$

$$= 0.897 d_{p1} = 0.897 \text{ mm}$$

The density of the drop of water at 293 K is $\rho_p = 995$ kg/m^3, the density of air at 323 K and 1 atm is $\rho = 1.094$ kg/m^3, its viscosity is $\mu = 195$ μP (Holman, 1990). By considering a force balance on a spherical particle falling in a fluid medium, we can show that the terminal velocity of the particle is

$$v_t = \sqrt{\frac{4d_p(\rho_p - \rho)g}{3C_D \rho}} = \sqrt{\frac{4d_p \rho_{pr} g}{3C_D \rho}} \qquad (2\text{-}40)$$

where $\rho_{pr} = (\rho_p - \rho)$, g is the acceleration of gravity, and C_D is the drag coefficient. The drag coefficient is a complicated function of the Reynolds number of the particle, which is proportional to the terminal velocity. To avoid having to resort to a trial-and-error calculation to estimate the terminal velocity, the following computational scheme has been proposed (Benítez, 1993):

(a) Define a new dimensionless number, the Galileo number, Ga

$$Ga = C_D \, Re^2 = C_D v_t^2 \left(\frac{d_p \rho}{\mu}\right)^2 \qquad (2\text{-}41)$$

Combining equations (2-40) and (2-41)

$$Ga = \frac{4 d_p^3 \rho \rho_{pr} g}{3 \mu^2} \qquad (2\text{-}42)$$

(b) Establish another useful relation between Re and C_D

$$\frac{Re}{C_D} = \frac{Re^3}{Ga} = \frac{3 \rho^2 v_t^3}{4 g \rho_{pr} \mu} \qquad (2\text{-}43)$$

(c) The following correlation is used to relate Re/C_D to Ga:

$$\ln\left(\frac{Re}{C_D}\right)^{1/3} = -3.194 + 2.153 \ln Ga^{1/3} - 0.238 \left[\ln Ga^{1/3}\right]^2$$

$$+ 0.01068 \left[\ln Ga^{1/3}\right]^3 \qquad (2\text{-}44)$$

(d) To calculate v_t for a particle of a given diameter, first calculate Ga from equation (2-42), calculate Re/C_D from equation (2-44), and then v_t from equation (2-43).

Applying the computational scheme to calculate the terminal velocity of our drop of water,

$$Ga^{1/3} = 8.97 \times 10^{-4} \left(\frac{4 \times 1.094 \times 994 \times 9.8}{3 \times \left(1.95 \times 10^{-5}\right)^2}\right)^{1/3} = 30.837$$

$$\ln Ga^{1/3} = 3.429$$

From equations (2-44) and (2-43)

$$\left(\frac{Re}{C_D}\right)^{1/3} = 6.178 = v_t \left[\frac{3 \rho^2}{4 g \rho_{pr} \mu}\right]^{1/3}$$

$$v_t = 3.56 \text{ m / s}$$

The vapor pressure of water at 293 K is 2.34 kPa (Smith, et al., 1996). Therefore, the water concentration at the inner edge of the gas film is 2.34/101.3 = 0.0231 mole fraction. The average molecular weight of the mixture is 28.75 kg/kmol, and the density is 1.195 kg/m³. The outer edge of the film is dry air at 323 K and 1 atm, with a density of 1.094 kg/m³. Then, $\Delta\rho = 0.101$ kg/m³. To determine whether natural convection effects are important, calculate Gr_D. The density and viscosity in the Grashof number are the properties at the average film conditions. Since the water vapor content of the film is so low, we may assume that the average film conditions are basically those of dry air at 1 atm and $(293 + 323)/2 = 308$ K. Under those conditios, $\rho = 1.14$ kg/m³ and $\mu = 191$ μP (Holman, 1990). Then,

$$Gr_D = \frac{gd_p^3 \rho \Delta \rho}{\mu^2} = \frac{9.8 \times \left(8.97 \times 10^{-4}\right)^3 \times 1.14 \times 0.101}{\left(1.91 \times 10^{-5}\right)^2} = 2.232$$

Estimate the diffusivity of water in air at 308 K and 1 atm: $D_{AB} = 0.242$ cm²/s. The Schmidt number is

$$Sc = \frac{\mu}{\rho D_{AB}} = \frac{1.91 \times 10^{-5}}{1.14 \times 2.42 \times 10^{-5}} = 0.692$$

The Reynolds number is

$$Re = \frac{\rho d_p v_t}{\mu} = \frac{1.14 \times 8.97 \times 10^{-4} \times 3.56}{1.91 \times 10^{-5}} = 191$$

In equation (2-37)

$$0.4 Gr_D^{1/2} Sc^{-1/6} = 0.4(2.232)^{1/2}(0.692)^{-1/6} = 0.635$$

This is much smaller than the Reynolds number, indicating that natural convection effects are negligible. The Froessling equation (2-36) can be used to evaluate the mass-transfer coefficient

$$Sh = 2 + 0.552(191)^{1/2}(0.692)^{1/3} = 8.748$$

Given that the solubility of air (component B) in water is negligible, $N_B = 0$. We have already shown that the gas phase is very dilute in water vapor (component A), therefore it is appropriate to use k-type mass-transfer coefficients. From Table 2.1

$$Sh = \frac{k_c p_{B,M} d_p}{P D_{AB}} \cong \frac{k_c d_p}{D_{AB}} = 8.748 \quad \text{since } p_{B,M} \cong P$$

$$k_c = \frac{Sh D_{AB}}{d_p} = \frac{8.748 \times 2.42 \times 10^{-5}}{8.97 \times 10^{-4}} = 0.236 \text{ m / s}$$

The average rate of evaporation is

$$w_A = \pi d_p^2 N_A M_A = \pi d_p^2 M_A k_c \left(c_{A1} - c_{A2}\right)$$

The dry-air concentration, $c_{A2} = 0$. The interface concentration is evaluated from the vapor pressure of water at 293 K:

$$c_{A1} = \frac{P_A}{RT_1} = \frac{2.34}{8.314 \times 293} = 9.61 \times 10^{-4} \text{ kmol / m}^3$$

Substituting the known values into the average rate of evaporation equation,

$$w_A = \pi \left(8.97 \times 10^{-4}\right)^2 \times 18 \times 0.236 \times 9.61 \times 10^{-4}$$

$$= 1.031 \times 10^{-8} \text{ kg / s} = 1.031 \times 10^{-5} \text{ g / s}$$

The amount of water evaporated, m, is given by

$$m = \rho \Delta V = \frac{\rho V_1}{2} = \frac{\rho \pi d_p^3}{12} = \frac{995 \times \pi \times \left(1.0 \times 10^{-3}\right)^3}{12}$$

$$= 2.61 \times 10^{-7} \text{ kg} = 2.61 \times 10^{-4} \text{ g}$$

The time necessary to reduce the volume by 50% is

$$t = \frac{m}{w_A} = \frac{2.61 \times 10^{-4}}{1.031 \times 10^{-5}} = 25.3 \text{ s}$$

The distance of fall is equal to $tv_t = 25.3 \times 3.56 = 90.0$ m.

2.5.3 Mass-Transfer Coefficients for Single Cylinders

Several investigators have studied the sublimation from a solid cylinder into air flowing normal to its axis. Additional results on the dissolution of solid cylinders into a turbulent water stream have been reported. Bedingfield and Drew (1950) correlated the available data by

$$\frac{k_G P \text{Sc}^{0.56}}{G_M} = 0.281(\text{Re})^{-0.4}$$

$$400 < \text{Re} < 25,000 \qquad\qquad (2\text{-}45)$$

$$0.6 < \text{Sc} < 2.6$$

where Re is the Reynolds number based on the diameter of the cylinder, $G_M = vc$ is the molar mass velocity of the gas, and P is the total pressure.

Example 2.10 Mass Transfer for Single Cylinder

Repeat Example 2.6 using equation (2-45), if it applies.

Solution

The values of the dimensionless parameters calculated in Example 2.6 (Re = 1,223 and Sc = 0.905) are within the limits of applicability of equation (2-45). The molar density calculated in that example was $c = 0.039$ kmol/m^3, and the gas velocity was specified as $v = 1.0$ m/s. Therefore, $G_M = vc = (1.0)(0.039) = 0.039$ kmol/m^2-s. Substituting in (2-45)

$$k_G P = \frac{0.281 G_M}{\text{Re}^{0.4} \, \text{Sc}^{0.56}} = \frac{0.281 \times 0.039}{(1,223)^{0.4} (0.905)^{0.56}}$$

$$= 6.75 \times 10^{-4} \text{ kmol} / \text{m}^2 - \text{s}$$

From equation (2-9), $k_G P = k_y$. However, in this problem we cannot use k-type coefficients because we are not dealing with dilute solutions. Therefore, we must try to relate our results so far with F. We know from equation (2-6) that $F = k_y \, y_{B,M}$. The fact that the correlation in equation (2-45) was developed in terms of k_G is evidence that the experimental data used were obtained under very dilute conditions for which $y_{B,M} \approx 1.0$, and $F \approx k_y$. Then, we may conclude that $F = 6.75 \times 10^{-4}$ kmol/m^2-s, which is only about 3% higher than the value estimated in Example 2.6 ($F = 6.52 \times 10^{-4}$ kmol/m^2-s) using an analogy between heat and mass transfer.

2.5.4 Turbulent Flow in Circular Pipes

Mass transfer from the inner wall of a tube to a moving fluid has been studied extensively, and most experimental data come from so-called wetted-wall towers. In Figure 2.2, a volatile pure liquid is allowed to flow down the inside surface of a circular pipe while a gas is blown upward or downward through the central core. Measurement of the rate of evaporation of the liquid into the gas stream over the known surface permits calculation of the mass-transfer coefficient for the gas phase. Use of different gases and liquids provides variations of Sc.

Gilliland and Sherwood (1934) studied the vaporization of nine different liquids into air. Their correlation is

$$\text{Sh} = 0.023 \, \text{Re}^{0.83} \, \text{Sc}^{0.44}$$
$$2,000 < \text{Re} < 35,000 \qquad\qquad (2\text{-}46)$$
$$0.6 < \text{Sc} < 2.5$$

where the characteristic length in Sh and Re is the tube diameter, and the physical properties of the gas are evaluated at the bulk conditions of the flowing gas stream.

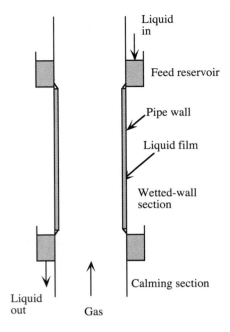

Figure 2.2 Wetted-wall tower.

In a subsequent study, Linton and Sherwood (1950) extended the range of Schmidt number when they investigated the rate of dissolution of benzoic acid, cinnamic acid, and β-naphthol. The combined results were correlated by

$$Sh = 0.023\,Re^{0.83}\,Sc^{1/3}$$
$$4,000 < Re < 70,000 \qquad\qquad (2\text{-}47)$$
$$0.6 < Sc < 3,000$$

Notice that equation (2-47) is the best correlation for all the data, including both gases and liquids. However, the data for gases only are best correlated by equation (2-46).

Example 2.11 Air Humidification in Wetted-Wall Column

Water flows down the inside wall of a wetted-wall tower of the design of Figure 2.2, while air flows upward through the core. The ID of the tower is 25.4 mm, and the length of the wetted section is 1.5 m. Dry air enters the wetted section at a mass velocity of 10.0 kg/m²-s, a temperature of 308 K, and a pressure of 1 atm. The water enters at 295 K. Estimate the partial pressure of water in the air leaving the tower. Assume that the temperature of the air and of the water remain constant.

Solution

In this problem, dry air enters the bottom of the tower. As it flows upward, water vapor is transferred from the gas-liquid interface toward the bulk of the gas phase due to a partial pressure driving force, $(p_{A1} - p_{A2})$, that changes continuously along the wetted section. The water vapor partial pressure at the interface remains constant at the vapor pressure of liquid water at 295 K, which is $p_{A1} = P_A = 2.64$ kPa. However, the water vapor partial pressure at the bulk of the gas phase increases from $p_{A2} = p_{Ain} = 0$, for the dry inlet air, to $p_{A2} = p_{Aout}$ for the air leaving the tower. Not only the driving force changes, but also the total gas flow rate increases as water evaporates along the wetted section. Since the mass-transfer coefficient is a function of gas flow rate, it also changes along the column.

For the particular case of the system air-water at near-ambient conditions, we are dealing with very dilute solutions of water vapor in air. For this reason, it is justified to assume that the total gas flow rate will remain relatively constant at its inlet value, and that the gas phase will behave as if it were dry air. If we assume that the temperature and pressure will remain constant, it follows that the dimensionless numbers Re and Sc will not change along the column, justifying the assumption of a constant value for the mass-transfer coefficient. Dilute gas solutions, and $N_B = 0$, also justify the use of a k-type coefficient coupled to a linear concentration driving force. These assumptions simplify considerably the analysis of the problem.

To take into account the fact that the driving force, hence, the mass-transfer rate, changes with position along the column, consider a segment of the wetted section of thickness Δz, as shown in Figure 2.3. A water (component A) balance, assuming steady-state and constant molar mass gas velocity (G_M), is

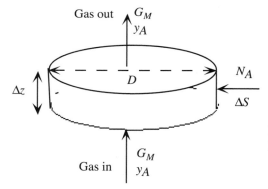

Figure 2.3 Differential section for Example 2.11

$$\frac{\pi D^2}{4} G_M y_A \Big|_{z+\Delta z} - \frac{\pi D^2}{4} G_M y_A \Big|_z = N_A \Delta S = \frac{\pi D^2}{4} \Delta z a N_A \quad (2\text{-}48)$$

where $a = \Delta S/\Delta V$ is the mass-transfer suface area per unit volume of column. Divide both sides by $\pi D^2 \Delta z/4$, rearrange, and take the limit as Δz tends to zero

$$G_M \frac{dy_A}{dz} = aN_A = k_G a(P_A - p_A) \qquad (2\text{-}49)$$

where P_A is the vapor pressure at the water temperature. But $y_A = p_A/P$, then

$$\frac{dp_A}{P_A - p_A} = \frac{k_G aP}{G_M} dz \qquad (2\text{-}50)$$

Integrating equation (2-50) within the limits $z = 0, p_A = 0$, and $z = Z, p_A = p_{Aout}$

$$\ln\left[\frac{P_A}{P_A - p_{Aout}}\right] = \frac{k_G aPZ}{G_M} \qquad (2\text{-}51)$$

where Z is the total height of the wetted section. Rearranging equation (2-51),

$$p_{Aout} = P_A\left[1 - \exp\left(\frac{-k_G aPZ}{G_M}\right)\right] \qquad (2\text{-}52)$$

From Table 2.1, for dilute solutions, $Sh \approx k_G RTD/D_{AB}$. Then,

$$k_G = \frac{Sh D_{AB}}{RTD}$$

$$\frac{k_G aPZ}{G_M} = \frac{Sh aPZD_{AB}}{RTG_M D} = \frac{Sh acZD_{AB}}{G_M D}$$

For the wetted-wall geometric arrangement,

$$a = \frac{\text{surface area}}{\text{volume}} = \frac{\pi DZ}{\pi D^2 Z/4} = \frac{4}{D} \qquad (2\text{-}53)$$

Substituting in equation (2-52)

$$p_{Aout} = P_A\left[1 - \exp\left(\frac{-4Sh cZD_{AB}}{G_M D^2}\right)\right] \qquad (2\text{-}54)$$

Equation (2-54) is the relation needed to estimate the water vapor concentration in the air leaving the tower. Assuming that the gas phase is basically dry air, $G_M = G_y/M_{av} = 10/29 = 0.345$ kmol/m²-s. The properties of dry air at 308 K and 1 atm are $\rho = 1.14$ kg/m³ and $\mu = 192$ μP, $D_{AB} = 0.242$ cm²/s, $Sc = 0.692$ (from Example 2.9). Calculate $Re = G_y D/\mu = (10)(0.0254)/1.91 \times 10^{-5} = 13,300$. From equation (2-46), $Sh = 51.8$. Calculate $c = P/RT = (101.3)/[(8.314)(308)] = 0.040$ kmol/m³. Substituting in equation (2-54),

$$p_{Aout} = 2.64\left[1 - \exp\left(\frac{-4 \times 51.8 \times 1.5 \times 0.04 \times 0.242}{0.345 \times 2.54^2}\right)\right]$$

$$= 1.96 \text{ kPa}$$

This is 74.2% of the maximum possible partial pressure of water in the air, which is $p_{Amax} = P_A = 2.64$ kPa.

2.5.5 Mass Transfer in Packed and Fluidized Beds

Packed and fluidized beds are commonly used in industrial mass-transfer operations, including adsorption, ion exchange, chromatography, and gaseous reactions that are catalyzed by solid surfaces because they offer a dramatical increase in the surface area available for heat and mass transfer for a given volume, compared to an empty tube. Numerous investigations have been conducted for measuring mass-transfer coefficients in packed beds and correlating the results. Sherwood et al. (1975) suggested the following correlation for gases

$$j_D = 1.17 \, \text{Re}^{-0.415}$$
$$10 < \text{Re} < 2,500 \qquad (2\text{-}55)$$

where $\text{Re} = d_p G_y / \mu$, where G_y is the gas mass velocity based on the total cross-sectional area of the tower, and d_p is the diameter of a sphere with the same surface area per unit volume as the particle.

Mass transfer in both gas and liquid fixed and fluidized beds of spheres has been correlated by Gupta and Thodos (1962) with the equation

$$\varepsilon j_D = 0.010 + \frac{0.863}{\text{Re}^{0.58} - 0.483} \qquad (2\text{-}56)$$

for Re between 1 and 2100, where ε is the void fraction of the bed.

For packed and fluidized beds, the area for mass transfer is usually expressed in terms of a, defined in this case as the area for mass transfer per unit volume of packed bed. It is easy to demonstrate that

$$a = \frac{6(1 - \varepsilon)}{d_p} \qquad (2\text{-}57)$$

Equation (2-57) suggests that the smaller the packing size, the higher the area available for mass transfer per unit of packed volume. However, pressure drop through the packed bed becomes a limiting factor as the size of the packing material diminishes.

Example 2.12 Air Humidification in a Packed Bed

The column of Example 2.11 is packed randomly with spherical glass beads, 3.5 mm in diameter. The water is now supplied as a fine mist that flows down through the packed bed at a flow rate just enough to replace the water lost by vaporization, and to keep the surface of the packing wet. Under these conditions, it may be assumed that equation (2-55) applies.

(a) Estimate the depth of packing required if the water partial pressure in the air leaving the bed is to be 99% of the water vapor pressure P_A.

(b) Estimate the gas pressure drop through the bed.

Solution

(a) Begin with the integrated mass-balance of equation (2-52) and specialize it for the given packed-bed geometry using

$$k_G \approx \frac{Sh D_{AB}}{RT d_p} \qquad a = \frac{6(1-\varepsilon)}{d_p}$$

Substituting in equation (2-52)

$$p_{Aout} = P_A\left[1 - \exp\left(\frac{-6(1-\varepsilon)Sh c Z D_{AB}}{G_M d_p^2}\right)\right] \qquad (2\text{-}58)$$

Calculate the Reynolds number, based on the particle size:

$$Re = \frac{G_y d_p}{\mu} = \frac{10 \times 3.5 \times 10^{-3}}{1.91 \times 10^{-5}} = 1,832$$

Substituting in equation (2-55), $j_D = 0.052$. From Example 2.11, Sc = 0.692. Then, $St_D = 0.052/0.692^{2/3} = 0.066$; Sh = St_D ReSc = 83.7. To estimate the bed porosity, use the following correlation developed for uniform spherical particles randomly packed in a cylindrical container of diameter D (Nandakumar, et al., 1999):

$$\varepsilon = 0.406 + 0.571\frac{d_p}{D} \qquad \text{for} \qquad \frac{d_p}{D} < 0.14 \qquad (2\text{-}59)$$

Substituting, ε = 0.406 + 0.571(3.5/25.4) = 0.485. To determine the bed depth required to achieve 99% relative humidity at the exit, substitute p_{Aout}= 0.99P_A in equation (2-58) and solve for Z.

$$Z = \frac{-0.345 \times (0.35)^2 \times \ln(1-0.99)}{6 \times (1-0.485) \times 83.7 \times 0.04 \times 0.242} = 0.078 \text{ m} = 7.8 \text{ cm}$$

Notice that, although the packed bed is 20 times smaller than the wetted-wall column, the air leaving it is virtually saturated with water, while the relative humidity achieved by the wetted-wall was only 74.2%. This explains why the wetted-wall arrangement is used only in experimental setups, and never for practical industrial applications.

(b) To estimate the gas pressure drop through the bed, assume that—for the given conditions of very low liquid flow—it is a problem of single-phase flow, reasonably well correlated by (Ergun, 1952):

$$\frac{\Delta P}{Z} \frac{\varepsilon^3 d_p \rho}{(1-\varepsilon)(G_y)^2} = \frac{150(1-\varepsilon)}{Re} + 1.75 \qquad (2\text{-}60)$$

Substituting in equation (2-60),

$$\Delta P = \left[\frac{150(1-0.485)}{1,832} + 1.75 \right] \frac{(1-0.485) \times 10^2 \times 0.078}{0.485^3 \times 0.0035 \times 1.14}$$
$$= 15,820 \text{ Pa}$$

This is a very high pressure drop, even though the packed bed is very shallow! For comparison purposes, let us estimate the pressure drop for the wetted-wall column of Example 2.11. To estimate the pressure drop in turbulent flow through a circular "smooth" pipe of diameter D and length Z (Welty et al., 1984)

$$\Delta P = 2f \frac{Z}{D} \frac{G_y^2}{\rho} \text{ , where}$$
$$f^{-1/2} = 4 \log_{10} \left(Re \, f^{1/2} \right) - 0.4 \qquad (2\text{-}61)$$

From Example 2.11, Re = 13,300, Z = 1.5 m, D = 0.0254 m, G_y = 10 kg/m²-s. Equation (2-60) yields f = 0.00717, ΔP = 74 Pa. Notice the dramatic difference between this result and the pressure drop estimated for the packed bed!

2.5.6 Mass Transfer in Hollow-Fiber Membrane Modules

Mass transfer in fiber bundles is a problem of great practical importance for membrane separation processes. Such processes commonly utilize a bundle of randomly packed hollow fibers enclosed in a case to contact two process streams. Ports on the case permit one to introduce and remove streams from the space inside the fibers, the *lumen*, and the space outside the fibers, the *shell*. Figure 2.4 illustrates the construction of a typical hollow-fiber membrane module (Bao, et al., 1999).

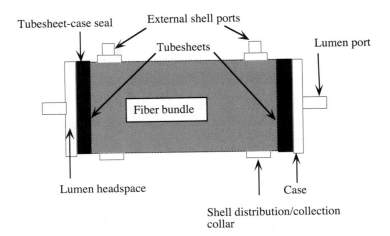

Figure 2.4 Construction of a typical hollow-fiber membrane module.

Hollow-fiber membrane modules are the mass-transfer equivalent of shell and tube heat exchangers. As fluids flow through the shell and lumen, mass is transferred from one stream to the other across the fiber wall. In contrast to heat exchangers, though, mass transfer may involve a combination of diffusion and convection, depending on the nature of the membrane. These modules are used for a wide range of membrane processes, including gas separation, reverse osmosis, filtration, and dialysis.

The literature contains numerous correlations for shell-side mass-transfer coefficients in hollow-fiber membrane modules (Bao, et al., 1999). A representative relationship for liquid flowing through the shell parallel to the fibers is (Costello, et al., 1993)

$$\text{Sh} = 0.53(1 - 1.1\phi)\left(\frac{1-\phi}{\phi}\right)^{-0.47}\text{Re}^{0.53}\,\text{Sc}^{0.33}$$

$$30 < 100\phi < 75$$

$$20 < \text{Re}\left(\frac{1-\phi}{\phi}\right) < 350$$

(2-62)

where $\text{Sh} = 2Rk_L/D_{AB}$ is the Sherwood number calculated using the effective mass-transfer coefficient for a module length L; R is the fiber radius; ϕ is the fiber packing fraction; the Reynolds number is defined as $\text{Re} = 2Rv_0\rho/\mu$, where v_0 is the superficial velocity based on an empty shell. Packing fraction refers to the fraction of the cross-sectional area that is occupied by the fibers.

Flow through the lumen usually corresponds to laminar flow inside a circular pipe. Theoretical solutions have been developed for this situation, with two different boundary conditions: (a) constant wall concentration, and (b) constant wall flux. Constant wall concentration is much more common than constant flux. The corresponding correlation is (Cussler, 1997)

$$Sh = 1.62 \left(Pe_D\right)^{1/3} \left(\frac{d_i}{L}\right)^{1/3} = 1.62 (Gz)^{1/3}$$
$$Re < 2,100$$

(2-63)

where $Sh = d_i k_L / D_{AB}$ is the Sherwood number calculated using the effective mass-transfer coefficient for a module length L; d_i is the fiber inside diameter; the Reynolds number is defined as $Re = d_i v \rho / \mu$, where v is the average fluid velocity; the Graetz number is defined as $Gz = Pe_D (d_i / L)$.

Example 2.13 Design of a Hollow-Fiber Boiler Feed Water (BFW) Deaerator

Boiler feed water (BFW) must be deareated to avoid corrosion problems in the boilers. Hollow fibers made of microporous polypropylene can be used for very fast removal of dissolved oxygen from water, therefore making a compact BFW deaerator possible (Yang and Cussler, 1986). For a given boiler, 40,000 kg/hr of BFW are needed. Design a hollow-fiber membrane module for that purpose, assuming that the unit is capable of removing 99% of the dissolved oxygen present in natural waters at 298 K.

Solution

To design the deaerator, we must first decide on the geometry for the hollow-fiber module. We will use commercially available microporous polypropylene hollow fibers in a module similar to that shown in Figure 2.4. Water at 298 K will flow through the shell side, parallel to the fibers, at a superficial velocity of 10 cm/s. Pure nitrogen at 298 K and 1 atm at the rate of 40 L/min will be used as a sweep gas in countercurrent flow through the lumen. The outside diameter of the available fibers is 290 μm; the packing factor is 40%; the surface area per unit volume is $a = 46.84$ cm^{-1} (Prasad and Sirkar, 1988).

First, calculate the volumetric water flow rate, Q_L, assuming that the density of liquid water at 298 K is approximately 1,000 kg/m^3:

$$Q_L = \frac{40,000}{3,600 \times 1,000} = 0.0111 \ \text{m}^3/\text{s}$$

Calculate the shell diameter, D, for the calculated water flow rate and the specified superficial velocity, $v_0 = 10$ cm/s.

$$D = \left(\frac{4Q_L}{\pi v_0}\right)^{1/2} = \left(\frac{4 \times 0.0111}{\pi \times .10}\right)^{0.5} = 0.376 \text{ m}$$

Estimate the properties of dilute mixtures of oxygen in water at 298 K: $\rho = 1{,}000 \text{ kg/m}^3$, $\mu = 0.9$ cP. The diffusivity is from equation (1-37), $D_{AB} = 1.93 \times 10^{-5}$ cm²/s. Calculate the dimensionless numbers, Sc = 467, Re = $dv_0\rho/\mu$ = 32.2. Substituting in equation (2-62),

$$\text{Sh} = 0.53 \times (1 - 1.1 \times 0.4) \times \left(\frac{1 - 0.4}{0.4}\right)^{-0.47} \times (32.2)^{0.53} \times (467)^{0.33}$$

$$= 11.74$$

Calculate the mass-transfer coefficient on the shell side, k_L = 11.74D_{AB}/d = 0.0078 cm/s.

The design equation is based on the observation by Yang and Cussler (1986) that, for this particular situation, all the resistance to mass transfer resides on the shell side. Then, the total volume of the module, V_T , for countercurrent flow is given by

$$V_T = \frac{L}{k_L ac(1 - A)} \ln\left[\frac{x_{in}}{x_{out}}(1 - A) + A\right] \qquad (2\text{-}64)$$

where

$$A = \frac{L}{mV} \qquad (2\text{-}65)$$

where L and V are the molar flow rates of liquid and gas, respectively, and m is the equilibrium ratio of gas concentration to liquid concentration. From the specified BFW flow rate,

$$L = 40{,}000\frac{\text{kg}}{\text{hr}} \times \frac{1 \text{ hr}}{3{,}600 \text{ s}} \times \frac{1 \text{ kmol}}{18 \text{ kg}} = 0.617 \text{ kmol / s}$$

From the ideal gas law,

$$V = 40\frac{\text{L}}{\text{min}} \times \frac{1 \text{ min}}{60 \text{ s}} \times \frac{1 \text{ m}^3}{1{,}000 \text{ L}} \times \frac{101.3 \text{ kPa}}{298 \text{ K}} \times \frac{\text{kmol - K}}{8.314 \text{ kPa - m}^3}$$

$$= 2.726 \times 10^{-5} \text{ kmol / s}$$

From the solubility of oxygen in water at 298 K, $m = 4.5 \times 10^4$ (Davis and Cornwell, 1998). Then, $A = 0.617/(4.5 \times 10^4 \times 2.726 \times 10^{-5}) = 0.503$. For 99% removal of the dissolved oxygen, x_{in}/x_{out} = 100. Substituting in equation (2-64),

$$V_T = \frac{0.617 \times \ln[100(1-0.503)+0.503]}{0.0078 \times 46.84 \times 55.5 \times (1-0.503)} = 0.24 \text{ m}^3$$

The module length, Z, is

$$Z = \frac{V_T}{\dfrac{\pi D^2}{4}} = \frac{0.24 \times 4}{\pi \times (0.376)^2} = 2.16 \text{ m}$$

2.6 ESTIMATION OF MULTICOMPONENT MASS-TRANSFER COEFFICIENTS

Your objectives in studying this section are to be able to:

1. Estimate convective mass-transfer coefficients for multicomponent problems based on binary correlations.
2. Write and solve the Maxwell-Stefan equations for multicomponent mixtures of ideal gases in terms of the corresponding binary mass-transfer coefficients.

Most published empirical correlations to estimate mass-transfer coefficients have concentrated on binary systems and there are no correlations available for the multicomponent case. The need to estimate multicomponent mass-transfer coefficients is very real, however. Various approximate methods have been proposed to estimate multicomponent mass-transfer coefficients based on binary correlations.

The simplest approach is to calculate binary mass-transfer coefficients F_{ij}, from the corresponding empirical correlation, substituting the MS diffusivity $\mathfrak{D}_{ij}$, for the Fick diffusivity in the Sc and Sh numbers. The Maxwell-Stefan equations are, then, written in terms of the binary mass-transfer coefficients. For ideal gas multicomponent mixtures, and one-dimensional fluxes, they become

$$\frac{dy_i}{d\eta} = \sum_{j=1}^{n} \frac{y_i N_j - y_j N_i}{F_{ij}} \qquad i = 1, 2, \cdots, n-1 \tag{2-66}$$

$\eta = z/\delta$, is the dimensionless distance along the gas film.

Example 2.14 Ternary Distillation in a Wetted-Wall Column.

Dribicka and Sandall (1979).distilled ternary mixtures of benzene (1), toluene (2), and ethylbenzene (3) in a wetted-wall column of 2.21-cm inside diameter (d). Samples of the vapor and liquid phases were taken from various points along the column. During one of their experiments the bulk of the vapor phase at a height of 300 mm from the bottom of the column had the following molar composition: 74.71% benzene, 20.72% toluene, and 4.57% ethylbenzene. From the experimental conditions, Taylor and Krishna (1993) estimated that the corresponding vapor composition at the interface was 89.06% benzene, 9.95% toluene, and 0.99% ethylbenzene. Estimate the individual molar fluxes, assuming that the total molar flux is zero. The following data apply:

Viscosity of the vapor, $\mu = 8.82 \times 10^{-6}$ kg/m-s

Density of the vapor, $\rho = 2.81$ kg/m^3

Total molar density, $c = 34.14$ mole/m^3

$\mathcal{D}_{12} = 2.228$ mm^2/s; $\mathcal{D}_{13} = 2.065$ mm^2/s; $\mathcal{D}_{23} = 1.832$ mm^2/s

Vapor mass velocity, $G_y = 8.68$ kg/m^2-s

Solution

The Sherwood-Gilliland correlation in the form of equation (2-46) may be used to estimate the binary mass-transfer coefficients in the vapor phase. Calculate the Reynolds number:

$$\text{Re} = \frac{G_y d}{\mu} = 21,750$$

Compute the Schmidt number for each binary pair using the MS diffusivities:

$$\text{Sc}_{ij} = \frac{\mu}{\rho \mathcal{D}_{ij}}$$

$$\text{Sc}_{12} = 1.409 \qquad \text{Sc}_{13} = 1.520 \qquad \text{Sc}_{23} = 1.713$$

The corresponding Sherwood numbers are from equation (2-46)

$$\text{Sh}_{ij} = 0.023 \, \text{Re}^{0.83} \, \text{Sc}_{ij}^{0.44}$$

$$\text{Sh}_{12} = 106.5 \qquad \text{Sh}_{13} = 110.1 \qquad \text{Sh}_{23} = 116.1$$

Calculate the binary mass-transfer coefficients from

$$F_{ij} = \frac{Sh_{ij}c\mathcal{D}_{ij}}{d}$$

$$F_{12} = 0.367 \text{ mole} / \text{m}^2\text{-s}$$
$$F_{13} = 0.351 \text{ mole} / \text{m}^2\text{-s}$$
$$F_{23} = 0.328 \text{ mole} / \text{m}^2\text{-s}$$

We have now all the information needed to modify the Mathcad program of Appendix C-1 to estimate the individual molar fluxes. In this case, there is an additional unknown value, namely N_3. The additional relation is the statement that the total flux is zero:

$$N_1 + N_2 + N_3 = 0$$

The program yields the following results:

$$N_1 = -0.0527 \text{ mole/m}^2\text{-s}$$
$$N_2 = 0.0395 \text{ mole/m}^2\text{-s}$$
$$N_3 = 0.0132 \text{ mole/m}^2\text{-s}$$

PROBLEMS

   **The problems at the end of each chapter have been
   grouped into four classes (designated by a superscript
   after the problem number).**

   Class a: Illustrates direct numerical application of the for-
   mulas in the text.
   Class b: Requires elementary analysis of physical situations,
   based on the subject material in the chapter.
   Class c: Requires somewhat more mature analysis.
   Class d: Requires computer solution.

2.1ª. Mass-transfer coefficients in a gas absorber.
   A gas absorber is used to remove benzene (C_6H_6) vapors from air by
scrubbing the gas mixture with a nonvolatile oil at 300 K and 1 atm. At a
certain point in the absorber, the benzene mole fraction in the bulk of the gas
phase is 0.02, while the corresponding interfacial benzene gas-phase concentration is 0.0158. The benzene flux at that point is measured as 0.62 g/m²-s.

(a) Calculate the mass-transfer coefficient in the gas phase at that point in the
equipment, expressing the driving force in terms of mole fractions.

Answer: $k_y = 1.9 \times 10^{-3}$ kmol/m²-s

(b) Calculate the mass-transfer coefficient in the gas phase at that point in
the equipment, expressing the driving force in terms of molar concentrations,
kmol/m³.

Answer: $k_c = 0.047$ m/s

(c) At the same place in the equipment, the benzene mole fraction in the bulk
of the liquid phase is 0.125, while the corresponding interfacial benzene liquid-phase concentration is 0.158. Calculate the mass-transfer coefficient in
the liquid phase, expressing the driving force in terms of mole fractions.

Answer: $F_L = 2.07 \times 10^{-4}$ kmol/m²-s

2.2ᵃ. Mass-transfer coefficients from naphthalene sublimation data.

In a laboratory experiment, air at 347 K and 1 atm is blown at high speed around a single naphthalene ($C_{10}H_8$) sphere, which sublimates partially. When the experiment begins, the diameter of the sphere is 2.0 cm. At the end of the experiment, 14.32 min later, the diameter of the sphere is 1.85 cm.

(a) Estimate the mass-transfer coefficient, based on the average surface area of the particle, expressing the driving force in terms of partial pressures. The density of solid naphthalene is 1.145 g/cm³, its vapor pressure at 347 K is 670 Pa (Perry and Chilton, 1973).

Answer: $k_G = 1.19 \times 10^{-6}$ mol/cm²-s-kPa

(b) Calculate the mass-transfer coefficient, for the driving force in terms of molar concentrations.

Answer: $k_c = 3.44$ cm/s

2.3ᵃ. Mass-transfer coefficients from acetone evaporation data.

In a laboratory experiment, air at 300 K and 1 atm is blown at high speed parallel to the surface of a rectangular shallow pan that contains liquid acetone (C_3H_6O), which evaporates partially. The pan is 1 m long and 50 cms wide. It is connected to a reservoir containing liquid acetone which automatically replaces the acetone evaporated, maintaining a constant liquid level in the pan. During an experimental run, it was observed that 2.0 L of acetone evaporated in 5 min. Estimate the mass-transfer coefficient. The density of liquid acetone at 300 K is 0.79 g/cm³; its vapor pressure is 27 kPa (Perry and Chilton, 1973).

Answer: $F_G = 5.86 \times 10^{-4}$ kmol/m²-s

2.4ᵇ. Mass-transfer coefficients from wetted-wall experimental data.

A wetted-wall experimental setup consists of a glass pipe, 50 mm in diameter and 1.0 m long. Water at 300 K flows down the inner wall. Dry air enters the bottom of the pipe at the rate of 1.04 m³/min, measured at 308 K and 1 atm. It leaves the wetted section at 308 K and with a relative humidity of 21.5%. With the help of equation (2-52), estimate the average mass-transfer coefficient, with the driving force in terms of molar fractions.

Answer: $k_y = 1.8 \times 10^{-3}$ kmol/m²-s

2.5[b]. Dimensional analysis: aeration.

Aeration is a common industrial process and yet one in which there is often serious disagreement about correlations (Cussler, 1997). This is especially true for deep-bed fermentors and for sewage treatment, where the rising bubbles can be the chief means of stirring. We expect that the rate of oxygen mass transfer through the bed can be calculated in terms of a mass-transfer coefficient, k_c, which will depend on the average bubble velocity, v, the solution's density ρ and viscosity μ, the bubble diameter, d, the depth of the bed, z, and the diffusivity, D_{AB}. Use the Buckingham method to determine the dimensionless groups to be formed from the variables significant to this problem. Choose $\{D_{AB}, d, \text{and } \rho\}$ as core variables.

2.6[b]. Dimensional analysis: the artificial kidney.

In the artificial kidney, blood flowing inside a tubular membrane is dialyzed against well-stirred saline solution. Toxins in the blood diffuse across the membrane into the saline solution, thus purifying the blood. This dialysis is often slow; it can take more than 40 hours per week (Cussler, 1997). Thus, increasing the mass transfer in this system would greatly improve its clinical value. It has been found that in modern designs of artificial kidneys (thin membranes and vigorous stirring of the saline solution) most of the resistance to toxin removal resides inside the membrane, in the blood. The mass-transfer coefficient, k_c, for a particular toxin varies with v, ρ, and μ (velocity, density, and viscosity of the blood), the diffusivity, D_{AB}, the diameter, d, and length, L, of the tube. Use the Buckingham method to determine the dimensionless groups to be formed from the variables significant to this problem. Choose $\{D_{AB}, d, \text{and } \rho\}$ as core variables.

2.7[c]. Mass transfer in an annular space.

(a) In studying rates of diffusion of naphthalene into air, an investigator replaced a 30.5-cm section of the inner pipe of an annulus with a naphthalene rod. The annulus was composed of a 51-mm-OD brass inner pipe surrounded by a 76-mm-ID brass pipe. While operating at a mass velocity within the annulus of 12.2 kg of air/m²-s at 273 K and 1 atm, the investigator determined that the partial pressure of naphthalene in the exiting gas stream was 0.041 Pa. Under the conditions of the investigation, the Schmidt number of the gas was 2.57, the viscosity was 175 µP, and the vapor pressure of naphthalene was 1.03 Pa. Estimate the mass-transfer coefficient from the inner wall for this set of conditions. Assume that equation (2-52) applies.

Answer: $k_y = 8.72 \times 10^{-4}$ kmol/m^2-s

(b) Monrad and Pelton (1942) presented the following correlation for heat-transfer coefficient in an annular space:

$$\frac{h_i}{C_p G_y} = 0.023 \left(\frac{d_o}{d_i}\right)^{0.5} \left(\frac{d_e G_y}{\mu}\right)^{-0.2} \left(\frac{C_p \mu}{k}\right)^{-2/3} \qquad (2\text{-}67)$$

where d_o and d_i are the outside and inside diameters of the annulus, d_e is the equivalent diameter defined as

$$d_e = 4 \frac{\text{cross-sectional area of flow}}{\text{wetted perimeter}} \qquad (2\text{-}68)$$

Write down the analogous expression for mass transfer and use it to estimate the mass-transfer coefficient for the conditions of part a. Compare both results.

2.8^c. The Chilton-Colburn analogy: flow across tube banks.

Winding and Cheney (1948) passed air at 310 K and 1 atm through a bank of rods of naphthalene. The rods were in a staggered arrangement, with the air flowing at right angles to the axes of the rods. The bank consisted of 10 rows containing alternately five and four 38-mm-OD tubes (d = 38 mm) spaced on 57-mm centers, with the rows 76 mm apart. The mass-transfer coefficient was determined by measuring the rate of sublimation of the naphthalene. The data could be correlated by:

$$k_G = 3.86 \times 10^{-9} \left(G_y\right)^{0.56}$$
$$5 < G_y < 40 \text{ kg / m}^2 - \text{s} \qquad (2\text{-}69)$$

where G_y is the maximum mass velocity through the tube bank, in kg/m^2-s, and k_G is in kmol/m^2-s-Pa.

(a) Rewrite equation (2-69) in terms of the Colburn j_D-factor. The diffusivity of naphthalene in air at 310 K and 1 atm is 0.074 cm^2/s.

Answer: $j_D = 0.554 \, \text{Re}^{-0.44}$; $\quad$ Re $= G_y d / \mu$

(b) Estimate the mass-transfer coefficient to be expected for evaporation of n-propyl alcohol into carbon dioxide for the same geometrical arrangement when the carbon dioxide flows at a maximum velocity of 10 m/s at 300 K and 1 atm. The vapor pressure of n-propyl alcohol at 300 K is 2.7 kPa.

Answer: $k_G = 1.87 \times 10^{-5}$ kmol/m^2-s-kPa

(c) Zakauskas (*Adv. Heat Transfer,* **8**, 93, 1972) proposed the following correlation for the heat-transfer coefficient in a staggered tube bank arrangement similar to that studied by Winding and Cheney:

$$Nu = 0.453 \, Re^{0.568} \, Pr^{0.36}$$

$$10 < Re < 10^6 \qquad\qquad (2\text{-}70)$$

$$0.7 < Pr < 500$$

Use the mass-transfer expression analogous to equation (2-69) to estimate the mass-transfer coefficient of part b. Compare the results.

2.9[b]. Mass transfer from a flat plate.

A 1-m square thin plate of solid naphthalene is oriented parallel to a stream of air flowing at 20 m/s. The air is at 310 K and 101.3 kPa. The naphthalene remains at 290 K; at this temperature the vapor pressure of naphthalene is 26 Pa. Estimate the moles of naphthalene lost from the plate per hour, if the end effects can be ignored.

Answer: 1.88 mol/hr

2.10[b]. Mass transfer from a flat plate.

A thin plate of solid salt, NaCl, measuring 15 by 15 cm, is to be dragged through seawater at a velocity of 0.6 m/s. The 291 K seawater has a salt concentration of 0.0309 g/cm^3 and a density of 1.022 g/cm^3. Estimate the rate at which the salt goes into solution if the edge effects can be ignored. Assume that the kinematic viscosity at the average liquid film conditions is 1.02×10^{-6} m^2/s, and the diffusivity is 1.25×10^{-9} m^2/s. The solubility of NaCl in water at 291 K is 0.35 g/cm^3, and the density of the saturated solution is 1.22 g/cm^3 (Perry and Chilton, 1973).

Answer: 0.857 kg/hr

2.11[b]. Mass transfer from a flat liquid surface.

During the experiment described in Problem 2.3, the air velocity was measured at 6 m/s, parallel to the longest side of the pan. Estimate the mass-transfer coefficient predicted by equation (2-28) or (2-29) and compare it to the value measured experimentally. Notice that, due to the high volatility of acetone, the average acetone concentration in the gas film is relatively high. Therefore, properties such as density and viscosity should be estimated carefully. The following data for acetone might be needed: T_c = 508.1 K, P_c = 47.0 bar, M = 58, V_c = 209 cm^3/mol, Z_c = 0.232 (Reid, et al., 1987).

Answer: $F_G = 6.17 \times 10^{-4}$ kmol/m^2-s

2.12[b]. Evaporation of a drop of water falling in air.

Repeat Example 2.9 for a drop of water which is originally 2 mm in diameter.

Answer: 390 m

2.13[b]. Dissolution of a solid sphere into a flowing liquid stream.

Estimate the mass-transfer coefficient for the dissolution of sodium chloride from a cast sphere, 1.5 cm in diameter, if placed in a flowing water stream. The velocity of the 291 K water stream is 1.0 m/s.

Assume that the kinematic viscosity at the average liquid film conditions is 1.02×10^{-6} m^2/s, and the mass diffusivity is 1.25×10^{-9} m^2/s. The solubility of NaCl in water at 291 K is 0.35 g/cm^3, and the density of the saturated solution is 1.22 g/cm^3 (Perry and Chilton, 1973) .

Answer: $F_L = 1.06 \times 10^{-3}$ kmol/m^2-s

2.14[b]. Sublimation of a solid sphere into a gas stream.

During the experiment described in Problem 2.2, the air velocity was measured at 10 m/s. Estimate the mass-transfer coefficient predicted by equation (2-36) and compare it to the value measured experimentally. The following data for naphthalene might be needed: T_b = 491.1 K, V_c = 413 cm^3/mol.

Answer: $k_G = 1.12 \times 10^{-6}$ mol/cm^2-s-kPa

2.15[b]. Dissolution of a solid sphere into a flowing liquid stream.

The crystal of Problem 1.26 is a sphere 2 cm in diameter. It is falling at terminal velocity under the influence of gravity into a big tank of water at

288 K. The density of the crystal is 1,464 kg/m^3 (Perry and Chilton, 1973).

(a) Estimate the crystal's terminal velocity.

Answer: 0.56 m/s

(b) Estimate the rate at which the crystal dissolves and compare it to the answer obtained in Problem 1.26.

2.16^c. Mass transfer inside a circular pipe.

Water flows through a thin tube, the walls of which are lightly coated with benzoic acid ($C_7H_6O_2$). The water flows slowly, at 298 K and 0.1 cm/s. The pipe is 1 cm in diameter. Under these conditions, equation (2-63) applies.

(a) Show that a material balance on a length of pipe Z leads to

$$c_{Aout} = c_A^* \left[1 - \exp\left(\frac{-k_L \, aZ}{v} \right) \right] \tag{2-71}$$

where v is the average fluid velocity, and c_A^* is the equilibrium solubility concentration.

(b) What is the average concentration of benzoic acid in the water after 2 m of pipe. The solubility of benzoic acid in water at 298 K is 0.003 g/cm^3, and the mass diffusivity is 1.0×10^{-5} cm^2/s (Cussler, 1997).

Answer: 0.00114 g/cm^3

2.17^b. Mass transfer in a wetted-wall tower.

Water flows down the inside wall of a 25-mm-ID wetted-wall tower of the design of Figure 2.2, while air flows upward through the core. Dry air enters at the rate of 7 kg/m^2-s. Assume the air is everywhere at its average conditions of 309 K and 1 atm, the water at 294 K, and the mass-transfer coefficient constant. Compute the average partial pressure of water in the air leaving if the tower is 1 m long.

Answer: 1.52 kPa

2.18^c. Mass transfer in an annular space.

In studying the sublimation of naphthalene into an airstream, an investigator constructed a 3-m-long annular duct. The inner pipe was made from a 25-mm-OD, solid naphthalene rod; this was surrounded by a 50-mm-ID naphthalene pipe.

Air at 289 K and 1 atm flowed through the annular space at an average velocity of 15 m/s. Estimate the partial pressure of naphthalene in the airstream exiting from the tube. At 289 K, naphthalene has a vapor pressure of 5.2 Pa, and a diffusivity in air of 0.06 cm^2/s. Use the results of Problem 2.7 to estimate the mass-transfer coefficient for the inner surface; and equation (2-46), using the equivalent diameter defined in Problem 2.7, to estimate the coefficient from the outer surface.

Answer: 3.59 Pa

2.19^c. Benzene evaporation on the outside surface of a single cylinder.

Benzene is evaporating at the rate of 20 kg/hr over the surface of a porous 10-cm-diameter cylinder. Dry air at 325 K and 1 atm flows at right angle to the axis of the cylinder at a velocity of 2 m/s. The liquid is at a temperature of 315 K where it exerts a vapor pressure of 26.7 kPa. Estimate the length of the cylinder. For benzene, T_c = 562.2 K, P_c = 48.9 bar, M = 78, V_c = 259 cm^3/mol, Z_c = 0.271 (Reid, et al., 1987).

Answer: 1.74 m

2.20^b. Mass transfer in a packed bed.

Wilke and Hougan (1945) reported the mass transfer in beds of granular solids. Air was blown through a bed of porous celite pellets wetted with water, and by evaporating this water under adiabatic conditions, they reported gas-film coefficients for packed beds. In one run, the following data were reported:

effective particle diameter	5.71 mm
gas stream mass velocity	0.816 kg/m^2-s
temperature at the surface	311 K
pressure	97.7 kPa
k_G	4.415 × 10^{-3} kmol/m^2-s-atm

With the assumption that the properties of the gas mixture are the same as those of air, calculate the gas-film mass-transfer coefficient using equation (2-55) and compare the result with the value reported by Wilke and Hougan.

Answer: 2.1% difference

2.21[b]. Mass transfer and pressure drop in a packed bed.

Air at 373 K and 2 atm is passed through a bed 10-cm in diameter composed of iodine spheres 0.7-cm in diameter. The air flows at a rate of 2 m/s, based on the empty cross section of the bed. The porosity of the bed is 40%.

(a) How much iodine will evaporate from a bed 0.1 m long? The vapor pressure of iodine at 373 K is 6 kPa.

Answer: 0.41 kg/min

(b) Estimate the pressure drop through the bed.

2.22[b]. Volumetric mass-transfer coefficients in industrial towers.

The interfacial surface area per unit volume, a, in many types of packing materials used in industrial towers is virtually impossible to measure. Both a and the mass-transfer coefficient depend on the physical geometry of the equipment and on the flow rates of the two contacting, immiscible streams. Accordingly, they are normally correlated together as the *volumetric mass-transfer coefficient, $k_c a$*.

Empirical equations for the volumetric coefficients must be obtained experimentally for each type of mass-transfer operation. Sherwood and Holloway (*Trans. AIChE*, **36**, 21, 39, 1940) obtained the following correlation for the liquid-film mass-transfer coefficient in packed absorption towers:

$$\frac{k_L a}{D_{AB}} = \alpha \left[\frac{G_x}{\mu}\right]^{1-n} \left[\frac{\mu}{\rho D_{AB}}\right]^{0.5} \tag{2-72}$$

The values of α and n to be used in equation (2-71) for various industrial packings are listed in the following table, *when SI units are used exclusively*.

Packing	α	n
50-mm Raschig ring	341	0.22
38-mm ring	384	0.22
25-mm ring	426	0.22
13-mm ring	1391	0.35
9.5-mm ring	3116	0.46
38-mm Berl saddle	731	0.28
25-mm saddle	777	0.28
13-mm saddle	686	0.28
76-mm spiral tiles	502	0.28

(a) Consider the absorption of SO_2 with water at 294 K in a tower packed with 25-mm Raschig rings. If the liquid mass velocity is $G_x = 2.04$ kg/m^2-s, estimate the liquid-film mass-transfer coefficient. The diffusivity of SO_2 in water at 294 K is 1.7×10^{-9} m^2/s.

Answer: 6.7×10^{-3} s^{-1}

(b) Whitney and Vivian (1949) measured rates of absorption of SO_2 in water and found the following expression for 25-mm Raschig rings at 294 K

$$k_x a = 0.152 (G_x)^{0.82} \qquad (2\text{-}73)$$

where $k_x a$ is in kmole/m^3-s, and G_x is in kg/m^2-s. For the conditions described in part a, estimate the liquid-film mass-transfer coefficient using equation (2-73). Compare the results.

2.23[b]. Mass transfer in fluidized beds.

Cavatorta, et al. (1999) studied the electrochemical reduction of ferri-cyanide ions, $\{Fe(CN)_6\}^{-3}$, to ferrocyanide, $\{Fe(CN)_6\}^{-4}$, in aqueous alka-line solutions. They studied different arrangements of packed columns, including fluidized beds. The fluidized bed experiments were performed in a 5-cm-ID circular column, 75-cm high. The bed was packed with 0.534-mm spherical glass beads, with a particle density of 2.612 g/cm^3. The properties of the aqueous solutions were: density = 1,083 kg/m^3, viscosity = 1.30 cP, diffusivity = 5.90×10^{-10} m^2/s. They found that the porosity of the fluidized bed, ε, could be correlated with the superficial liquid velocity based on the empty tube, v_s, through

$$v_s = 8.88\,\varepsilon^{3.17}$$
$$0.35 < \varepsilon < 1.0$$

where v_s is in cm/s.

(a) Using equation (2-56), estimate the mass-transfer coefficient, k_L, if the porosity of the bed is 60%.

Answer: $k_L = 5.78 \times 10^{-5}$ m/s.

(b) Cavatorta, et al. (1999) proposed the following correlation to estimate the mass-transfer coefficient for their fluidized bed experimental runs:

$$\varepsilon j_D = 0.333 \left[\frac{Re}{\varepsilon(1-\varepsilon)} \right]^{-0.364}$$

$$j_D = \frac{k_L}{v_s} Sc^{2/3}$$

where Re is based on the empty tube velocity. Using this correlation, estimate the mass-transfer coefficient, k_L, if the porosity of the bed is 60%. Compare your result to that of part a.

2.24[b]. Mass transfer in a hollow-fiber boiler feedwater deaerator.

Consider the hollow-fiber BFW deaerator described in Example 2-13. If the water flow rate increases to 60,000 kg/hr while everything else remains constant, calculate the fraction of the entering dissolved oxygen that can be removed.

Answer: 94.2%

2.25[b]. Mass transfer in a hollow-fiber boiler feedwater deaerator.

(a) Consider the hollow-fiber BFW deaerator described in Example 2.13. Assuming that only oxygen diffuses across the membrane, calculate the gas volume flow rate and composition at the lumen outlet. The water enters the shell side at 298 K saturated with atmospheric oxygen, which means a dissolved oxygen concentration of 8.38 mg/L.

Answer: 44.2 L/min, 9.6% O_2

(b) Calculate the mass-transfer coefficient at the average conditions inside the lumen. Neglect the thickness of the fiber walls when estimating the gas velocity inside the lumen.

Answer: $k_c = 0.37$ cm/s

2.26$^{b, d}$. Ternary condensation inside a vertical tube.

Isopropanol (1) and water (2) are condensing in the presence of nitrogen (3) inside a vertical tube, 2.54 cm inside diameter. At the vapor inlet, the bulk composition of the gas phase is 11.23 mole% isopropanol, 42.46% water, and 46.31% nitrogen. The interfacial vapor composition is 14.57 mole% isopropanol, 16.40% water, and 69.03% nitrogen. Estimate the molar rates of condensation of isopropanol and water at that point. The following data apply:

viscosity of the vapor, $\mu = 1.602 \times 10^{-5}$ kg/m-s

density of the vapor, $\rho = 0.882$ kg/m^3

$\mathcal{D}_{12} = 16.00$ mm^2/s; $\mathcal{D}_{13} = 14.44$ mm^2/s; $\mathcal{D}_{23} = 38.73$ mm^2/s

vapor-phase Reynolds number, Re = 9,574

Answer: $N_2 = 0.575$ mole/m^2-s

REFERENCES

Bao, L., B. Liu, and G. G. Lipscomb, *AIChE J.*, **45**, 2346 (1999).

Bedingfield, C. H., and T. B. Drew, *Ind. Eng. Chem.*, **42**, 1164 (1950).

Benítez, J., *Process Engineering and Design for Air Pollution Control*, Prentice Hall, Englewood Cliffs, NJ (1993).

Bird, R. B., W. E. Stewart, and E. N. Lightfoot, *Transport Phenomena*, Wiley, New York, (1960).

Brian, P. L. T., and H. B. Hales, *AIChE J.*, **15**, 419 (1969).

Buckingham, E., *Phys. Rev.*, **2**, 345 (1914).

Cavatorta, O. N., et al., *AIChE J.*, **45**, 938 (1999)

Costello, M. J., A. G. Fane, P. A. Hogan, and R. W. Schofield, *J. Memb. Sci.*, **80**, 1 (1993).

Chilton, T. H., and A. P. Colburn, *Ind. Eng. Chem.*, **26**, 1183 (1934).

Cussler, E. L., *Diffusion*, 2nd ed., Cambridge University Press, New York (1997).

Davis, M. L., and D. A. Cornwell, *Introduction to Environmental Engineering*, 3rd ed., McGraw-Hill, New York (1998).

Dribicka, M. M., and O. C. Sandall, *Chem. Eng. Sci.*, **34**, 733 (1979).

Eckert, E. R. G., and R. M. Drake, *Heat and Mass Transfer*, 2nd ed., McGraw-Hill, New York (1959).

Ergun, S., *Chem. Eng. Prog.*, **48**, 89 (1952).

Evnochides, S., and G. Thodos, *AIChE J.*, **5**, 178 (1959).

Froessling, N., *Gerlands Beitr. Geophys.*, **52**, 170 (1939).

Gilliland, E. R., and T. K. Sherwood, *Ind. Eng. Chem.*, **26**, 516 (1934).

Gupta, A. S., and G. Thodos, *AIChE J*, **8**, 608 (1962).

Himmelblau, D. M., *Basic Principles and Calculations in Chemical Engineering*, 5th ed., Prentice Hall, Englewood Cliffs, NJ (1989).

Holman, J. P., *Heat Transfer*, 7th ed., McGraw-Hill, New York, NY (1990).

Levich, V. G., *Physicochemical Hydrodynamics*, Prentice Hall, Englewood Cliffs, NJ (1962).

Linton, W. H., and T. K. Sherwood, *Chem. Eng. Prog.*, **46**, 258 (1950).

Lucas, K., *Phase Equilibria and Fluid Properties in the Chemical Industry*, p. 573, Dechema, Frankfurt (1980).

Monrad, C. C., and J. F. Pelton, *Trans. AIChE*, **38**, 593 (1942).

Nandakumar, K., Y. Shu, and K. T. Chuang, *AIChE J.*, **45**, 2286 (1999).

Perry, R. H., and C. H. Chilton (eds), *Chemical Engineers' Handbook*, 5th ed., McGraw-Hill, New York (1973).

Prasad, R., and K. K. Sirkar, *AIChE J.*, **34**, 177 (1988).

Reid, R. C., J. M. Prausnitz, and B. E. Poling, *The Properties of Gases and Liquids*, 4th ed., McGraw-Hill, Boston (1987).

Sherwood, T. K., and F. A. Holloway, *Trans. AIChE,* **36**, 21, 39 (1940).

Sherwood, T. K., R. L. Pigford, and C. R. Wilke, *Mass Transfer,* McGraw-Hill, New York (1975).

Smith, J. M., H. C. Van Ness, and M. M. Abbott, *Introduction to Chemical Engineering Thermodynamics,* 5th ed., McGraw-Hill, New York (1996).

Steinberger, R. L., and R. E. Treybal, *AIChE J.,* **6**, 227 (1960).

Taylor, R., and R. Krishna, *Multicomponent Mass Transfer,* Wiley, New York (1993).

Treybal, R. E., *Mass-Transfer Operations*, 3rd ed., McGraw-Hill, New York (1980).

Welty, J. R., C. E. Wicks, and R. E. Wilson, *Fundamentals of Momentum, Heat, and Mass Transfer*, 3rd ed., Wiley, New York (1984).

Whitney, R. P., and J. E. Vivian, *Chem. Eng. Progr.,* 45, 323 (1949).

Wilke, C. R., and O. A. Hougan, *Trans. AIChE,* **41**, 445 (1945).

Winding, C. C., and A. J. Cheney, *Ind. Eng. Chem.,* **40**, 1087 (1948).

Yang, M. C., and E. L. Cussler, *AIChE J.,* **32**, 1910 (1986).

3

Interphase
Mass Transfer

3.1 INTRODUCTION

Thus far, we have considered only the diffusion of substances within a single phase. In most of the mass-transfer operations, however, two insoluble phases are brought into contact in order to permit transfer of constituent substances between them. Therefore we are now concerned with the simultaneous application of the diffusional mechanism for each phase to the combined system. We have seen that the rate of diffusion within each phase is dependent upon the concentration gradient existing within it. At the same time, the concentration gradients of the two-phase system are indicative of the departure from equilibrium which exists between the phases. Should equilibrium be established, the concentration gradients and hence the rate of diffusion will fall to zero. It is necessary, therefore, to consider both the diffusional phenomena and the equilibria in order to describe the various situations fully.

3.2 EQUILIBRIUM

Your objectives in studying this section are to be able to:

1. Understand the concept of equilibrium between two insoluble phases, and its importance from the point of view of mass transfer.
2. Represent interphase equilibrium data in the form of an equilibrium-distribution curve.
3. Review and use in phase equilibria calculations the concepts of Raoult's law, modified Raoult's law, and Henry's law.

120

It is convenient first to consider the equilibrium characteristics of a particular operation and then to generalize the result for others. As an example, consider the gas-absorption operation which occurs when ammonia is dissolved from an ammonia-air mixture by liquid water. Suppose a fixed amount of liquid water is placed in a closed container together with a gaseous mixture of ammonia and air, the whole arranged so that the system can be maintained at constant temperature and pressure. Since ammonia is very soluble in water, some ammonia molecules will instantly transfer from the gas into the liquid, crossing the interphase. A portion of the ammonia molecules escapes back into the gas, at a rate proportional to their concentration in the liquid. As more ammonia moves into the liquid, with the consequent increase in concentration within the liquid, the rate at which ammonia returns to the gas increases until eventually the rate at which it enters the liquid exactly matches that at which it leaves. At the same time, through the mechanism of diffusion, the concentrations throughout each phase become uniform. A dynamic equilibrium develops, and while ammonia molecules continue to transfer back and forth from one phase to the other, the net transfer falls to zero. The concentrations within each phase no longer change. To the observer who cannot see the individual molecules the diffusion has apparently stopped.

If we now inject additional ammonia into the container, a new set of equilibrium concentrations will be eventually established, with higher concentrations in each phase than were initially obtained. In this manner, we can eventually obtain the complete relationship between the equilibrium concentrations in both phases. If the ammonia is designated as substance A, the equilibrium mole fractions in the gas (y_A) and liquid (x_A) give rise to an *equilibrium-distribution curve* as shown in Figure 3.1. This curve results irrespective of the amounts of air and water that we start with, and is influenced only by the temperature and pressure of the system. It is important to note that at equilibrium the concentrations in the two phases are not equal; instead, the chemical potential of the ammonia is the same in both phases. It is this equality of chemical potentials which causes the net transfer of ammonia to stop. The curve of Figure 3.1 does not, of course, show all the equilibrium concentrations existing within the system. For example, water will partially vaporize into the gas phase, the components of the air will also dissolve to a small extent into the liquid, and equilibrium concentrations for these substances will also be established. For the moment, we need not consider these equilibria, since they are of minor importance.

Equations relating the equilibrium concentrations in the two phases have been developed and are presented in textbooks on thermodynamics. In cases involving ideal gas and liquid phases, the fairly simple, yet useful relation known as Raoult's law applies,

$$y_A P = x_A P_A \qquad (3\text{-}1)$$

where P_A is the vapor pressure of pure A at the equilibrium temperature, and P is the equilibrium pressure. If the liquid phase does not behave ideally, a modified form of

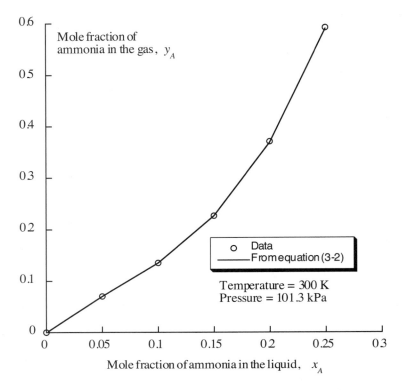

Figure 3.1 Equilibrium distribution of ammonia between air and water at constant temperature and pressure (data are from Perry and Chilton, 1973).

Raoult's law is

$$y_A P = x_A \gamma_A P_A \qquad (3\text{-}2)$$

where γ_A is the *activity coefficient* of species A in solution. Another equilibrium relation which is found to be true for dilute solutions is Henry's law, expressed by

$$p_A = y_A P = H x_A \qquad (3\text{-}3)$$

where p_A is the equilibrium partial pressure of component A in the vapor phase, H is the Henry's law constant. An equation similar to Henry's law relation describes the distribution of a solute between two immiscible liquids. This equation, the "distribution-law" equation, is

$$c_{A,liquid1} = K c_{A,liquid2} \qquad (3\text{-}4)$$

where K is the distribution coefficient.

A detailed discussion of the characteristic shapes of the equilibrium curves for the various situations and the influence of conditions such as temperature and pressure must be left for the studies of the individual unit operations. Nevertheless, the following principles are common to all systems involving the distribution of a component between two phases:

1. At a fixed set of conditions, such as temperature and pressure, Gibbs' phase rule stipulates that a set of equilibrium relations exists which may be shown in the form of an equilibrium-distribution curve.
2. When a system is in equilibrium, there is no net mass transfer between the phases.
3. When a system is not in equilibrium, diffusion of the components between the phases will occur in such a manner as to cause the system composition to shift toward equilibrium. If sufficient time is allowed, the system will eventually reach equilibrium.

The following examples illustrate the application of equilibrium relations for determining equilibrium concentrations.

Example 3.1 Application of Raoult's Law to a Binary System

Raoult's law may be used to determine phase compositions for the binary system benzene-toluene, at low temperatures and pressures. Determine the composition of the vapor in equilibrium with a liquid containing 0.4 mole fraction of benzene at 300 K, and calculate the total equilibrium pressure. Estimate the vapor pressure of benzene and toluene at 300 K from Antoine equation

$$\ln P_i = A_i - \frac{B_i}{C_i + T} \tag{3-5}$$

where P_i is the vapor pressure of component i, in mm Hg, and T is the temperature in K. The Antoine constants for benzene and toluene are given in the following table (Himmelblau, 1989):

Component	A	B	C
Benzene	15.9008	2788.51	− 52.36
Toluene	16.0137	3096.52	− 53.67

Solution

Benzene is more volatile than toluene, and will be designated component A. From equation (3-5), at 300 K, the vapor pressures of benzene and toluene are $P_A = 103.5$ mm Hg = 13.8 kPa and $P_B = 31.3$ mm Hg = 4.17 kPa. The partial pressures are $p_A = x_A P_A = (0.4)(13.8) = 5.52$ kPa; $p_B = x_B P_B = (0.6)(4.17) = 2.50$ kPa. The total pressure is $P = p_A + p_B = 8.02$ kPa. The molar fraction of benzene in the gas phase is $y_A = p_A/P = 5.52/8.02 = 0.688$.

Example 3.2 Henry's Law: Saturation of Water with Oxygen

The Henry's law constant for oxygen dissolved in water at 298 K is 4.5×10^4 atm/mole fraction. Determine the saturation concentration of oxygen in water exposed to dry air at 298 K and 1 atm.

Solution
Dry air contains 21% oxygen, then $p_A = y_A P = 0.21$ atm. The equilibrium liquid concentration is from Henry's law:

$$x_A = \frac{p_A}{H} = \frac{0.21}{4.5 \times 10^4} = 4.67 \times 10^{-6}$$

Basis: 1L of saturated solution

For one liter of very dilute solution of oxygen in water, the total moles of solution, n_T, will be approximately equal to the moles of water. The density is around 1kg/L, then $n_T = 1000/18 = 55.6$ mole. The moles of oxygen in 1 L of solution are $n_O = 55.6 \times 4.67 \times 10^{-6} = 2.6 \times 10^{-4}$ mole. The saturation concentration is $(2.6 \times 10^{-4}$ mole/L)(32 g/mole)(1000 mg/g) = 8.32 mg/L.

Example 3.3 Material Balances Combined with Equilibrium Relations: Algebraic Solution

Ten kilograms of dry gaseous ammonia, NH_3, and 15 m^3 of dry air measured at 300 K and 1 atm are mixed together and then brought into contact with 45 kg of water at 300 K in a piston/cylinder device. After a long period of time, the system reaches equilibrium. Assuming that the temperature and pressure remain constant, calculate the equilibrium concentrations of ammonia in the liquid and gas phases. Assume that the amount of water that evaporates and the amount of air that dissolves in the water are negligible. At 300 K and 1 atm, the equilibrium solubility of ammonia in air can be described by equation (3-2) with $P_A = 10.51$ atm, and the activity coefficient of ammonia given by

$$\gamma_A = 0.156 + 0.622 x_A (5.765 x_A - 1)$$
$$x_A \leq 0.3 \tag{3-6}$$

Solution
Convert the given quantities of ammonia, air, and water to moles. From the ideal gas law, $n_{air} = 15 \times 101.3/(8.314 \times 300) = 0.609$ kmol; $n_{water} = 45/18 = 2.5$ kmol. The

total amount of ammonia in the system is $n_A = 10/17 = 0.588$ kmol; in equilibrium, part of it will be in the liquid phase, and the rest in the gas phase. Define L_A as the number of kmol of ammonia in the liquid phase when equilibrium is achieved. Then, at equilibrium, $0.588 - L_A$ kmol of ammonia will remain in the gas phase. Assuming that all the air will remain in the gas phase, and all the water will remain in the liquid phase, the equilibrium concentrations can be written in terms of L_A :

$$x_A = \frac{L_A}{2.5 + L_A} \qquad y_A = \frac{0.588 - L_A}{0.609 + 0.588 - L_A} = \frac{0.588 - L_A}{1.197 - L_A} \qquad (3\text{-}7)$$

The equilibrium relation for 300 K and 1 atm is

$$y_A = 10.51 \gamma_A x_A \qquad (3\text{-}8)$$

Equations (3-6), (3-7), and (3-8) must be solved simultaneously for L_A, x_A, γ_A, and y_A.

This system of simultaneous nonlinear algebraic equations is easily solved using the "solve block" capabilities of Mathcad, as shown in Figure 3.2. Initial guesses of the values of the four variables must be supplied. However, the computational algorithm used by Mathcad is very robust and convergence to the true solution is virtually independent of the initial guesses. Figure 3.2 shows that, at equilibrium, the ammonia content of the liquid phase will be 14.4% by mole, and that of the gas phase will be 21.4% by mole. The amount of ammonia absorbed by the water will be 0.422 kmol, which is 71.8% of all the ammonia present in the system.

3.3 DIFFUSION BETWEEN PHASES

Your objectives in studying this section are to be able to:

1. Calculate interfacial mass-transfer rates in terms of the local mass-transfer coefficients for each phase.
2. Define and use, where appropriate, overall mass-transfer coefficients.

Having established that departure from equilibrium provides the driving force for diffusion, we can now study the rates of diffusion in terms of the driving forces. Many of the mass-transfer operations are carried out in steady-flow fashion, with continuous and invariant flow of the contacted phases and under circumstances such that concentrations at any position in the equipment used do not change with time. It will be convenient at this point to use one of these operations as an example with which to establish the principles, and to generalize with respect to other operations later.

Initial guesses:

$L_A := 0.5$ $\gamma_A := 1.0$ $x_A := 0.5$ $y_A := 0.5$

Given

$$x_A = \frac{L_A}{L_A + 2.5} \qquad y_A = \frac{0.588 - L_A}{1.197 - L_A}$$

$$\gamma_A = 0.156 + 0.622 \cdot x_A \cdot (5.765 \cdot x_A - 1)$$

$$y_A = 10.51 \cdot \gamma_A \cdot x_A$$

$$\begin{bmatrix} x_A \\ y_A \\ \gamma_A \\ L_A \end{bmatrix} := \mathrm{Find}(x_A, y_A, \gamma_A, L_A)$$

$$\begin{bmatrix} x_A \\ y_A \\ \gamma_A \\ L_A \end{bmatrix} = \begin{bmatrix} 0.144 \\ 0.214 \\ 0.141 \\ 0.422 \end{bmatrix}$$

Figure 3.2 Mathcad solution of Example 3.3.

For this purpose, let us consider the absorption of a soluble gas such as ammonia (substance A) from a mixture with air, by liquid water, in a wetted-wall tower. The ammonia-air mixture may enter at the bottom of the tower and flow upward, while the water flows downward around the inside of the pipe. The ammonia concentration in the gas mixture diminishes as it flows upward, while the water absorbs the ammonia as it flows downward and leaves at the bottom as an aqueous ammonia solution. Under steady-state conditions, the concentrations at any point in the apparatus do not change with passage of time.

3.3.1 Two-Resistance Theory

Consider the situation at a particular level along the tower. Since the solute is diffusing from the gas phase into the liquid, there must be a concentration gradient in the direction of mass transfer within each phase. This can be shown graphically in terms of the distance through the phases, as in Figure 3.3, where a section through the two phases in contact is shown. The concentration of A in the main body of the gas is $y_{A,G}$ mole fraction and it falls to $y_{A,i}$ at the interface. In the liquid, the concentration falls from $x_{A,i}$ at the interface to $x_{A,L}$ in the bulk liquid. The bulk concentrations $y_{A,G}$ and $x_{A,L}$ are clearly not equilibrium values, since otherwise diffusion of the solute would not occur. At the same time, these bulk concentrations cannot be used directly with a mass-transfer coefficient to describe the rate of interphase mass transfer since the two concentrations are differently related to the chemical potential, which is the real "driving force" of mass transfer.

To get around this problem, Lewis and Whitman (1924) assumed that the only diffusional resistances are those residing in the fluids themselves. There is then no resistance to solute transfer across the interface separating the phases, and as a result the concentrations $y_{A,i}$ and $x_{A,i}$ are equilibrium values given by the system's equilibrium-distribution curve. This concept has been called the "two-resistance theory."

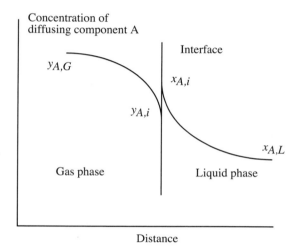

Figure 3.3 The two-resistance concept.

The reliability of this theory has been the subject of a great amount of study. A careful review of the results indicates that departure from concentration equilibrium at the interface must be a rarity (Treybal, 1980). Consequently, in most situations the interfacial concentrations in Figure 3.3 are those corresponding to a point on the equilibrium-distribution curve. Notice that the concentration rise at the interface, from $y_{A,i}$ to $x_{A,i}$, is not a barrier to diffusion in the direction gas to liquid. They are equilibrium concentrations and hence correspond to equal chemical potentials of substance A in both phases at the interface.

The various concentrations can also be shown graphically, as in Figure 3.4, whose coordinates are those of the equilibrium-distribution curve. Point P represents the two bulk-phase concentrations, and point M those at the interface. For steady-state mass transfer, the rate at which A reaches the interface from the gas must be equal to the rate at which it diffuses to the bulk liquid, so that no accumulation or depletion of A at the interface occurs. We can, therefore, write the flux of A in terms of the mass-transfer coefficients for each phase and the concentration changes appropriate to each. Let us assume at this point that we are dealing with dilute solutions in both the liquid and gas phases. Furthermore, in both phases, substance A (ammonia) diffuses through nondiffusing B (air in the gas phase, and water in the liquid phase). Therefore, the development that follows will be done in terms of the k-type coefficients. The general results, in terms of the F-type coefficients will be presented later. Thus, when k_x and k_y are the locally applicable coefficients:

$$N_A = k_y\left(y_{A,G} - y_{A,i}\right) = k_x\left(x_{A,i} - x_{A,L}\right) \tag{3-9}$$

Rearrangement as

$$\frac{y_{A,G} - y_{A,i}}{x_{A,L} - x_{A,i}} = -\frac{k_x}{k_y} \tag{3-10}$$

provides the slope of the line PM. If the mass-transfer coefficients are known, the interfacial concentrations and hence the flux N_A can be determined, either graphically by plotting the line PM on the equilibrium-distribution diagram or analytically by solving equation (3-10) simultaneously with an algebraic expression for the equilibrium-distribution curve,

$$y_{A,i} = f\left(x_{A,i}\right) \tag{3-11}$$

3.3.2 Overall Mass-Transfer Coefficients

In experimental determination of the rate of mass transfer, it is usually possible to determine the solute concentrations in the bulk of the fluids by sampling and analyzing. Successful sampling of the fluids at the interface, however, is ordinarily quite difficult since the greatest part of the concentration gradients takes place over

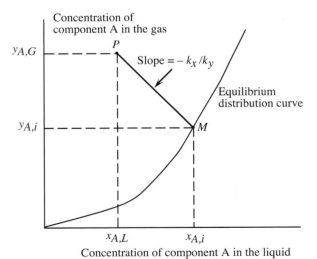

Figure 3.4 Interfacial concentrations as predicted by the two-resistance theory.

extremely small distances. Any ordinary sampling device will be so large in comparison that it would be practically impossible to get close enough to the interface. Sampling and analyzing, then, will yield $y_{A,G}$ and $x_{A,L}$, but not $y_{A,i}$ and $x_{A,i}$. Under these circumstances, only an overall effect, in terms of the bulk concentrations, can be determined. The bulk concentrations, however, are not by themselves on the same basis in terms of chemical potential.

Consider the situation shown in Figure 3.5. Since the equilibrium-distribution curve for the system is unique at fixed temperature and pressure, then $y_A{}^*$, in equilibrium with $x_{A,L}$, is as good a measure of $x_{A,L}$ as $x_{A,L}$ itself, and moreover it is on the same thermodynamic basis as $y_{A,G}$. The entire two-phase mass-transfer effect can then be measured in terms of an overall mass-transfer coefficient, K_y, which includes the resistance to diffusion in both phases in terms of a gas phase molar fraction driving force

$$N_A = K_y\left(y_{A,G} - y_A{}^*\right) \qquad (3-12)$$

In similar fashion, $x_A{}^*$ is a measure of $y_{A,G}$ and can be used to define another overall mass-transfer coefficient, K_x

$$N_A = K_x\left(x_A{}^* - x_{A,L}\right) \qquad (3-13)$$

A relation between the overall coefficients and the individual phase coefficients can be obtained when the equilibrium relation is linear as expressed by

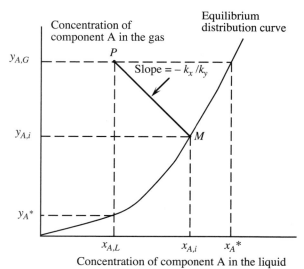

Figure 3.5 Overall concentration differences.

$$y_{A,i} = mx_{A,i} \tag{3-14}$$

This condition is always encountered at low concentrations, where Henry's law is obeyed; the proportionality constant is then $m = H/P$. Utilizing equation (3-14), we may relate the gas- and liquid-phase concentrations by

$$y_A{}^* = mx_{A,L} \tag{3-15}$$

$$y_{A,G} = mx_A{}^* \tag{3-16}$$

Rearranging equation (3-12), we obtain

$$\frac{1}{K_y} = \frac{y_{A,G} - y_A{}^*}{N_A} \tag{3-17}$$

From the geometry of Figure 3.5

$$y_{A,G} - y_A{}^* = (y_{A,G} - y_{A,i}) + (y_{A,i} - y_A{}^*) \tag{3-18}$$

Substituting equation (3-18) in (3-17)

$$\frac{1}{K_y} = \frac{y_{A,G} - y_A{}^*}{N_A} = \frac{y_{A,G} - y_{A,i}}{N_A} + \frac{y_{A,i} - y_A{}^*}{N_A} \tag{3-19}$$

or, in terms of m,

$$\frac{1}{K_y} = \frac{y_{A,G} - y_A{}^*}{N_A} = \frac{y_{A,G} - y_{A,i}}{N_A} + \frac{m(x_{A,i} - x_{A,L})}{N_A} \tag{3-20}$$

The substitution of equation (3-9) into the above relation relates K_y to the individual phase coefficients by

$$\frac{1}{K_y} = \frac{1}{k_y} + \frac{m}{k_x} \tag{3-21}$$

A similar expression for K_x may be derived as follows:

$$\frac{1}{K_x} = \frac{x_A{}^* - x_{A,L}}{N_A} = \frac{y_{A,G} - y_{A,i}}{mN_A} + \frac{(x_{A,i} - x_{A,L})}{N_A} \tag{3-22}$$

or

$$\frac{1}{K_x} = \frac{1}{mk_y} + \frac{1}{k_x} \tag{3-23}$$

Equations (3-21) and (3-23) lead to the following relationships between the mass-transfer resistances:

$$\frac{\text{Resistance in gas phase}}{\text{Total resistance in both phases}} = \frac{1/k_y}{1/K_y} \tag{3-24}$$

$$\frac{\text{Resistance in liquid phase}}{\text{Total resistance in both phases}} = \frac{1/k_x}{1/K_x} \qquad (3\text{-}25)$$

Assuming that the numerical values of k_x and k_y are roughly the same, the importance of the solubility of the gas—as indicated by the slope of the equilibrium curve, m—can readily be demonstrated. If m is small (solute A is very soluble in the liquid), the term m/k_x in equation (3-21) becomes minor, the major resistance is represented by $1/k_y$, and it is said that the rate of mass transfer is gas-phase-controlled. In the extreme, this becomes

$$\frac{1}{K_y} \approx \frac{1}{k_y} \qquad (3\text{-}26)$$

Under such circumstances, even fairly large percentage changes in k_x will not significantly affect K_y, and efforts to increase the rate of mass transfer would best be directed toward decreasing the gas-phase resistance. Conversely, when m is very large (solute A relatively insoluble in the liquid), with k_x and k_y nearly equal, the first term on the right of equation (3-23) becomes minor and the major resistance to mass transfer resides within the liquid, which is then said to control the rate. Ultimately, this becomes

$$\frac{1}{K_x} \approx \frac{1}{k_x} \qquad (3\text{-}27)$$

Remember that the relationships derived between the individual coefficients and the overall coefficients are valid only for a straight equilibrium-distribution line. Under such circumstances, the use of overall coefficients will eliminate the need to calculate the concentrations at the interface. If the distribution line is not straight, the overall mass-transfer coefficients will change with changing concentrations of the two phases, and the local coefficients should be used instead.

Example 3.4 Mass-Transfer Resistances During Absorption of Ammonia by Water

In an experimental study of the absorption of ammonia by water in a wetted-wall column, the value of K_G was found to be 2.75×10^{-6} kmol/m²-s-kPa. At one point in the column, the composition of the gas and liquid phases were 8.0 and 0.115 mole % NH_3, respectively. The temperature was 300 K, and the total pressure was 1 atm. Eighty-five percent of the total resistance to mass transfer was found to be in the gas phase. At 300 K, ammonia-water solutions follow Henry's law up to 5 mole%

ammonia in the liquid, with $m = 1.64$ when the total pressure is 1 atm. Calculate the individual film coefficients and the interfacial concentrations.

Solution

The first step in the solution is to convert the given overall coefficient from K_G to K_y. The relations among the overall coefficients are the same as those among the individual coefficients, presented in Chapter 2. Therefore, $K_y = K_G P = 2.75 \times 10^{-6} \times 101.3 = 2.786 \times 10^{-4}$ kmol/m²-s. From equation (3-24), for a gas-phase resistance that accounts for 85% of the total resistance,

$$k_y = \frac{K_y}{0.85} = 3.28 \times 10^{-4} \text{ kmol / m}^2\text{-s}$$

From equation (3-21)

$$\frac{m}{k_x} = \frac{1}{K_y} - \frac{1}{k_y} = \frac{0.15}{K_y}$$

or

$$k_x = \frac{mK_y}{0.15} = \frac{1.64 \times 2.786 \times 10^{-4}}{0.15} = 3.05 \times 10^{-3} \text{ kmol / m}^2\text{-s}$$

To estimate the ammonia flux and the interfacial concentrations at this particular point in the column, use equation (3-15) to calculate

$$y_A^* = mx_{A,L} = 1.64 \times 1.15 \times 10^{-3} = 1.886 \times 10^{-3}$$

The flux is from equation (3-12)

$$N_A = 2.768 \times 10^{-4} \left(0.080 - 1.866 \times 10^{-3}\right) = 2.18 \times 10^{-5} \text{ kmol / m}^2\text{-s}$$

Calculate the gas-phase interfacial concentration from equation (3-9)

$$y_{A,i} = y_{A,G} - \frac{N_A}{k_y} = 0.080 - \frac{2.18 \times 10^{-5}}{3.28 \times 10^{-4}} = 0.01362$$

Since the interfacial concentrations lie on the equilibrium line,

$$x_{A,i} = \frac{y_{A,i}}{m} = \frac{0.01362}{1.64} = 8.305 \times 10^{-3}$$

As a double check on the calculations so far, calculate the ammonia flux based on the conditions in the liquid phase,

$$N_A = 3.05 \times 10^{-3} \left(8.305 \times 10^{-3} - 1.15 \times 10^{-3}\right) = 2.18 \times 10^{-5} \text{ kmol / m}^2\text{-s}$$

The two estimates of the flux agree within the round-off error. Notice that, in this example, it was not necessary to calculate the interfacial concentrations to estimate the ammonia flux since the use of overall coefficients was appropriate. They were calculated just for illustration purposes. Usually, for the dilute conditions inherent to the use of k-type coefficients, the equilibrium relation follows Henry's law, and overall coefficients are justified.

3.3.3 Local Mass-Transfer Coefficients—General Case

When we deal with situations which do not involve either diffusion of only one substance or equimolar counterdiffusion, or if mass-transfer rates are large, the F-type coefficients should be used. The general approach is the same, although the resulting expressions are more cumbersome than those developed above. Thus, in a situation like that shown in Figures 3.3 to 3.5, the mass-transfer flux is

$$N_A = \Psi_{A,G} F_G \ln \frac{\Psi_{A,G} - y_{A,i}}{\Psi_{A,G} - y_{A,G}} = \Psi_{A,L} F_L \ln \frac{\Psi_{A,L} - x_{A,L}}{\Psi_{A,L} - x_{A,i}} \qquad (3\text{-}28)$$

Rearranging equation (3-28)

$$\frac{\Psi_{A,G} - y_{A,i}}{\Psi_{A,G} - y_{A,G}} = \left[\frac{\Psi_{A,L} - x_{A,L}}{\Psi_{A,L} - x_{A,i}} \right]^{F_L \Psi_{A,L} / F_G \Psi_{A,G}} \qquad (3\text{-}29)$$

The interfacial compositions $y_{A,i}$ and $x_{A,i}$ must satisfy simultaneously equation (3-29) and the equilibrium-distribution relation. If the equilibrium relation is in the form of an algebraic expression, then the interfacial compositions will be obtained solving simultaneously the equilibrium expression and equation (3-29). If the equilibrium relation is in the form of a distribution diagram, plot equation (3-29)—with $y_{A,i}$ replaced by y_A and $x_{A,i}$ by x_A—on the distribution diagram and determine graphically the intersection of the resulting curve with the distribution curve.

Example 3.5 Absorption of Ammonia by Water: Use of F-Type Mass-Transfer Coefficients

A wetted-wall absorption tower is fed with water as the wall liquid and an ammonia-air mixture as the central-core gas. At a particular point in the tower, the ammonia concentration in the bulk gas is 0.60 mole fraction, that in the bulk liquid is 0.12 mole fraction. The temperature is 300 K and the pressure is 1 atm. Ignoring the vaporization of water, calculate the local ammonia mass-transfer flux. The rates of flow are such that $F_L = 0.0035$ kmol/m²-s, and $F_G = 0.0020$ kmol/m²-s. The equilibrium-distribution data for the system at 300 K and 1 atm are those shown graphically in Figure 3.1, and algebraically in Example 3.3.

Solution

In this case, the ammonia concentrations are too high ($y_{A,G} = 0.60$, $x_{A,L} = 0.12$), particularly in the gas phase, to justify the use of k-type coefficients. Therefore, equations (3-28) and (3-29) will be used, together with the equilibrium-distribution data, to calculate the interfacial ammonia compositions and the local flux. In the gas phase, substance A is ammonia and substance B is air. Since the solubility of air in water is so low, me may assume that $N_{B,G} = 0$, and $\Psi_{A,G} = 1.0$. In the liquid phase, substance B is water. Ignoring the vaporization of water (the vapor pressure of water at 300 K is only 3.5 kPa), we may assume that $N_{B,L} = 0$, and $\Psi_{A,L} = 1.0$. Then, equation (3-29) reduces to

$$\frac{1 - y_{A,i}}{1 - y_{A,G}} = \left[\frac{1 - x_{A,L}}{1 - x_{A,i}}\right]^{F_L / F_G} \tag{3-30}$$

Substituting into equation (3-30) the numerical values given, and rearranging:

$$y_{A,i} = 1 - \left(1 - y_{A,G}\right)\left[\frac{1 - x_{A,L}}{1 - x_{A,i}}\right]^{F_L / F_G} = 1 - 0.4\left[\frac{0.88}{1 - x_{A,i}}\right]^{1.75} \tag{3-31}$$

Equation (3-31) must be combined with the equilibrium data to calculate the interfacial concentrations. Next, we will illustrate both the algebraic and graphical solutions.

(a) Algebraic solution:

The equilibrium relations for this case are equations (3-8) and (3-6), which become

$$y_{A,i} = 10.51\gamma_A x_{A,i} \tag{3-8a}$$

$$\gamma_A = 0.156 + 0.622 x_{A,i}(5.765 x_{A,i} - 1) \tag{3-6a}$$

Equations (3-31), (3-6a), and (3-8a) constitute a system of simultaneous nonlinear algebraic equations which is easily solved using the "solve block" capabilities of Mathcad, as shown in Figure 3.6. Initial guesses of the values of the three variables must be supplied. Figure 3.6 shows that the local interfacial concentrations are $x_{A,i} = 0.231$ and $y_{A,i} = 0.494$.

(b) Graphical solution:

Equation (3-31) relates the interfacial concentrations to each other. So, it is valid only at that point on the equilibrium-distribution curve which represents the local interfacial compositions. For the purpose of locating that point, rewrite equation (3-31) as a continuous relation between the variables in the equilibrium-distribution diagram, namely x_A and y_A:

Initial guesses:

$\gamma_A := 1.0 \qquad x_A := 0.2 \qquad y_A := 0.5$

Given

$$y_A = 1 - 0.4 \cdot \left(\frac{0.88}{1 - x_A} \right)^{1.75}$$

$$\gamma_A = 0.156 + 0.622 \cdot x_A \cdot (5.765 \cdot x_A - 1)$$

$$y_A = 10.51 \cdot \gamma_A \cdot x_A$$

$$\begin{bmatrix} x_A \\ y_A \\ \gamma_A \end{bmatrix} := \text{Find}(x_A, y_A, \gamma_A)$$

$$\begin{bmatrix} x_A \\ y_A \\ \gamma_A \end{bmatrix} = \begin{bmatrix} 0.231 \\ 0.494 \\ 0.204 \end{bmatrix}$$

Figure 3.6 Mathcad solution of Example 3.5 a.

$$y_A = 1 - 0.4 \left[\frac{0.88}{1 - x_A} \right]^{1.75} \qquad (3\text{-}32)$$

Equation (3-32) represents a continuous curve on the xy diagram; its intersection with the equilibrium curve will yield the interfacial concentrations. To plot equation (3-32), generate the following table of (x_A, y_A) values:

x_A	0.12	0.15	0.20	0.25	0.30
y_A	0.600	0.575	0.527	0.471	0.403

These values are plotted on the equilibrium-distribution diagram, as shown in Figure 3.7. The resulting curve intersects the equilibrium curve to give the interface compositions $x_{A,i} = 0.231$ and $y_{A,i} = 0.494$. This solution agrees with the solution obtained previously by the algebraic method.

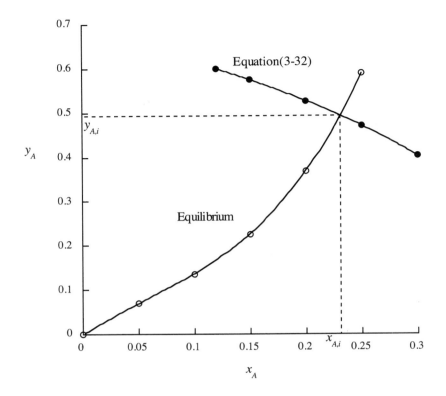

Figure 3.7 Graphical solution of Example 3.5 b.

Once the interfacial compositions have been calculated by either method, the local flux is from equation (3-28), which, in this case, becomes

$$N_A = F_G \ln \frac{1 - y_{A,i}}{1 - y_{A,G}} = 2 \times 10^{-3} \ln \frac{1 - 0.494}{1 - 0.600}$$

$$N_A = 4.7 \times 10^{-4} \text{ kmol / m}^2\text{- s}$$

Example 3.6 Distillation of a Mixture of Methanol and Water in a Packed Tower: Use of F-Type Mass-Transfer Coefficients

A mixture of methanol (substance A, the more volatile) and water (substance B) is being distilled in a packed tower at constant pressure of 1 atm. At a point along the tower, the methanol content of the bulk of the gas phase is 36 mole%; that of the bulk of the liquid phase is 20 mole%. The temperature at that point in the tower is around 360 K. At that temperature, the molar latent heat of vaporization of methanol is $\lambda_A = 33.3$ MJ/kmol, and that of water is $\lambda_B = 41.3$ MJ/kmol. The packing characteristics and flow conditions at that point are such that $F_G = 0.0017$ kmol/m²-s, and $F_L = 0.0149$ kmol/m²-s. Estimate the local flux of methanol from the liquid to the gas phase.

The VLE for this system is adequately described by the modified Raoult's law, equation (3-2), using the Wilson equation to estimate the activity coefficients (Smith, et al., 1996)

$$\ln \gamma_A = -\ln\left(x_A + x_B \Delta_{12}\right) + x_B\left[\frac{\Delta_{12}}{x_A + x_B \Delta_{12}} - \frac{\Delta_{21}}{x_B + x_A \Delta_{21}}\right] \qquad (3\text{-}33)$$

$$\ln \gamma_B = -\ln\left(x_B + x_A \Delta_{21}\right) - x_A\left[\frac{\Delta_{12}}{x_A + x_B \Delta_{12}} - \frac{\Delta_{21}}{x_B + x_A \Delta_{21}}\right] \qquad (3\text{-}34)$$

where

$$\Delta_{i,j} = \frac{V_j}{V_i}\exp\left[\frac{-a_{i,j}}{RT}\right] \qquad (i \neq j) \qquad (3\text{-}35)$$

where V_j and V_i are the molar volumes at temperature T of pure liquids j and i, and $a_{i,j}$ is a constant independent of composition and temperature. Recommended values of the parameters for this system are $V_A = 40.73$ cm³/mole, $V_B = 18.07$ cm³/mole, $a_{A,B} = 107.38$ cal/mole, $a_{B,A} = 469.55$ cal/mole (Smith, et al., 1996).

Solution

Assume that the distillation operation takes place inside the tower adiabatically, in such a way that the energy required to evaporate methanol is supplied by condensation of water vapor. Therefore, from an energy balance

$$N_B = -\frac{\lambda_A}{\lambda_B} N_A = -\frac{33.3}{41.3} N_A = -0.806 N_A$$

$$\Psi_{A,G} = \Psi_{A,L} = \frac{1}{1 - 0.806} = 5.155$$

Equation (3-29) becomes

$$\frac{5.155 - y_{A,i}}{5.155 - 0.360} = \left[\frac{5.155 - 0.200}{5.155 - x_{A,i}}\right]^{0.0149/0.0017}$$

Rearranging,

$$y_{A,i} = 5.155 - 4.795 \left[\frac{4.955}{5.155 - x_{A,i}} \right]^{8.765} \tag{3-36}$$

The modified form of Raoult's law is written for both components:

$$y_{A,i} = \frac{x_{A,i}\gamma_A P_A}{P}, \qquad y_{B,i} = \frac{x_{B,i}\gamma_B P_B}{P} \tag{3-37}$$

For a binary system, such as this

$$y_{A,i} + y_{B,i} = 1 , \qquad x_{A,i} + x_{B,i} = 1 \tag{3-38}$$

Antoine equations for methanol and water are (Smith, et al., 1996):

$$\ln P_A = 16.5938 - \frac{3644.3}{T - 33}, \qquad \ln P_B = 16.2620 - \frac{3800.0}{T - 47} \tag{3-39}$$

where P_i is the vapor pressure in kPa, and T is in K. Equations (3-36) to (3-38) constitute a system of five equations in five unknowns: the equilibrium temperature, and the four interfacial mole fractions. They are solved, with the help of equations (3-33) to (3-35), and (3-39), using Mathcad as shown in Figure 3.8. The solution is $x_{A,i} = 0.177$, $y_{A,i} = 0.549$, $T = 356.5$ K. The local methanol flux is from equation (3-28)

$$N_A = 5.155 \times 0.0017 \times \ln \frac{5.155 - 0.549}{5.155 - 0.360}$$

$$= -3.52 \times 10^{-4} \text{ kmol / m}^2\text{-s}$$

The negative sign of the flux implies that the methanol flows from the liquid to the gas phase, which is in the opposite direction to that assumed in the derivation of equation (3-28).

An alternative approximate solution to this problem can be obtained through the use of k'-type coefficients. As mentioned in Example 2.3, the McCabe-Thiele method of analysis of distillation columns assumes equimolar counterdiffusion, which would justify the use of k'-type coefficients regardless of the concentration levels. This would be exactly so only if the molar latent heats of vaporization of both components were equal. Although in this case there is a significant difference between the two heats of vaporization, we will show that the approximate solution is fairly close to the rigorous solution obtained above. It is easily shown that in this case, assuming steady-state equimolar counterdiffusion:

$$\frac{y_{A,G} - y_{A,i}}{x_{A,L} - x_{A,i}} = -\frac{k'_x}{k'_y} = -\frac{F_L}{F_G} \tag{3-40}$$

This is the equation of a straight line on the xy diagram with slope $= -F_L/F_G$.

Parameters:

a12:= 107.38   a21:= 469.55   V1:= 40.73

R:= 1.987   P:= 101.3   V2:= 18.07

Initial estimates:

x1:= 0.1   x2:= 0.9   y1:= 0.2   y2:= 0.8   T:= 360

Correlations:

$$\Delta 12(T):= \frac{V2}{V1} \cdot e^{\frac{-a12}{R \cdot T}} \qquad \Delta 21(T):= \frac{V1}{V2} \cdot e^{\frac{-a21}{R \cdot T}}$$

$$P1(T):= e^{\left[16.5938 - \frac{3644.3}{T-33}\right]} \qquad P2(T):= e^{\left[16.2620 - \frac{3800.0}{T-47}\right]}$$

$$\gamma 1(T,x1,x2):= e^{-\ln(x1+x2 \cdot \Delta 12(T))+x2 \cdot \left[\frac{\Delta 12(T)}{x1+x2 \cdot \Delta 12(T)} - \frac{\Delta 21(T)}{x2+x1 \cdot \Delta 21(T)}\right]}$$

$$\gamma 2(T,x1,x2):= e^{-\ln(x2+x1 \cdot \Delta 21(T))-x1 \cdot \left[\frac{\Delta 12(T)}{x1+x2 \cdot \Delta 12(T)} - \frac{\Delta 21(T)}{x2+x1 \cdot \Delta 21(T)}\right]}$$

Solve block:

Given

x1 + x2 = 1   y1 + y2 = 1

$$y1 = \frac{x1 \cdot \gamma 1(T,x1,x2) \cdot P1(T)}{P} \qquad y2 = \frac{x2 \cdot \gamma 2(T,x1,x2) \cdot P2(T)}{P}$$

$$y1 = 5.155 - 4.795 \cdot \left[\frac{4.955}{5.155 - x1}\right]^{8.765}$$

$$\begin{bmatrix} x1 \\ y1 \\ x2 \\ y2 \\ T \end{bmatrix} := \text{Find}(x1,y1,x2,y2,T) \qquad \begin{bmatrix} x1 \\ y1 \\ x2 \\ y2 \\ T \end{bmatrix} = \begin{bmatrix} 0.177 \\ 0.549 \\ 0.823 \\ 0.451 \\ 356.464 \end{bmatrix}$$

Figure 3.8 Mathcad solution of Example 3.6.

Figure 3.9 is the equilibrium-distribution diagram for this system at a pressure of 1 atm. Point P ($y_{A,G}$, $x_{A,L}$) represents the bulk concentration of both phases, and point M ($y_{A,i}$, $x_{A,i}$) on the equilibrium curve represents the interfacial concentrations. The straight line joining those two points has a slope of $-F_L/F_G = -8.765$.

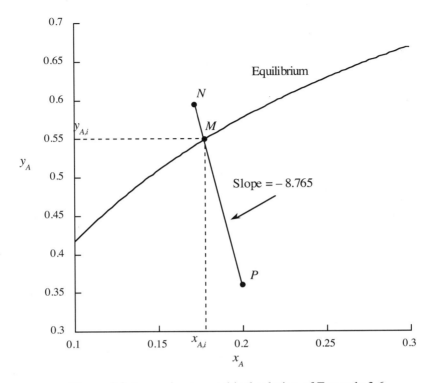

Figure 3.9 Approximate graphical solution of Example 3.6.

To draw the straight line passing through point P $(y_{A,G}$, $x_{A,L})$ with the given slope of -8.765, choose arbitrarily a value of y_A on the line, for example, $y_A = 0.6$, and calculate the corresponding value of x_A from the expression for the slope:

$$\frac{0.36 - 0.6}{0.2 - x_A} = -8.765 \quad \Rightarrow \quad x_A = 0.172$$

Therefore, the points N (0.172, 0.600) and P (0.200, 0.360) completely define the straight line that contains the segment PM. The intersection of line PN with the equilibrium curve yields the interfacial concentrations at point M. They are very similar to those obtained by the rigorous method using the F-type coefficients, namely, $x_{A,i} = 0.178$, $y_{A,i} = 0.550$. The local methanol flux is

$$N_A = k'_y\left(y_{A,i} - y_{A,G}\right) = F_G\left(y_{A,i} - y_{A,G}\right)$$
$$= 1.7 \times 10^{-3}(0.550 - 0.360)$$
$$= 3.23 \times 10^{-4} \text{ kmol / m}^2\text{- s}$$

which is about 8% lower than the flux value obtained through the rigorous solution method. This was to be expected since assuming equimolar counterdiffusion implies that there is no net bulk flow contribution, which underestimates the actual flow in the present situation where $N_A + N_B = 0.194 \, N_A$.

3.4 MATERIAL BALANCES

Your objectives in studying this section are to be able to:

1. Understand the importance of material balances in interphase mass-transfer calculations.
2. Define and generate minimum and actual operating lines for batch operations, and for steady-state cocurrent and countercurrent processes.

The concentration-difference driving forces discussed so far are those existing at a given position in the equipment used to contact the immiscible phases. In the case of a steady-state flow process, because of the transfer of solute from one phase to the other, the concentration within each phase changes as it moves through the equipment. Similarly, in the case of a batch process, the concentration in each phase changes with time. These changes produce corresponding variations in the driving forces, and these can be followed with the help of material balances.

3.4.1 Countercurrent Flow

Consider any steady-state mass-transfer operation which involves the countercurrent contact of two immiscible phases as shown schematically in Figure 3.10. The two phases will be identified as phase V and phase L, and for the present consider only the case where a single substance A diffuses from phase V to phase L during their contact.

At the bottom of the mass-transfer device, or plane z_1, the flow rates and concentrations are defined as follows:

V_1 total moles of phase V entering the tower per unit time
L_1 total moles of phase L entering the tower per unit time
y_1 mole fraction of component A in V_1;
x_1 mole fraction of component A in L_1;

Similarly, at the top of the device, or plane z_2, the total moles of each phase will be V_2 and L_2, and the compositions of each stream will be y_2 and x_2.

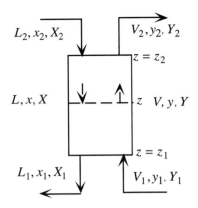

Figure 3.10 Steady-state countercurrent process.

An overall macroscopic mass balance for component A around the steady-state device (from plane $z = z_1$ to plane $z = z_2$), in which there is no chemical production or disappearance of A, requires that

$$\begin{bmatrix} \text{moles of A entering} \\ \text{the device} \end{bmatrix} = \begin{bmatrix} \text{moles of A leaving} \\ \text{the device} \end{bmatrix} \qquad (3\text{-}41)$$

or

$$V_1 y_1 + L_2 x_2 = V_2 y_2 + L_1 x_1 \qquad (3\text{-}42)$$

A mass balance for component A around plane $z = z_1$ and the arbitrary plane at z stipulates

$$V_1 y_1 + Lx = Vy + L_1 x_1 \qquad (3\text{-}43)$$

Simpler relations, and certainly easier equations to use, may be expressed in terms of *solute-free concentration units or mole ratios*. The concentration of each phase will be defined as follows:

Y moles of A in phase V per mole of A-free V; that is,

$$Y = \frac{y}{1-y} \qquad\qquad y = \frac{Y}{1+Y} \qquad (3\text{-}44)$$

X moles of A in phase L per mole of A-free L; that is,

$$X = \frac{x}{1-x} \qquad\qquad x = \frac{X}{1+X} \qquad (3\text{-}45)$$

When using mole ratios as concentration units, it is convenient to express the flow rates in terms of L_S and V_S. These are defined as:

L_S moles of phase L on a solute-free basis, per unit time

V_S moles of phase V on a solute-free basis, per unit time.

The overall balance on component A may be written, using the solute-free terms as

$$V_S Y_1 + L_S X_2 = V_S Y_2 + L_S X_1 \qquad (3\text{-}46)$$

Rearranging, we obtain

$$\frac{L_S}{V_S} = \frac{Y_1 - Y_2}{X_1 - X_2} \qquad (3\text{-}47)$$

Equation (3-47) is the equation of a straight line on the XY diagram which passes through the points (X_1, Y_1) and (X_2, Y_2) with a slope of L_S/V_S. A mass balance on component A around plane z_1 and the arbitrary plane $z = z$ in solute-free terms is

$$\frac{L_S}{V_S} = \frac{Y_1 - Y}{X_1 - X} \qquad (3\text{-}48)$$

As before, equation (3-48) is the equation of a straight line which passes through the points (X_1, Y_1) and (X, Y) with a slope of L_S/V_S. It is a general expression relating the bulk compositions of the two phases at any plane in the device. Since it defines operating conditions within the equipment, it is designated the *operating-line equation for countercurrent operation*. Figures 3.11 and 3.12 illustrate the location of the operating line relative to the equilibrium line when the transfer is from phase V to phase L, and from phase L to phase V.

It is important to notice the difference between equations (3-43) and (3-48). Although both equations describe the mass balance for component A, only (3-48) is

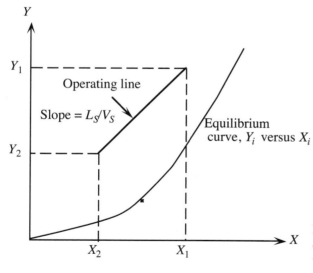

Figure 3.11 Steady-state countercurrent process, transfer from phase V to L.

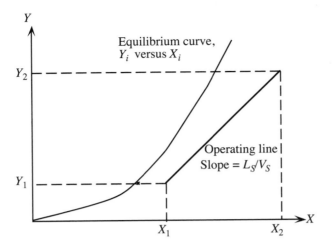

Figure 3.12 Steady-state countercurrent process, transfer from phase L to V.

the equation of a straight line. When written in terms of the solute-free units X and Y, the operating line is straight because the mole-ratio concentrations are based on the constant quantities L_S and V_S. On the other hand, when written in terms of mole-fraction units—x and y—the total moles in a phase—L or V—change as the solute is transferred into or out of the phase. This produces a curved operating line on x-y coordinates.

In the design of mass-transfer equipment, the flow rate of one phase, and three of the four entering and exiting compositions, must be fixed by the process requirements. The necessary flow rate of the second phase is a design variable. For example, consider the case in which phase V, with a known value of V_S, changes in composition from Y_1 to Y_2 by transferring solute to a second phase which enters the tower with composition X_2. According to equation (3-48), the operating line must pass through point (X_2, Y_2) with a slope L_S/V_S, and must end at the ordinate Y_1. Recall that V_S, is fixed, but L_S is subject to choice. If such a quantity L_S is used to give operating line DE on Figure 3.13 (a), the exiting phase L will have the composition X_1. If less L_S is used, the exit-L phase composition will clearly be greater, as at point F, but, since the operating line will now be closer to the equilibrium curve, the driving forces for diffusion are less and the solute transfer is more difficult. The time of contact between the phases must be greater, and the mass-transfer device must be correspondingly bigger.

The minimum value of L_S which can be used while still achieving the desired amount of solute transfer (a reduction in concentration of the V phase from Y_1 to Y_2) corresponds to the operating line DM, which is tangent to the equilibrium curve at P. Under these conditions, the exiting phase L will have the composition X_1 (max). Any value of L_S smaller than L_S (min) would result in an operating line that would cross and go under the equilibrium curve, a physical impossibility if the solute transfer is to remain from phase V to phase L. At P, the diffusional driving force is zero, the required time of contact for the concentration change desired is infinite, and an infinitely big mass-transfer device results. This then represents a limiting condition, the *minimum L_S/V_S ratio for mass transfer*. The actual L_S/V_S ratio must be higher than the minimum to ensure a finite size of the device; it is usually specified as a multiple of L_S (min)/V_S. Therein lies the importance of estimating accurately the ratio $L_S(\text{min})/V_S$.

The equilibrium curve is frequently concave upward, as in Figure 3.13 (b), and $L_S(\text{min})/V_S$ corresponds to the phase L_1 leaving in equilibrium with the entering phase G_1; that is, point M [$X_1(\text{max})$, Y_1] lies on the equilibrium curve.

Figure 3.14 illustrates the location of the minimum and actual operating lines when solute transfer is from phase L to phase V. In this case, L_S is specified and V_s is subject to change. The value of V_s to be used must be higher than the minimum.

Example 3.7 Recovery of Benzene Vapors from a Mixture with Air

A waste airstream from a chemical process flows at the rate of 1.0 m³/s at 300 K and 1 atm, containing 7.4% by volume of benzene vapors. It is desired to recover 85% of the benzene in the gas by a three-step process. First, the gas is scrubbed using a nonvolatile wash oil to absorb the benzene vapors. Then, the wash oil leaving the absorber is stripped of the benzene by contact with steam at 1 atm and 373 K. The

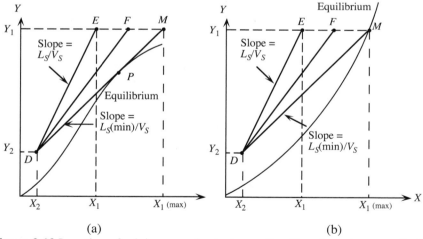

(a)   (b)

Figure 3.13 Location of minimum and actual operating lines for solute transfer from phase V to phase L.

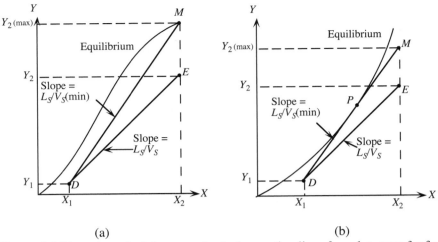

(a)   (b)

Figure 3.14 Location of minimum and actual operating lines for solute transfer from phase L to phase V.

mixture of benzene vapor and steam leaving the stripper will then be condensed. Because of the low solubility of benzene in water, two distinct liquid phases will form and the benzene layer will be recovered by decantation. The aqueous layer will be purified and returned to the process as boiler feedwater. The oil leaving the stripper will be cooled to 300 K and returned to the absorber. Figure 3.15 is a schematic diagram of the process.

The wash oil entering the absorber will contain 0.0476 mole fraction of benzene; the pure oil has an average molecular weight of 198. An oil circulation rate of twice the minimum will be used. In the stripper, a steam rate of 1.5 times the minimum will be used.

Compute the oil-circulation rate and the steam rate required for the operation. Wash oil–benzene solutions are ideal. The vapor pressure of benzene at 300 K is 0.136 atm, and 1.77 atm at 373 K.

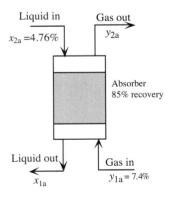

Liquid in   Gas out

$x_{2a} = 4.76\%$ y_{2a}

Absorber
85% recovery

Liquid out   Gas in

x_{1a} $y_{1a} = 7.4\%$

Solution

Consider first the conditions in the absorber. From the flow conditions of the entering gas and the given tower diameter, using the ideal gas, calculate the molar gas velocity entering at the bottom of the tower. (In all calculations in this example, we will use a subscript "a" to indicate flows and concentrations in the absorber, and a subscript "s" to refer to conditions in the stripper.)

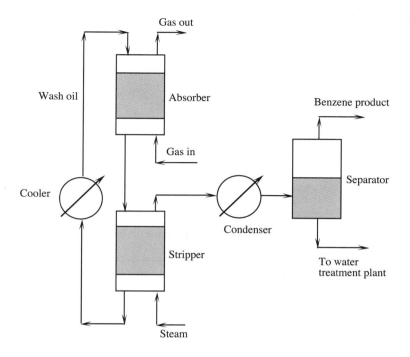

Gas out

Wash oil

Absorber

Gas in

Cooler

Benzene product

Separator

Condenser

Stripper

To water
treatment plant

Steam

Figure 3.15 Schematic diagram of the benzene-recovery process of Example 3.7.

$$V_{1a} = \frac{101.3 \times 1.0}{8.314 \times 300} = 0.041 \text{ kmol / s}$$

Calculate the inert-gas molar velocity

$$V_{sa} = V_{1a}(1 - y_{1a}) = 0.041 \times (1 - 0.074) = 0.038 \text{ kmol / s}$$

Convert the entering-gas mole fraction to a mole ratio using equation (3-44)

$$Y_{1a} = \frac{y_{1a}}{1 - y_{1a}} = \frac{0.074}{1 - 0.074} = 0.08 \text{ kmol of benzene / kmol dry gas}$$

Convert the entering-liquid mole fraction to a mole ratio using equation (3-45)

$$X_{2a} = \frac{x_{2a}}{1 - x_{2a}} = \frac{0.0476}{1 - 0.0476} = 0.05 \text{ kmol of benzene / kmol oil}$$

Since the absorber will recover 85% of the benzene in the entering gas, the concentration of the gas leaving it will be

$$Y_{2a} = (1 - 0.85) \times 0.08 = 0.012 \text{ kmol of benzene / kmol dry gas}$$

To determine the minimum amount of wash oil required, we need the equilibrium distribution curve for this system, expressed in terms of molar ratios. The benzene–wash oil solutions are ideal, and the pressure is low, therefore Raoult's law applies. From equations (3-1), (3-44), and (3-45)

$$y_{ia} = 0.136 x_{ia}$$

$$\frac{Y_{ia}}{1 + Y_{ia}} = 0.136 \frac{X_{ia}}{1 + X_{ia}}$$

The following table of equilibrium values, for the conditions prevailing in the absorber, is generated from the equation above:

X_{ia}, kmol benzene/kmol oil	Y_{ia}, kmol benzene/kmol dry gas
0.00	0.000
0.10	0.013
0.20	0.023
0.30	0.032
0.40	0.040
0.60	0.054
0.80	0.064
1.00	0.073
1.20	0.080
1.40	0.086

Figure 3.16 is the resulting equilibrium distribution curve. Notice that, although Raoult's law is always represented as a straight line on the xy diagram, it is not necessarily so on the XY diagram. We are now ready to locate the minimum operating line. Locate point D (X_{2a}, Y_{2a}) on the XY diagram. This is one end of the operating line and its location is fixed, independent of the amount of liquid used. Starting with any arbitrary operating line, such as DE, rotate it toward the equilibrium curve using D as a pivot point until the operating line touches the equilibrium curve for the first time. In this case, because of the shape of the equilibrium curve, the first contact occurs at point P where the operating line is tangent to the curve. Extending this line all the way to $Y_a = Y_{1a}$, we generate operating line DM, corresponding to the minimum oil rate. From the diagram, we read that X_{1a}(max) = 0.91 kmol benzene/kmol oil. Then,

$$\frac{L_{Sa}(\min)}{V_{Sa}} = \frac{Y_{1a} - Y_{2a}}{X_{1a}(\max) - X_{2a}} = \frac{0.080 - 0.012}{0.91 - 0.05} = 0.079 \text{ kmol oil / kmol air}$$

$$L_{Sa}(\text{min}) = 0.079 \times 0.038 = 3.0 \times 10^{-3} \text{ kmol oil / s}$$

For an actual oil flow rate which is twice the minimum, $L_{Sa} = 0.006$ kmol/s. The molecular weight of the oil is 198. The mass rate of flow of the oil, w_{Sa}, is

$$w_{Sa} = L_{Sa}M_{Sa} = 0.006 \times 198 = 1.19 \text{ kg oil/s}$$

The next step is to calculate the actual concentration of the liquid phase leaving the absorber. Rearranging equation (3-47)

$$X_{1a} = X_{2a} + \frac{V_{Sa}(Y_{1a} - Y_{2a})}{L_{Sa}} = 0.05 + \frac{0.038(0.080 - 0.012)}{0.006}$$

$$= 0.48 \text{ kmol benzene / kmol oil}$$

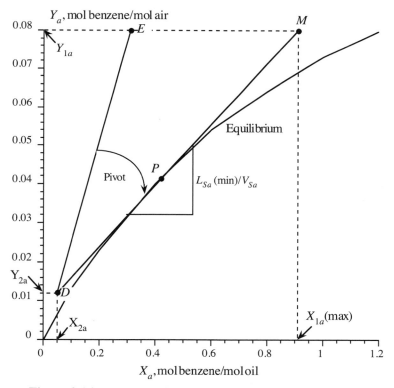

Figure 3.16 *XY* diagram for the absorber of Example 3.7.

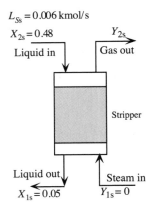

$L_{Ss} = 0.006\ \text{kmol/s}$
$X_{2s} = 0.48$
Y_{2s}
Liquid in   Gas out

Stripper

Liquid out
$X_{1s} = 0.05$ Steam in $Y_{1s} = 0$

We turn our attention now to conditions in the stripper. Figure 3.15 shows that the wash oil cycles continuously from the absorber to the stripper, and through the cooler back to the absorber. Therefore, $L_{Ss} = L_{Sa} = 0.006$ kmol/s. The concentration of the liquid entering the stripper is the same as that of the liquid leaving the absorber ($X_{2s} = X_{1a} = 0.48$ kmol benzene/kmol oil), and the concentration of the liquid leaving the stripper is the same as that of the liquid entering the absorber ($X_{1s} = X_{2a} = 0.05$ kmol benzene/kmol oil). The gaseous phase entering the stripper is pure steam, therefore $Y_{1s} = 0$. To determine the minimum amount of steam needed, we need to generate the equilibrium distribution curve for the system at 373 K. Applying Raoult's law at this temperature

$$y_{is} = 1.770 x_{is}$$

$$\frac{Y_{is}}{1 + Y_{is}} = 1.770 \frac{X_{is}}{1 + X_{is}}$$

The following table of equilibrium values, for the conditions prevailing in the stripper, is generated from the equation above:

X_{is}, kmol benzene/kmol oil	Y_{is}, kmol benzene/kmol steam
0.00	0.000
0.05	0.092
0.10	0.192
0.15	0.300
0.20	0.418
0.25	0.548
0.30	0.691
0.35	0.848
0.40	1.023
0.45	1.219
0.50	1.439

Figure 3.17 is the resulting equilibrium distribution curve. We are now ready to locate the minimum operating line. Locate point D (X_{1s}, Y_{1s}) on the XY diagram. This is one end of the operating line and its location is fixed, independent of the amount of steam used. Starting with any arbitrary operating line, such as DE, rotate

it toward the equilibrium curve using D as a pivot point until the operating line touches the equilibrium curve for the first time. In this case, because of the shape of the equilibrium curve, the first contact occurs at point P where the operating line is tangent to the curve. Extending this line all the way to $X_s = X_{2s}$, we generate operating line DM, corresponding to the minimum oil rate. From the diagram, we read that $Y_{2s}(\text{max}) = 1.175$ kmol benzene/kmol steam. Then,

$$\frac{L_{Ss}}{V_{Ss}(\text{min})} = \frac{Y_{2s}(\text{max}) - Y_{1s}}{X_{2s} - X_{1s}} = \frac{1.175 - 0.0}{0.48 - 0.05} = 2.733 \text{ kmol oil / kmol steam}$$

$$V_{Ss}(\text{min}) = \frac{0.006}{2.733} = 2.2 \times 10^{-3} \text{ kmol / s}$$

$$V_{Ss} = 1.5 V_{Ss}(\text{min}) = 3.292 \times 10^{-3} \text{ kmol / s}$$

The mass flow rate of steam used in the stripper, w_{Ss}, is

$$w_{Ss} = V_{Ss} M_{Ss} = 3.292 \times 10^{-3} \times 18 = 0.059 \text{ kg steam / s}$$

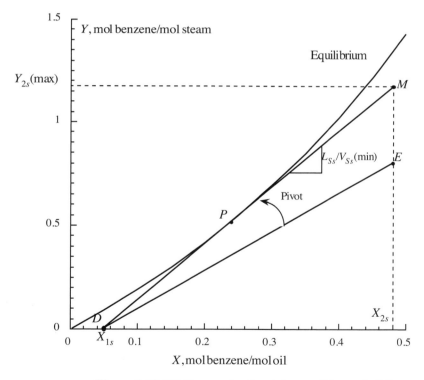

Figure 3.17 *XY* diagram for the stripper of Example 3.7.

The next step is to calculate the actual concentration of the gas phase leaving the stripper. Rearranging equation (3-47)

$$Y_{2s} = Y_{1s} + \frac{L_{Ss}(X_{2s} - X_{1s})}{V_{Ss}} = 0.00 + \frac{0.006(0.48 - 0.05)}{0.003292}$$

$$= 0.784 \text{ kmol benzene / kmol steam}$$

> Masses, mass fractions, and mass ratios can be substituted consistently for moles, mole fractions, and mole ratios in the mass balances of equations (3-41) to (3-48). The following example illustrates this concept.

Example 3.8 Adsorption of Nitrogen Dioxide on Silica Gel.

Nitrogen dioxide, NO_2, produced by a thermal process for fixation of nitrogen, is to be removed from a dilute mixture with air by adsorption on silica gel in a continuous countercurrent adsorber. The mass flow rate of the gas entering the adsorber is 0.50 kg/s; it contains 1.5% NO_2 by volume, and 85% of the NO_2 is to be removed. Operation is to be isothermal at 298 K and 1 atm. The entering gel will be free of NO_2. If twice the minimum gel rate is to be used, calculate the gel mass flow rate and the composition of the gel leaving the process. The equilibrium adsorption data at this temperature are (Treybal, 1980):

Partial pressure NO_2, p_A, mm Hg	0	2	4	6	8	10	12
Solid concentration, m, kg NO_2/100 kg gel	0	0.4	0.9	1.65	2.60	3.65	4.85

Solution
The material balances in this problem are easier to establish in terms of mass flow rates and mass ratios. Define:

$$L_S' \quad = \text{mass flow rate of the "dry" gel, kg/s}$$
$$V_S' \quad = \text{mass flow rate of the "dry" air, kg/s}$$

X' = mass ratio in the gel, kg NO_2/kg gel

Y' = mass ratio in the gas, kg NO_2/kg air

The mass ratio in the entering gas is

$$Y_1' = \frac{y_1}{1-y_1} \times \frac{M_{NO_2}}{M_{air}} = \frac{0.015}{1-0.015} \times \frac{46}{29} = 0.0242 \text{ kg } NO_2 / \text{kg air}$$

For 85% removal of the NO_2, $Y_2' = 0.15 \times 0.0242 = 0.0036$ kg NO_2/kg air. Since the entering gel is free of NO_2, $X_2' = 0$. The equilibrium data are converted to mass ratios as follows

$$Y_i' = \frac{P_{NO_2}}{760 - P_{NO_2}} \times \frac{46}{29} , \qquad X_i' = \frac{m}{100}$$

to yield:

$X_i' \times 100$, kg NO_2/kg gel	$Y_i' \times 100$, kg NO_2/kg air
0.00	0.00
0.40	0.42
0.90	0.83
1.65	1.26
2.60	1.69
3.65	2.11
4.85	2.54

Figure 3.18 shows the corresponding equilibrium-distribution curve, together with the minimum operating line DM. From Figure 3.18, X_1' (max) = 0.0375 kg NO_2/kg gel. Then,

$$L_S'(\text{min}) = \frac{Y_1' - Y_2'}{X_1'(\text{max}) - X_2'} \times V_S' = \frac{0.0242 - 0.0036}{0.0375 - 0.0} \times V_S' = 0.550 \times V_S'$$

To calculate the mass velocity of the inert gas, air in this case, $V_S' = V_1' \times \omega_{B1}$. It is easy to show that

$$\omega_B = \frac{1}{1 + Y'}$$

Then, $V_S' = 0.5 \times (1 + 0.0242)^{-1} = 0.488$ kg air/s; L_S' (min) = $0.55 \times 0.488 = 0.268$ kg gel/s. The actual mass velocity of the gel to be used is $L_S' = 2L_S'$ (min) = 0.536 kg gel/s. The composition of the gel leaving the adsorber is

$$X_1' = X_2' + \frac{V_S'(Y_1' - Y_2')}{L_S \copyright} = 0.00 + \frac{0.488 \times (0.0242 - 0.0036)}{0.536}$$

$$= 0.0188 \text{ kg NO}_2 / \text{kg gel}$$

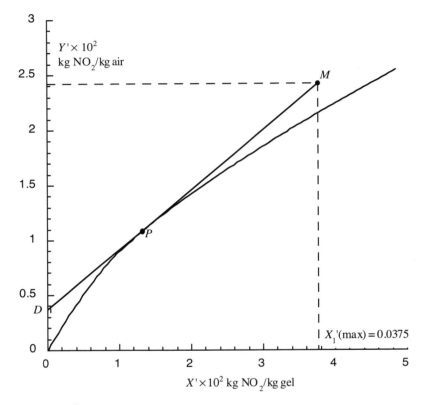

Figure 3.18 $X'Y'$ diagram for the adsorber of Example 3.8.

3.4.2 Cocurrent Flow

For steady-state mass-transfer operations involving cocurrent contact of two insoluble phases, as shown in Figure 3.19, the overall mass balance for component A is

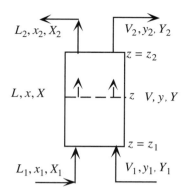

Figure 3.19 Steady-state cocurrent process.

$$V_S Y_1 + L_S X_1 = V_S Y + L_S X = V_S Y_2 + L_S X_2 \tag{3-49}$$

Rearranging equation (3-49),

$$\frac{Y_1 - Y_2}{X_1 - X_2} = -\frac{L_S}{V_S} = \frac{Y_1 - Y}{X_1 - X} \tag{3-50}$$

This is the equation of a straight line on the XY diagram through the points (X_1, Y_1) and (X_2, Y_2) with slope $- L_S/V_S$, the operating line for cocurrent operation. Figures 3.20 and 3.21 illustrate the location of the operating line relative to the equilibrium line when the transfer is from phase V to phase L, and from phase L to phase V.

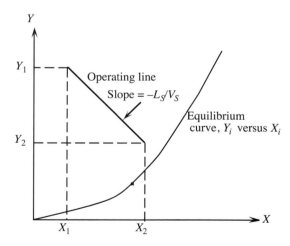

Figure 3.20 Steady-state cocurrent process, transfer from phase V to L.

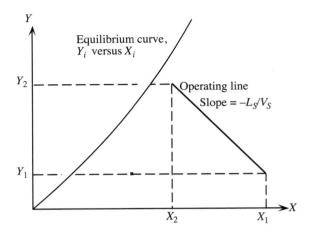

Figure 3.21 Steady-state cocurrent process, transfer from phase L to V.

As in the case for countercurrent flow, there is a minimum L_S/V_S ratio for cocurrent mass-transfer operations established from the fixed process compositions: Y_1, Y_2, and X_1 for transfer from V to L; X_1, X_2, and Y_1 for transfer from L to V.

Example 3.9 Adsorption of NO_2 on Silica Gel: Cocurrent Operation

Repeat Example 3.8, but for cocurrent contact between the gas mixture and the silica gel.

Solution

Figure 3.22 is the $X'Y'$ diagram for cocurrent operation of the adsorber. In this case, as before, $Y_1' = 0.0242$ kg NO_2/kg air, $Y_2' = 0.0036$ kg NO_2/kg air; however, now $X_1' = 0.0$. The minimum operating line goes from D (X_1', Y_1') to point M $[X_2(max)', Y_2']$ on the equilibrium curve, with a slope of $- L_S(min)/V_S$. From Figure 3.21, $X_2(max)' = 0.0034$ kg NO_2/kg gel. From equation (3-50),

$$-\frac{L_S'(min)}{V_S'} = \frac{0.0242 - 0.0036}{0.0 - 0.0034} = -6.09 \text{ kg gel / kg air}$$

Therefore, L_S' (min) $= 6.09 \times 0.488 = 2.957$ kg/s; $L_S' = 2.0 \times 2.957 = 5.920$ kg/s. Notice that the mass velocity of the silica gel required for cocurrent operation is about 11 times that required for countercurrent operation. This example dramatically illustrates the fact that the driving force for mass transfer is used much more efficiently in countercurrent than in cocurrent operation.

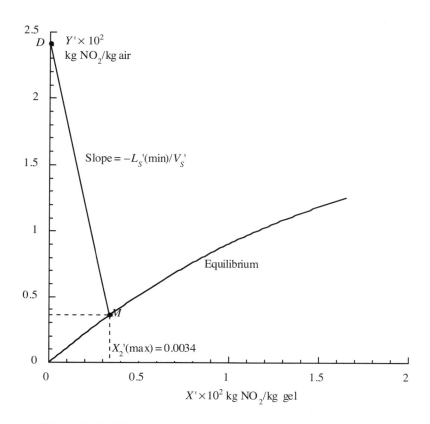

Figure 3.22 $X'Y'$ diagram for the cocurrent adsorber of Example 3.9.

3.4.3 Batch Processes

It is characteristic of batch processes that while there is no flow into and out of the equipment used, the concentrations within each phase change with time. When initially brought into contact, the phases will not be in equilibrium, but they will approach equilibrium with passage of time. The material-balance equation (3-50), developed for cocurrent steady-state operation, also shows the relation between the concentrations X and Y in the phases which coexist at any time after the start of the operation. Figures 3.20 and 3.21 give the graphical representation of these composition changes. If the phases are kept in contact long enough to reach equilibrium, the end of the operating line—point (X_2, Y_2)—lies on the equilibrium curve.

3.5 EQUILIBRIUM-STAGE OPERATIONS

Your objectives in studying this section are to be able to:

1. Define the concepts: stage, ideal stage, and cascade.
2. Calculate the number of ideal stages in a cascade required for a given separation and ratio of flow rates.

One class of mass-transfer devices consists of assemblies of individual units, or *stages*, interconnected so that the materials being processed pass through each stage in turn. The two streams move countercurrently through the assembly; in each stage they are brought into contact, mixed, and then separated. Such multistage systems are called *cascades*. For mass transfer to take place the streams entering each stage must not be in equilibrium with each other, for it is precisely the departure from equilibrium conditions that provides the driving force for mass transfer. The leaving streams are usually not in equilibrium either, but are much closer to being so than are the entering streams. If, as may actually happen in practice, the mass transfer in a given stage is so effective that the leaving streams are in fact in equilibrium, the stage is, by definition, an *ideal stage*. Obviously, the cocurrent process and apparatus of Figure 3.19 is that of a single stage, and if the stage were ideal, the effluent compositions would be in equilibrium with each other. A batch mass-transfer operation is also a single stage.

Two or more stages connected so that flow is cocurrent between stages will, of course, never be equivalent to more than one equilibrium stage, although the overall stage efficiency can thereby be increased. For effects greater than one equilibrium stage, the stages in a cascade must be connected for countercurrent flow. This is the most efficient arrangement requiring the fewest stages for a given change of composition and ratio of flow rates, and is always used in multistage equipment. (Foust, et al., 1980). Figure 3.23 shows a countercurrent cascade of N equilibrium stages.

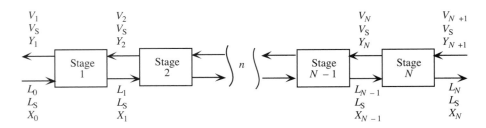

Figure 3.23 Countercurrent cascade of ideal stages.

In Figure 3.23, the flow rates and compositions are numbered corresponding to the effluent from a stage, so that, for example, X_2 is the concentration in the L phase leaving stage 2, etc. Since the stages are ideal, the effluents from each stage are in equilibrium with each other (Y_2 in equilibrium with X_2, etc.). The cascade as a whole has the characteristics of the countercurrent process of Figure 3.12, with an operating line on the XY diagram that goes through the points (X_0, Y_1) and (X_N, Y_{N+1}) with slope L_S/V_S. The most generally satisfactory methods of determining the number of ideal stages in a cascade are graphical ones. The simplest graphical method, and one that is sufficiently precise for most two-component situations, is based on the use of operating line in conjunction with the equilibrium line. The graphical relations are shown in Figure 3.24 for transfer of solute from phase L to phase V.

The operating line ST may be plotted either by knowing all four of the compositions at both ends of the cascade, or by knowing three compositions and the slope (L_S/V_S) of the operating line. The concentration of the V phase leaving stage 1 is Y_1. Since the stage is ideal, X_1, the concentration of the L phase leaving this stage, is such that the point (X_1, Y_1) must lie on the equilibrium curve. This fact fixes point Q, found by moving horizontally from point T to the equilibrium curve. The operating line is now used to determine Y_2. The operating line passes through all points having coordinates of the type (X_n, Y_{n+1}), and since X_1 is known, Y_2 is found by moving vertically from point Q to the operating line at point B, the coordinates of which are (X_1, Y_2). The step, or triangle, defined by the points T, Q, and B represents one ideal stage, the first one in this cascade. The second stage is located graphically on the diagram by repeating the same construction, passing horizontally to the equilibrium curve at point N, having coordinates (X_2, Y_2), and vertically to the operating line again at point C, having coordinates (X_2, Y_3). This stepwise construction is repeated until point S (X_N, Y_{N+1}) is reached.

We can therefore determine the number of ideal stages required for a countercurrent process by drawing the stairlike construction $TQBNC \ldots S$ and counting the number of steps. If anywhere the equilibrium curve and the cascade operating line should touch, the stages become *pinched* and an infinite number are required to bring about the desired composition change. If the solute transfer is from V to L, the entire construction falls above the equilibrium curve of Figure 3.24.

For most cases, because of either a curved operating line or equilibrium curve, the relation between number of stages, compositions, and flow ratio must be determined graphically, as shown. For the special case where both are straight, however, with the equilibrium curve continuing straight to the origin ($Y_i = mX_i$), an analytical solution was developed by Kremser (1930) which is most useful. Define the *absorption factor A* as

$$A = \frac{L_S}{mV_s} \tag{3-51}$$

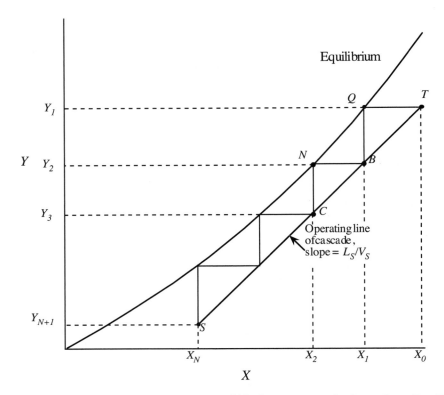

Figure 3.24 Countercurrent cascade of ideal stages, transfer from phase L to V.

Notice from equation (3-51) that the absorption factor is really the ratio of the slope of the operating line to that of the equilibrium relation. The Kremser equations are as follows:

For transfer from L to V (stripping of L)

$A \neq 1$:

$$N = \frac{\ln\left[\dfrac{X_0 - Y_{N+1}/m}{X_N - Y_{N+1}/m}(1-A)+A\right]}{\ln(1/A)} \tag{3-52}$$

$A = 1$

$$N = \frac{X_0 - X_N}{X_N - Y_{N+1}/m} \tag{3-53}$$

For transfer from V to L (absorption into L)

$A \neq 1$:

$$N = \frac{\ln\left[\frac{Y_{N+1} - mX_0}{Y_1 - mX_0}\left(1 - \frac{1}{A}\right) + \frac{1}{A}\right]}{\ln A} \tag{3-54}$$

$A = 1$

$$N = \frac{Y_{N+1} - Y_1}{Y_1 - mX_0} \tag{3-55}$$

Example 3.10 Benzene Recovery System: Number of Ideal Stages.

Estimate the number of ideal stages in the absorber and stripper of the benzene recovery system of Example 3.7.

Solution

Absorber: Using the nomenclature introduced for multistage cascades, and the results of Example 3.7:

$$
\begin{aligned}
X_0 \quad &= X_{2a} = 0.050 \text{ kmol benzene/kmol oil} \\
Y_1 \quad &= Y_{2a} = 0.012 \text{ kmol benzene/kmol dry gas} \\
X_N \quad &= X_{1a} = 0.480 \text{ kmol benzene/kmol oil} \\
Y_{N+1} \quad &= Y_{1a} = 0.080 \text{ kmol benzene/kmol dry gas}
\end{aligned}
$$

Figure 3.25 shows the operating line for the cascade, DE, located using the four compositions listed above. Since A is transferred in this case from phase V to L, the stairlike construction to determine the number of ideal stages begins at point D moving horizontally toward the equilibrium curve, and so on. Figure 3.25 shows that almost 4 equilibrium stages are required. A more precise estimate is obtained from

$$N \cong 3 + \frac{X_B - X_A}{X_C - X_A} = 3 + \frac{0.480 - 0.283}{0.530 - 0.283} = 3.80 \text{ stages}$$

For comparison purposes, we tried to estimate the number of ideal stages using the Kremser equations. By least-squares analysis, a straight line through the

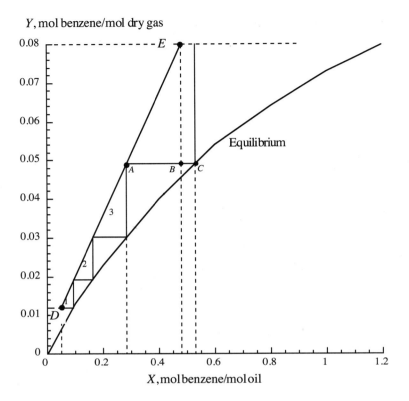

Figure 3.25 Ideal stages for the absorber of Example 3.7.

origin was fitted to the equilibrium data in the range of X_i between 0 and 0.6, giving a slope $m = 0.097$ mole oil/mole dry gas. The corresponding value of the absorption factor is $A = 0.006/(0.097 \times 0.038) = 1.628$. Substituting in equation (3-54), $N = 3.16$ stages. Therefore, in this case the Kremser equation underestimates the number of ideal stages by approximately 17%. This was to be expected since the curvature of the equilibrium line is significant.

Stripper: Using the nomenclature introduced for multistage cascades, and the results of Example 3.7:

$$
\begin{aligned}
X_0 &= X_{2s} = 0.480 \text{ kmol benzene/kmol oil} \\
Y_1 &= Y_{2s} = 0.784 \text{ kmol benzene/kmol steam} \\
X_N &= X_{1s} = 0.050 \text{ kmol benzene/kmol oil} \\
Y_{N+1} &= Y_{1s} = 0.000 \text{ kmol benzene/kmol steam}
\end{aligned}
$$

Figure 3.26 shows the operating line for the cascade, *DE,* located using the four compositions listed above. Since A is transferred in this case from phase *L* to *V,* the stairlike construction to determine the number of ideal stages begins at point *E* moving horizontally towards the equilibrium curve, and so on. Figure 3.26 shows that almost 6 equilibrium stages are required. A more precise estimate is obtained from

$$N \cong 5 + \frac{0.070 - 0.050}{0.070 - 0.028} = 5.50 \text{ stages}$$

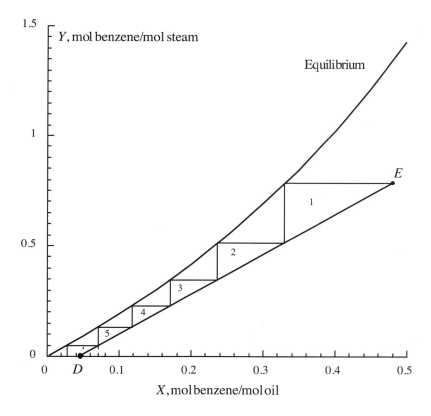

Figure 3.26 Ideal stages for the stripper of Example 3.7.

PROBLEMS

The problems at the end of each chapter have been grouped into four classes (designated by a superscript after the problem number).

Class a: Illustrates direct numerical application of the formulas in the text.
Class b: Requires elementary analysis of physical situations, based on the subject material in the chapter.
Class c: Requires somewhat more mature analysis.
Class d: Requires computer solution.

3.1[a]. Application of Raoult's law to a binary system.

Repeat Example 3.1, but for a liquid concentration of 0.6 mole fraction of benzene and a temperature of 320 K.

Answer: $y_A = 0.818$

3.2[b]. Application of Raoult's law to a binary system.

(a) Determine the composition of the liquid in equilibrium with a vapor containing 60 mole percent benzene–40 mole percent toluene if the system exists in a vessel under 1 atm pressure. Predict the equilibrium temperature.

Answer: $x_A = 0.379$

(b) Determine the composition of the vapor in equilibrium with a liquid containing 60 mole percent benzene–40 mole percent toluene if the system exists in a vessel under 1 atm pressure. Predict the equilibrium temperature.

Answer: $y_A = 0.791$

3.3[a]. Application of Raoult's law to a binary system.

Normal heptane, $n\text{-}C_7H_{16}$, and normal octane, $n\text{-}C_8H_{18}$, form ideal solutions. At 373 K, normal heptane has a vapor pressure of 106 kPa and normal octane of 47.1 kPa.

(a) What would be the composition of a heptane-octane solution that boils at

373 K under a 93 kPa pressure?

Answer: $x_A = 0.779$

(b) What would be the composition of the vapor in equilibrium with the solution that is described in part a?

Answer: $y_A = 0.888$

3.4[a]. Henry's law: saturation of water with oxygen.

A solution with oxygen dissolved in water containing 0.5 mg O_2/100 g of H_2O is brought in contact with a large volume of atmospheric air at 283 K and a total pressure of 1 atm. The Henry's law constant for the oxygen-water system at 283 K equals 3.27×10^4 atm/mole fraction.

(a) Will the solution gain or lose oxygen?

(b) What will be the concentration of oxygen in the final equilibrium solution?

Answer: 11.4 mg O_2/L

3.5[d]. Material balances combined with equilibrium relations.

Repeat Example 3.3, but assuming that the ammonia, air, and water are brought into contact in a closed container. There is 10 m^3 of gas space over the liquid. Assuming that the gas-space volume and the temperature remain constant until equilibrium is achieved, modify the Mathcad program in Figure 3.2 to calculate:

(a) the total pressure at equilibrium

Answer: 1.754 atm

(b) the equilibrium ammonia concentration in the gas and liquid phases

3.6[b]. Mass-transfer resistances during absorption.

In the absorption of component A (molecular weight = 60) from an airstream into an aqueous solution, the bulk compositions of the two adjacent streams at a point in the apparatus were analyzed to be $p_{A,G} = 0.1$ atm, and $c_{A,L} = 1.0$ kmol of A/m^3 of solution. The total pressure was 2.0 atm; the density of the solution was 1,100 kg/m^3. The Henry's constant for these conditions was 0.85 atm/mole fraction. The overall gas coefficient was $K_G =$ 0.27 kmol/m²-hr-atm. If 57% of the total resistance to mass transfer resides in the gas film, determine

(a) the gas-film coefficient, k_G

(b) the liquid-film coefficient, k_L

Answer: 0.91 cm/hr

(c) the concentration on the liquid side of the interface, $x_{A,i}$

Answer: 0.06 mole fraction

(d) the mass flux of A

3.7ᵇ. Mass-transfer resistances during absorption.

For a system in which component A is transferring from the gas phase to the liquid phase, the equilibrium relation is given by

$$p_{A,i} = 0.80 x_{A,i}$$

where $p_{A,i}$ is the equilibrium partial pressure in atm and $x_{A,i}$ is the equilibrium liquid concentration in molar fraction. At one point in the apparatus, the liquid stream contains 4.5 mole% and the gas stream contains 9.0 mole% A. The total pressure is 1 atm. The individual gas-film coefficient at this point is $k_G = 3.0$ mole/m²-s-atm. Fifty percent of the overall resistance to mass transfer is known to be encountered in the liquid phase. Evaluate

(a) the overall mass-transfer coefficient, K_y

Answer: 1.5 mole/m²-s

(b) the molar flux of A

(c) the liquid interfacial concentration of A

Answer: 7.88 mole %

3.8ᵈ. Absorption of ammonia by water: use of F-type mass-transfer coefficients.

Modify the Mathcad program in Figure 3.6 to repeat Example 3.5, but with $y_{A,G} = 0.70$ and $x_{A,L} = 0.10$. Everything else remains constant.

Answer: $y_{A,i} = 0.588$

3.9$^{d.}$ Absorption of ammonia by water: use of F-type mass-transfer coefficients.

Modify the Mathcad program in Figure 3.6 to repeat Example 3.5, but with $F_L = 0.0050$ kmol/m^2-s. Everything else remains constant.

Answer: $N_A = 0.62$ mole/m^2-s

3.10^b. Mass-transfer resistances during absorption of ammonia.

In the absorption of ammonia into water from an air-ammonia mixture at 300 K and 1 atm, the individual film coefficients were estimated to be $k_L = 6.3$ cm/hr and $k_G = 1.17$ kmol/m^2-hr-atm. The equilibrium relationship for very dilute solutions of ammonia in water at 300 K and 1 atm is

$$y_{A,i} = 1.64 x_{A,i}$$

Determine the following mass-transfer coefficients:

(a) k_y

(b) k_x

(c) K_y

Answer: 0.21 moles/m^2-s

(d) Fraction of the total resistance to mass transfer that resides in the gas phase.

Answer: 65%

3.11^b. Mass-transfer resistances in hollow-fiber membrane contactors.

For mass transfer across the hollow-fiber membrane contactors described in Example 2.13, the overall mass-transfer coefficient based on the liquid concentrations, K_L, is given by (Yang and Cussler, 1986):

$$\frac{1}{K_L} = \frac{1}{k_L} + \frac{1}{Hk_M} + \frac{1}{Hk_c} \tag{3-56}$$

where k_L, k_M, and k_c are the individual mass-transfer coefficients in the liquid, across the membrane, and in the gas, respectively; and H is Henry's law constant, the gas equilibrium concentration divided by that in the liquid. The mass-transfer coefficient across a hydrophobic membrane is estimated from (Prasad and Sirkar, 1988):

$$k_M = \frac{D_{AB}\varepsilon_M}{\tau_M \delta} \tag{3-57}$$

where

D_{AB}	= molecular diffusion coefficient in the gas filling the pores
ε_M	= membrane porosity
τ_M	= membrane tortuosity
δ	= membrane thickness

For the membrane modules of Example 2.13, $\varepsilon_M = 0.4$, $\tau_M = 2.2$, and $\delta = 25 \times 10^{-6}$ m (Prasad and Sirkar, 1988).

(a) Calculate the corresponding value of k_M.

Answer: 15.9 cm/s

(b) Using the results of part a, Example 2.13, and Problem 2.25, calculate K_L, and estimate what fraction of the total resistance to mass transfer resides in the liquid film.

3.12[c]. Combined use of F- and k-type coefficients: absorption of low-solubility gases.

During absorption of low-solubility gases, mass transfer from a highly concentrated gas mixture to a very dilute liquid solution frequently takes place. In that case, although it is appropriate to use a k-type mass-transfer coefficient in the liquid phase, an F-type coefficient must be used in the gas phase. Since dilute liquid solutions usually obey Henry's law, the interfacial concentrations during absorption of low-solubility gases are related through $y_{A,i} = mx_{A,i}$.

(a) Show that, under the conditions described above, the gas interfacial concentration satisfies the equation

$$\ln\left(1 - y_{A,i}\right) - \frac{k_x y_{A,i}}{m F_G} + \frac{k_x x_{A,L}}{F_G} - \ln\left(1 - y_{A,G}\right) = 0 \qquad (3\text{-}58)$$

(b) In a certain apparatus used for the absorption of SO_2 from air by means of water, at one point in the equipment the gas contained 30% SO_2 by volume and was in contact with a liquid containing 0.2% SO_2 by mole. The temperature was 303 K and the total pressure 1 atm. Estimate the interfacial concentrations and the local SO_2 molar flux. The mass-transfer coefficients were calculated as $F_G = 0.002$ kmol/m²-s, $k_x = 0.160$ kmol/m²-s. The equilibrium SO_2 solubility data at 303 K are (Perry and Chilton, 1973):

kg SO_2/100 kg water	Partial pressure of SO_2, mm Hg
0.0	0
0.5	42
1.0	85
1.5	129
2.0	176
2.5	224

Answer: $N_A = 0.34$ moles/m²-s

3.13[d]. Distillation of a mixture of methanol and water in a packed tower: use of F-type mass-transfer coefficients.

At a different point in the packed distillation column of Example 3.6, the methanol content of the bulk of the gas phase is 76.2 mole %; that of the bulk of the liquid phase is 60 mole %. The temperature at that point in the tower is around 343 K. The packing characteristics and flow rates at that point are such that $F_G = 1.542 \times 10^{-3}$ kmol/m²-s, and $F_L = 8.650 \times 10^{-3}$ kmol/m²-s. Calculate the interfacial compositions and the local methanol flux. To calculate the latent heat of vaporization at the new temperature, modify the values given in Example 3.6 using Watson's method (Smith, et al., 1996):

$$\frac{\lambda_2}{\lambda_1} = \left[\frac{1 - T_{r2}}{1 - T_{r1}}\right]^{0.38}$$ (3-59)

For water, $T_c = 647.1$ K; for methanol, $T_c = 512.6$ K.

Answer: $N_A = 1.24 \times 10^{-4}$ kmol/m²-s

3.14[b]. Material balances: adsorption of benzene vapor on activated carbon.

Activated carbon is used to recover benzene from a nitrogen-benzene vapor mixture. A nitrogen-benzene mixture at 306 K and 1 atm containing 1% benzene by volume is to be passed countercurrently at the rate of 1.0 m³/s to a moving stream of activated carbon so as to remove 85% of the benzene from the gas in a continuous process. The entering activated carbon contains 15 cm³ benzene vapor (at STP) adsorbed per gram of the carbon. The temperature and total pressure are maintained at 306 K and 1 atm. Nitrogen is not adsorbed. The equilibrium adsorption of benzene on this activated carbon at 306 K is reported as follows:

Benzene vapor adsorbed cm³ (STP)/g carbon	Partial pressure benzene, mm Hg
15	0.55
25	0.95
40	1.63
50	2.18
65	3.26
80	4.88
90	6.22
100	7.83

(a) Plot the equilibrium data as X' = kg benzene/kg dry carbon, Y' = kg benzene/kg nitrogen for a total pressure of 1 atm.

(b) Calculate the minimum flow rate required of the entering activated carbon (remember that the entering carbon contains some adsorbed benzene).

Answer: 0.096 kg of activated carbon/s

(c) If the carbon flow rate is 20% above the minimum, what will be the concentration of benzene adsorbed on the carbon leaving?

Answer: 0.295 kg benzene/kg activated carbon

d) For the conditions of part c, calculate the number of ideal stages required.

Answer: 3.31 stages

3.15[b]. Material balances: desorption of benzene vapor from activated carbon.

The activated carbon leaving the adsorber of Problem 3.14 is regenerated by countercurrent contact with steam at 380 K and 1 atm. The regenerated carbon is returned to the adsorber, while the mixture of steam and desorbed benzene vapors is condensed. The condensate separates into an organic and an aqueous phase and the two phases are separated by decantation. Due to the low solubility of benzene in water, most of the benzene will be concentrated in the organic phase, while the aqueous phase will contain only traces of benzene. The equilibrium adsorption data at 380 K are as follows:

Benzene vapor adsorbed kg benzene/100 kg carbon	Partial pressure benzene, kPa
2.9	1.0
5.5	2.0
12.0	5.0
17.1	8.0
20.0	10.0
25.7	15.0
30.0	20.0

(a) Calculate the minimum steam flow rate required.

Answer: 0.035 kg/s

(b) For a steam flow rate 50% above the minimum, calculate the benzene concentration in the gas mixture leaving the desorber, and the number of ideal stages required.

Answer: 4.52 stages

3.16[b]. Material balances: adsorption of benzene vapor on activated carbon; cocurrent operation.

If the adsorption process described in Problem 3.14 took place cocurrently, calculate the minimum flow rate of activated carbon required.

Answer: 0.54 kg/s

3.17[b]. Material balances in batch processes: drying of soap with air.

It is desired to dry 10 kg of soap from 20% moisture by weight to no more than 6% moisture by contact with hot air. The wet soap is placed in a vessel containing 8.06 m^3 of air at 350 K, 1 atm, and a water-vapor partial pressure of 1.6 kPa. The system is allowed to reach equilibrium, and then the air in the vessel is entirely replaced by fresh air of the original moisture content and temperature. How many times must the process be repeated in order to reach the specified soap moisture content of no more than 6%? When this soap is exposed to air at 350 K and 1 atm, the equilibrium distribution of moisture between the air and the soap is as follows:

Wt % moisture in soap	Partial pressure of water, kPa
2.40	1.29
3.76	2.56
4.76	3.79
6.10	4.96
7.83	6.19
9.90	7.33
12.63	8.42
15.40	9.58
19.02	10.60

Answer: 5 times

3.18[b]. Material balances in batch processes: extraction of an aqueous nicotine solution with kerosene.

Nicotine in a water solution containing 2% nicotine is to be extracted with kerosene at 293 K. Water and kerosene are essentially insoluble. Determine the percentage extraction of nicotine if 100 kg of the feed solution is extracted in a sequence of four batch ideal extractions using 49.0 kg of fresh, pure kerosene each. The equilibrium data are as follows (Claffey, et al., 1950):

X', $\times 10^3$ kg nicotine/kg water	Y', $\times 10^3$ kg nicotine/kg kerosene
1.01	0.81
2.46	1.96
5.02	4.56
7.51	6.86
9.98	9.13
20.4	18.70

Answer: 79.3%

3.19[b]. Cross-flow cascade of ideal stages.

The drying and liquid–liquid extraction operations described in Problems 3.17 and 3.18, respectively, are examples of a flow configuration called a *cross-flow cascade*. Figure 3.27 is a schematic diagram of a cross-flow cascade of ideal stages. Each stage is represented by a circle, and within each stage mass transfer occurs as if in cocurrent flow. The L phase flows from one stage to the next, being contacted in each stage by a fresh V phase. If the equilibrium-distribution curve of the cross-flow cascade is everywhere straight and of slope m, it can be shown that (Treybal, 1980)

$$N = \frac{\ln\left[(X_0 - Y_0 / m) / (X_N - Y_0 / m)\right]}{\ln(S+1)} \qquad (3\text{-}60)$$

where S is the stripping factor, mV_S/L_S, constant for all stages, and N is the total number of stages.

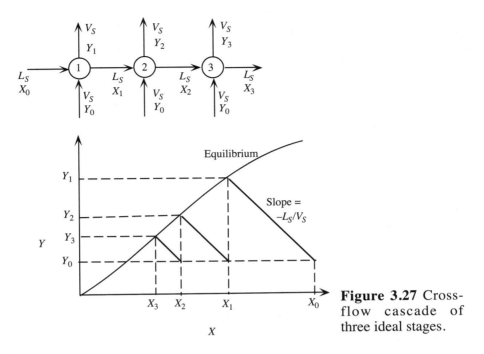

Figure 3.27 Cross-flow cascade of three ideal stages.

Solve Problem 3.18 using equation (3-60), and compare the results obtained by the two methods.

Answer: 78%

3.20[a]. Cross-flow cascade of ideal stages: nicotine extraction.

Consider the nicotine extraction of Problems 3.18 and 3.19. Calculate the number of ideal stages required to achieve at least 95% extraction efficiency.

3.21[b]. Kremser equations: absorption of hydrogen sulfide.

A scheme for the removal of H_2S from a flow of 1.0 std m^3/s of natural gas by scrubbing with water at 298 K and 10 atm is being considered. The initial composition of the feed gas is 2.5 mole percent H_2S. A final gas stream containing only 0.1 mole percent H_2S is desired. The absorbing water will enter the system free of H_2S. At the given temperature and pressure, the system will follow Henry's law, according to $Y_i = 48.3X_i$, where X_i = moles

H_2S/mole of water; Y_i = moles H_2S/mole of air.

(a) For a countercurrent absorber, determine the flow rate of water that is required if 1.5 times the minimum flow rate is used.

Answer: 54.5 kg water/s

(b) Determine the composition of the exiting liquid.

Answer: 0.67 g H_2S/L of water

(c) Calculate the number of ideal stages required.

Answer: 5.86 stages

3.22[b]. Absorption with chemical reaction: H_2S scrubbing with MEA.

As shown in Problem 3.21, scrubbing of hydrogen sulfide from natural gas using water is not practical since it requires large amounts of water due to the low solubility of H_2S in water. If a 2N solution of monoethanolamine (MEA) in water is used as the absorbent, however, the required liquid flow rate is reduced dramatically because the MEA reacts with the absorbed H_2S in the liquid phase, effectively increasing its solubility.

For this solution strength and a temperature of 298 K, the solubility of H_2S can be approximated by (de Nevers, 2000):

$$p_{H_2S,i} = (19.4 \text{ Pa})\exp\left(195 x_{H_2S,i}\right)$$

Repeat the calculations of Problem 3.21, but using a 2N monoethanolamine solution as absorbent.

Answer: 0.76 kg MEA solution/s

3.23[b]. Kremser equations: absorption of sulfur dioxide.

A flue gas flows at the rate of 10 kmol/s at 298 K and 1 atm with a SO_2 content of 0.15 mole%. Ninety percent of the sulfur dioxide is to be removed by absorption with pure water at 298 K. The design water flow rate will be 50% higher than the minimum. Under these conditions, the equilibrium line is (Benítez, 1993):

$$Y_i = 10 X_i$$

where X_i = moles SO_2/mole of water; Y_i = moles SO_2/mole of air.

(a) Calculate the water flow rate and the SO_2 concentration in the water leaving the absorber.

Answer: 2,426 kg/s

(b) Calculate the number of ideal stages required for the specified flow rates and percentage SO_2 removal.

Answer: 4.01 stages

3.24[b]. Kremser equations: absorption of sulfur dioxide.

An absorber is available to treat the flue gas of Problem 3.23 which is equivalent to 8.5 equilibrium stages.

(a) Calculate the water flow rate to be used in this absorber if 90% of the SO_2 is to be removed. Calculate also the SO_2 concentration in the water leaving the absorber.

Answer: 1,819 kg/s

(b) What is the percentage removal of SO_2 that can be achieved with this absorber if the water flow rate used is the same that was calculated in Problem 3.23 a?

Answer: 97.9%

3.25[b]. Kremser equations: extraction of acetic acid with 3-heptanol.

An aqueous acetic acid solution flows at the rate of 1,000 kg/hr. The solution is 1.1% (by weight) acetic acid. It is desired to reduce the concentration of this solution to 0.037% (by weight) acetic acid by extraction with 3-heptanol at 298 K. For practical purposes, water and 3-heptanol are inmiscible. The inlet 3-heptanol contains 0.02% (by weight) acetic acid. An extraction column is available which is equivalent to a countercurrent cascade of 35 equilibrium stages. What solvent flow rate is required? Calculate the composition of the solvent phase leaving the column. For this system and range of concentrations, equilibrium is given by (Wankat,1988):

Wt ratio acetic acid in solvent = 0.828 × Wt ratio acetic acid in water

Answer: 1,248 kg/hr

3.26^c. Countercurrent versus cross-flow extraction.

A 1-butanol acid solution is to be extracted with pure water. The butanol solution contains 4.5% (by weight) of acetic acid and flows at the rate of 400 kg/hr. A total water flow rate of 1005 kg/hr is used. Operation is at 298 K and 1 atm. For practical purposes, 1-butanol and water are immiscible. At 298 K, the equilibrium data can be represented by $Y_{Ai} = 0.62 \, X_{Ai}$, where Y_{Ai} is the weight ratio of acid in the aqueous phase and X_{Ai} is the weight ratio of acid in the organic phase.

(a) If the outlet butanol stream is to contain 0.10% (by weight) acid, how many equilibrium stages are required for a countercurrent cascade?

(b) If the water is split up equally among the same number of stages, but in a cross-flow cascade, what is the outlet toluene concentration (see Problem 3.19)?

Answer: 1.1%

3.27^c. Glucose sorption on an ion exchange resin.

Ching and Ruthven (1985) found that the equilibrium of glucose on an ion exchange resin in the calcium form was linear for concentrations below 50 g/L. Their equilibrium expression at 303 K is $Y_{Ai} = 1.961 \, X_{Ai}$, where X_{Ai} is the glucose concentration in the resin.(g of glucose per liter of resin) and Y_{Ai} is the glucose concentration in solution.(g of glucose per liter of solution).

(a) We wish to sorb glucose onto this ion exchange resin at 303 K in a countercurrent cascade of ideal stages. The concentration of the feed solution is 15 g/L. We want an outlet concentration of 1.0 g/L. The inlet resin contains 0.25 g of glucose/L. The feed solution flows at the rate of 100 L/min, while the resin flows at the rate of 250 L/min. Find the number of equilibrium stages required.

(b) If 5 equilibrium stages are added to the cascade of part a, calculate the resin flow required to maintain the same degree of glucose sorption.

Answer: 216.3 L/min

REFERENCES

Benítez, J., *Process Engineering and Design for Air Pollution Control,* Prentice Hall, Englewood Cliffs, NJ (1993).

Bird, R. B., W. E. Stewart, and E. N. Lightfoot, *Transport Phenomena*, Wiley, New York (1960).

Ching, and Ruthven, *AIChE Symp. Ser.*, **81**, 242 (1985).

Claffey, et al., *Ind. Eng. Chem.,* **42**, 166 (1950).

Cussler, E. L., *Diffusion,* 2nd ed., Cambridge University Press, New York (1997).

Davis, M. L., and D. A. Cornwell, *Introduction to Environmental Engineering,* 3rd ed., McGraw-Hill, New York (1998).

Foust, A. S., L. A. Wenzel, C. W. Clump, L. Maus, and L. B. Andersen, *Principles of Unit Operations,* 2nd ed., Wiley, New York (1980).

Himmelblau, D. M., *Basic Principles and Calculations in Chemical Engineering,* 5th ed., Prentice Hall, Englewood Cliffs, NJ (1989).

Kremser, A., *Natl. Petrol. News,* **22**, 42 (1930).

Lewis, W. K., and W. G. Whitman, *Ind. Eng. Chem.,* **16**, 1215 (1924).

de Nevers, N., *Air Pollution Control Engineering,* 2nd ed., McGraw-Hill, Boston (2000)

Perry, R. H. and C. H. Chilton (eds.), *Chemical Engineers' Handbook*, 5th ed., McGraw-Hill, New York (1973).

Prasad, R., and K. K. Sirkar, *AIChE J.*, **34**, 177 (1988).

Smith, J. M., H. C. Van Ness, and M. M. Abbott, *Introduction to Chemical Engineering Thermodynamics,* 5th ed., McGraw-Hill, New York (1996).

Treybal, R. E., *Mass-Transfer Operations*, 3rd ed., McGraw-Hill, New York (1980).

Wankat, P. C., *Equilibrium Staged Separations,* Elsevier, New York (1988).

Welty, J. R., C. E. Wicks, and R. E. Wilson, *Fundamentals of Momentum, Heat, and Mass Transfer*, 3rd ed., Wiley, New York (1984).

Yang, M. C., and E. L. Cussler, *AIChE J.*, **32**, 1910 (1986).

4

Equipment for Gas–Liquid Mass-Transfer Operations

4.1 INTRODUCTION

The purpose of the equipment used for mass-transfer operations is to provide intimate contact of the immiscible phases in order to permit interphase diffusion of the constituents. The rate of mass transfer is directly dependent upon the interfacial area exposed between the phases, and the nature and degree of dispersion of one phase into the other are therefore of prime importance.

The operations which include humidification and dehumidification, gas absorption and desorption, and distillation all have in common the requirement that a gas and a liquid phase be brought into contact for the purpose of diffusional interchange between them. The equipment for gas–liquid contact can be broadly classified according to whether its principal action is to disperse the gas or the liquid, although in many devices both phases become dispersed. In principle, at least, any type of equipment satisfactory for one of these operations is suitable for the others, and the major types are indeed used for all. For this reason, the main emphasis of this chapter is on equipment for gas-liquid operations.

4.2 GAS–LIQUID OPERATIONS: LIQUID DISPERSED

Your objectives in studying this section are to be able to:

1. Identify the most common types of tower packings.
2. Estimate gas-pressure drop in packed towers.
3. Estimate volumetric mass-transfer coefficients in packed towers.

This group includes devices in which the liquid is dispersed into thin films or drops, such as wetted-wall towers, sprays and spray towers, and packed towers.

A thin film of liquid flowing down the inside of a vertical pipe, with gas flowing cocurrently or countercurrently, constitutes a *wetted-wall tower*. Such devices have been used for theoretical studies of mass transfer, as described in Chapter 2, because the interfacial surface is readily kept under control and is measurable. Industrially, they have been used as absorbers for hydrochloric acid, where absorption is accompanied by a very large evolution of heat (Hulswitt and Mraz, 1972). In this case, the wetted-wall tower is surrounded with rapidly flowing cooling water. Multitube devices have also been used for distillation, where the liquid film is generated at the top by partial condensation of the rising vapor. Gas-pressure drop in these towers is probably lower than in any other gas–liquid contacting device, for a given set of operating conditions.

The liquid can be sprayed into a gas stream by means of a nozzle which disperses the liquid into a fine spray of drops. The flow may be countercurrent, as in vertical *spray towers* with the liquid sprayed downward, or parallel, as in horizontal *spray chambers*. They are frequently used for adiabatic humidification-cooling operations with recirculating liquid. These devices have the advantage of low pressure drop for the gas, but also have a number of disadvantages. There is a relatively high pumping cost for the liquid, owing to the pressure drop through the spray nozzle. The tendency for entrainment of liquid by the gas leaving is considerable, and mist eliminators will almost always be necessary. Unless the diameter/length ratio is very small, the gas will be fairly thoroughly mixed by the spray and full advantage of countercurrent flow cannot be taken. Ordinarily, however, the diameter/length ratio cannot be made very small since then the spray would quickly reach the walls of the tower and become ineffective as a spray.

Packed towers, used for continuous contact between liquid and gas in both countercurrent and cocurrent flow, are vertical columns which have been filled with packings or devices of large surface, as in Figure 4.1. The liquid is distributed over, and trickles down through, the packed bed, exposing a large surface to contact the gas. The tower *packing*, or *fill*, should provide for large interfacial surface between liquid and gas per unit volume of packed space. It should possess desirable fluid-flow characteristics. This means that the fractional void volume ε in the packed bed should be large enough to permit passage of the volumes of gas and liquid to be processed at relatively high velocity, with low pressure drop for the gas. The packing should be chemically inert to the fluids being processed, and should possess structural strength to permit easy handling and installation.

4.2.1 Types of Packing

Packings are of two major types, random and regular. Random packings are simply dumped into the tower during installation and allowed to fall at random. The Raschig ring, first patented by Dr. Fritz Raschig in Germany in 1907, was the first

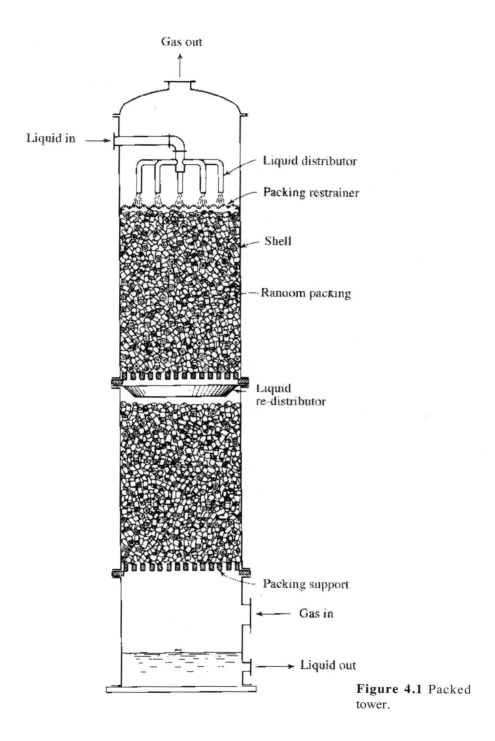

Figure 4.1 Packed tower.

standardized packing. Until then coke or broken glass and pottery were used as packings. Until the sixties, packed columns were mostly filled with Raschig rings or Berl saddles, known as *first-generation* packings. Then development of more advanced packings with higher separation efficiency at low pressure drop accelerated. Today, Pall rings (*second-generation*) and exotically shaped saddles made of ceramics, metals, or plastics (*third generation*) are widely used as packings. Figure 4.2 shows some of the most common types of random packings manufactured at present.

Metal packings are lighter and resist breakage better than ceramic packings, making metal the choice for deep beds. Metal also lends itself to packing geometries that yield higher efficiencies than ceramic or plastic packing shapes. Compared to standard plastic packings, metal packings withstand higher temperatures and provide better wettability. Ceramic packings manufactured in chemical porcelain offer optimal corrosion resistance for applications such as SO_2 and SO_3 absorption, mercaptan removal, natural gas or LPG sweetening, and corrosive distillation. Plastic packings offer the advantage of lightness in weight, but they must be carefully chosen since they may deteriorate rapidly with certain organic solvents and with oxygen-bearing gases at only slightly elevated temperatures.

Raschig rings are hollow cylinders, as shown, of diameters ranging from 6 to 100 mm or more. They may be made of metal, ceramic, or plastic. Pall rings (and as a variant, Hy-Pak), cylinders with partitions, are available in metal and plastic in five nominal sizes from 16 to 89 mm. The saddle-shaped packings, Intalox and Super Intalox saddles, are available in sizes from 6 to 75 mm, made of ceramic or plastic. The Super Intalox saddle has scalloped edges and holes for improved mass transfer, and reduced settling and channeling. The IMTP packing is ideal in a wide range of mass-transfer services. It is used extensively in distillation towers, and in absorbers and strippers. It is available in most metals and in six nominal sizes ranging from 15 to 70 mm. Other, more exotic packing shapes are available, such as the plastic 38-mm Snowflakes.

As a rough guide, packing sizes of 25 mm or larger are ordinarily used for gas rates of 0.25 m^3/s, and 50 mm or larger for gas rates of 1.0 m^3/s or more. During installation, the packings are poured into the tower to fall at random, and in order to prevent breakage of ceramic packings, the tower may first be filled with water to reduce the velocity of fall (Treybal, 1980).

Regular or structured packings are of great variety. They offer the advantages of low pressure drop for the gas and greater possible flow rates, usually at the expense of more costly installation than random packing. Figure 4.3 shows an example of the frequently used Intalox high-performance structured packing. Wood grids, or *hurdles*, are inexpensive and frequently used where large void volumes are required, as with cooling towers.

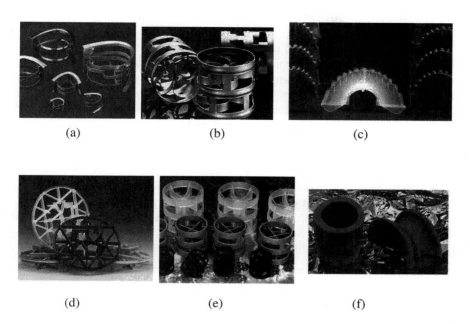

(a)   (b)   (c)

(d)   (e)   (f)

Figure 4.2 Some random tower packings: (a) IMTP, (b) Hy-Pak, (c) Super Intalox saddle, (d) Snowflake, (e) Pall rings, (f) Raschig ring and Intalox saddle (*Chemical Processing Products Division, Norton, Co.*).

Figure 4.3 Intalox high-performance structured packing (*Chemical Processing Products Division, Norton, Co.*).

Packed tower shells may be of wood, metal, chemical stoneware, acidproof brick, plastic- or glass-lined metal, or other material depending on corrosion conditions. For ease of construction and strength, they are usually circular in cross section. An open space at the bottom of the tower is necessary for ensuring good distribution of the gas into the packing. The support must, of course, be strong enough to carry the weight of a reasonable height of packing, and it must have ample free area to allow for flow of liquid and gas with a minimum of restriction. Specially designed supports which provide separate passageways for gas and liquid are available.

4.2.2 Liquid Distribution

Adequate initial distribution of the liquid at the top of the packing is of paramount importance. Otherwise, a significant portion of the packing near the top of the tower will remain dry. Dry packing is of course completely ineffective for mass transfer, and various devices are used for liquid distribution. The arrangement shown in Figure 4.1, or a ring of perforated pipe, can be used in small towers. For large diameters, special liquid distributors are available.

In the case of random packings, the packing density, (i.e., the number of packing pieces per unit volume) is ordinarily less in the immediate vicinity of the tower walls, and this leads to a tendency of the liquid to segregate towards the walls and the gas to flow in the center of the tower (channeling). This tendency is much less pronounced when the diameter of the individual packing pieces (d_p) is smaller than one-eighth the tower diameter (D). It is recommended that the ratio $d_p/D = 1/15$ (Treybal, 1980). Even so, it is customary to provide for redistribution of the liquid at intervals varying from 3 to 10 times the tower diameter, but at least every 6 or 7 m.

Example 4.1 Oscillatory Variation of Void Fraction Near the Walls of Packed Beds

It is well known that the wall in a packed bed affects the packing density resulting in void fraction variations in the radial direction. Mueller (1992) developed the following equation to predict the radial void fraction distribution in a cylindrical tower packed with equal-sized spheres:

$$\varepsilon = \varepsilon_b + (1 - \varepsilon_b) J_0(\alpha r^*) \exp(-\beta r^*) \qquad (4\text{-}1)$$

where

$$\alpha = 7.45 - \frac{3.15}{D/d_p} \qquad \text{for } 2.02 \leq D/d_p \leq 13.0$$

$$\alpha = 7.45 - \frac{11.25}{D/d_p} \qquad \text{for } D/d_p \geq 13.0$$

$$\beta = 0.315 - \frac{0.725}{D/d_p} \qquad \varepsilon_b = 0.365 + \frac{0.220}{D/d_p} \qquad (4\text{-}2)$$

$$r^* = r/d_p \qquad 0 \leq r^* \leq D/2d_p$$

J_0 = Bessel function of the first kind of order zero
r = radial distance measured from the wall

Consider a cylindrical vessel with a diameter of 305 mm packed with solid spheres with a diameter of 20 mm. Plot the radial void fraction fluctuations near the walls for this packed bed.

Solution

For this case, d_p = 20 mm, D = 305 mm, D/d_p = 15.25. Substituting this ratio in the definitions of equation (4-2), α = 6.712, β = 0.267, and ε_b = 0.379. Substituting in equation (4-1),

$$\varepsilon = 0.379 + 0.621\, J_0(6.712r*)\exp(-0.267r*)$$

Figure 4.4 shows the result of the analysis. Notice the wide fluctuations of void fraction near the wall, decaying gradually as we move toward the center of the bed. According to Govindarao and Froment (1986), when $D/d_p > 10$, the radial void fraction approaches a constant asymptotic value (ε_b) at a distance from the wall of approximately 5 particle diameters, consistent with the results shown in Figure 4.4.

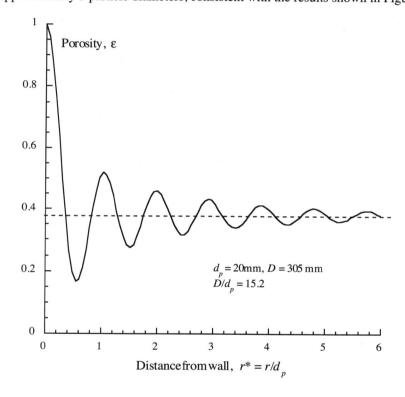

Figure 4.4 Oscillatory variation of void fraction near the walls of a packed bed.

4.2.3 Liquid Holdup

For most random packings, the pressure drop suffered by the gas is influenced by the gas and liquid flow rates. At a fixed gas velocity, the gas-pressure drop increases with increased liquid rate, principally because of the reduced free cross section available for flow of gas resulting from the presence of the liquid. Below a certain limiting gas velocity, the *liquid holdup* (i. e., the quantity of liquid contained in the packed bed) is reasonably constant with changing gas velocity, although it increases with liquid rate. For a given liquid velocity, the upper limit to the gas velocity for a constant liquid holdup is termed the *loading point*. Below this point, the gas phase is the continuous phase. Above this point, liquid begins to accumulate or load the bed replacing gas holdup and causing a sharp increase in pressure drop. Finally, a gas velocity is reached at which the liquid surface is continuous across the top of the packing and the column is flooded. At the *flooding point*, the pressure drop increases infinitely with increasing gas velocity.

The region between the loading point and the flooding point is the *loading region*; significant liquid entrainment is observed, liquid holdup increases sharply, mass-transfer efficiency decreases, and column operation is unstable. Typically, according to Billet (1989), the superficial gas velocity at the loading point is approximately 70% of that at the flooding point. Although a packed column can operate in the loading region, most packed columns are designed to operate below the loading point, in the *preloading region*.

The specific liquid holdup (i. e., volume of liquid holdup/volume of packed bed) in the preloading region has been found from extensive experiments by Billet and Schultes (1995) for a wide variety of random and structured packings and for a number of gas–liquid systems to depend on packing characteristics, and the viscosity, density, and superficial velocity of the liquid according to the dimensionless expression

$$h_L = \left[12 \frac{\mathrm{Fr}_L}{\mathrm{Re}_L} \right]^{1/3} \left[\frac{a_h}{a} \right]^{2/3} \tag{4-3}$$

where:

h_L = specific liquid holdup, m³ holdup/m³ packed be

Re_L = liquid Reynolds number = $v_L \rho_L / a \mu_L$

v_L = superficial liquid velocity, m/s

a = specific surface area of packing, m²/m³

Fr_L = liquid Froude number = $v_L^2 a / g$

g = acceleration of gravity

a_h = hydraulic, or effective, specific area of packing

The ratio of specific areas is given by

$$\frac{a_h}{a} = C_h \, \mathrm{Re}_L^{0.5} \, \mathrm{Fr}_L^{0.1} \qquad \text{for } \mathrm{Re}_L < 5 \tag{4-4}$$

$$\frac{a_h}{a} = 0.85 C_h \, \mathrm{Re}_L^{0.25} \, \mathrm{Fr}_L^{0.1} \qquad \text{for } \mathrm{Re}_L \geq 5 \tag{4-5}$$

Values of a and C_h are characteristic of the particular type and size of packing, as listed together with packing void fraction, ε, and other packing constants in Table 4.1. Because the specific liquid holdup in the preloading region is constant, equation (4-3) does not involve gas-phase properties or gas velocity.

At low liquid velocities, liquid holdup can be so small that the packing is no longer completely wetted. When this happens, packing mass-transfer efficiency decreases dramatically, particularly for aqueous systems of high surface tension. To ensure complete wetting of packing, proven liquid distributors and redistributors should be used and superficial liquid velocities should exceed the following values (Seader and Henley, 1998):

Type of packing material	$v_{L, min}$, mm/s
Ceramic	0.15
Oxidized or etched metal	0.30
Bright metal	0.90
Plastic	1.20

Example 4.2 Specific Liquid Holdup and Void Fraction in Second- and Third-Generation Random Packings

An absorption column is to be designed using an absorbent oil with a kinematic viscosity of 3.0×10^{-6} m^2/s. The superficial liquid velocity will be 0.01 m/s, which is safely above the minimum value for good wetting. The superficial gas velocity will be such that operation will be in the preloading region. Two random packings are being considered: (1) 50-mm metal Pall rings and (2) 50-mm metal Hiflow rings (resembling the Pall ring, but with wider openings). Estimate the specific liquid holdup for each of these two packings.

Solution
From Table 4.1,

Packing	a, m^2/m^3	ε	C_h
50-mm metal Pall rings	112.6	0.951	0.784
50-mm Hiflow rings	92.3	0.977	0.876

Substituting the information given on kinematic viscosity and liquid velocity into the definitions of the Reynolds and Froude numbers,

$$\mathrm{Re}_L = \frac{0.01}{3 \times 10^{-6}\, a} \qquad \mathrm{Fr}_L = \frac{(0.01)^2\, a}{9.8}$$

Table 4.1 Hydraulic Characteristics of Random Packings

Packing	Size	$F_p{}^a$ ft²/ft³	a m²/m³	ε	C_h	C_p
Berl saddle						
Ceramic	25 mm	110	260.0	0.680	0.620	
Ceramic	13 mm	240	545.0	0.65	0.833	
Bialecki ring						
Metal	50 mm		121.0	0.966	0.798	0.719
Metal	35 mm		155.0	0.967	0.787	1.011
Metal	25 mm		210.0	0.956	0.692	0.891
DINPAC ring						
Plastic	70 mm		110.7	0.938	0.991	
Plastic	45 mm		182.9	0.922	1.173	
Envi Pac ring						
Plastic	80, no. 3		60.0	0.955	0.641	0.358
Plastic	60, no. 2		98.4	0.961	0.794	0.338
Plastic	32, no. 1		138.9	0.936	1.039	0.549
Cascade miniring						
Metal	1.5" CMR;T		188.0	0.972	0.870	
Metal	1.5" CMR	29	174.9	0.974	0.935	
Metal	1.0" CMR	40	232.5	0.971	1.040	
Metal	0.5" CMR		356.0	0.955	1.338	
Hiflow ring						
Ceramic	75 mm	15	54.1	0.868		0.435
Ceramic	50 mm	29	89.7	0.809		0.538
Ceramic	35 mm	37	108.3	0.833		0.621
Ceramic	20 mm, 6stg.		265.8	0.776	0.958	
Ceramic	20 mm, 4stg.		261.2	0.779	1.167	0.628
Metal	50 mm	16	92.3	0.977	0.876	0.421
Metal	25 mm	42	202.9	0.962	0.799	0.689
Plastic	90 mm	9	69.7	0.968		0.276
Plastic	50 mm	20	117.1	0.924	1.038	0.327
Plastic	25 mm		194.5	0.918		0.741
IMTP						
Metal	# 25	41	245.0	0.967		
Metal	# 40	24	169.0	0.973		
Metal	# 50	18	81.0	0.978		
Metal	# 70	12	48.0	0.981		
Intalox saddle						
Ceramic	50 mm	40	114.6	0.761		0.747
Plastic	50 mm	28	122.1	0.908		0.758

Table 4.1 (Cont.)

Packing	Size	$F_p{}^a$	a	ε	C_h	C_p
		ft²/ft³	m²/m³			
NORPAC ring						
Plastic	50 mm	14	86.8	0.947	0.651	0.350
Plastic	35 mm	21	141.8	0.944	0.587	0.371
Plastic	15 mm		311.4	0.918	0.343	0.365
Nutter ring						
Metal	# 0.7	39	226.0	0.978		
Metal	# 1.0	27	167.0	0.978		
Metal	# 1.5	20	125.0	0.978		
Metal	# 2.0	17	95.0	0.979		
Metal	# 2.5	15	82.0	0.982		
Metal	# 3.0	11	66.0	0.984		
Plastic	# 2.0	17	82.0	0.920		
Pall ring						
Ceramic	50 mm	43	116.5	0.783	1.335	0.662
Metal	50 mm	27	112.6	0.951	0.784	0.763
Metal	35 mm	40	139.4	0.965	0.644	0.967
Metal	25 mm	56	223.5	0.954	0.719	0.957
Metal	15 mm	70	368.4	0.933	0.590	0.990
Plastic	50 mm	26	111.1	0.919	0.593	0.698
Plastic	35 mm	40	151.1	0.906	0.718	0.927
Plastic	25 mm	55	225.0	0.887	0.528	0.865
Raschig ring						
Ceramic	25 mm	179	190.0	0.680	0.577	1.329
Ceramic	15 mm	380	312.0	0.690	0.648	
Ceramic	10 mm	1,000	440.0	0.650	0.791	
Ceramic	6 mm	1,600	771.9	0.620	1.094	
Metal	15 mm	170	378.4	0.917	0.455	
Tellerette						
Plastic	25 mm	40	190.0	0.930	0.588	0.538
Top-Pak ring						
Aluminum	50 mm		105.5	0.956	0.881	0.604
VSP ring						
Metal	50 mm, no. 2		104.6	0.980	1.135	0.773
Metal	25 mm, no. 1		199.6	0.975	1.369	0.782

[a] When no value of F_p is given, it can be estimated from $F_p = a/\varepsilon^3$.
Sources: Seader and Henley (1998); Katmar Software (1998).

Therefore,

Packing	Re_L	Fr_L
50-mm metal Pall rings	29.6	0.00115
50-mm Hiflow rings	36.1	0.000942

From equation (4-5), since $Re_L > 5$, for the Pall rings,

$$\frac{a_h}{a} = 0.85 \times 0.784 \times (29.6)^{0.25} (0.00115)^{0.1} = 0.79$$

For the Hiflow rings,

$$\frac{a_h}{a} = 0.85 \times 0.876 \times (36.1)^{0.25} (0.000942)^{0.1} = 0.909$$

From equation (4-3), for the Pall rings,

$$h_L = \left[\frac{12 \times 0.00115}{29.6}\right]^{1/3} (0.79)^{2/3} = 0.066$$

For the Hiflow rings,

$$h_L = \left[\frac{12 \times 0.000942}{36.1}\right]^{1/3} (0.909)^{2/3} = 0.064$$

For the Pall rings, the void fraction available for gas flow is reduced by the liquid flow from $\varepsilon = 0.951$ to $0.951 - 0.066 = 0.885$ m³/m³. For the Hiflow packing, the reduction is from $\varepsilon = 0.977$ to $0.977 - 0.064 = 0.913$ m³/m³.

4.2.4 Pressure Drop

Most packed columns consist of cylindrical vertical vessels. The column diameter is determined so as to safely avoid flooding and operate in the preloading region with a pressure drop of no greater than 1.2 kPa/m of packed height (equivalent to 1.5 in. of water head per foot of packed height).

Flooding data for packed columns with countercurrent flow of gas and liquid were first correlated successfully by Sherwood et al. (1938) in terms of the *flow parameter*, $X = (L'/V')(\rho_G/\rho_L)^{0.5}$. The superficial gas velocity at flooding, v_{GF}, was embedded in the dimensionless term $v_{GF}^2 a/g\varepsilon^3 = v_{GF}^2 F_p/g$. The ratio a/ε^3 is a function of the packing only, and is known as the *packing factor*, F_p. Values of a, ε, and F_p are listed in Table 4.1. In some cases, F_p is a modified packing factor, treated as an empirical constant evaluated from experimental data. Leva (1954) used experimental data on ring and saddle packings to extend the Sherwood et al. (1938) flooding correlation to include lines of constant pressure, with the resulting chart becoming known as the *generalized pressure drop correlation* (GPDC). A modern version of the GPDC is shown in Figure 4.5 where the ordinate is given by

$$Y = F_p C_S^2 \mu_L^{0.1} \tag{4-6}$$

$$C_S = v_G \left[\frac{\rho_G}{\rho_L - \rho_G} \right]^{1/2} \tag{4-7}$$

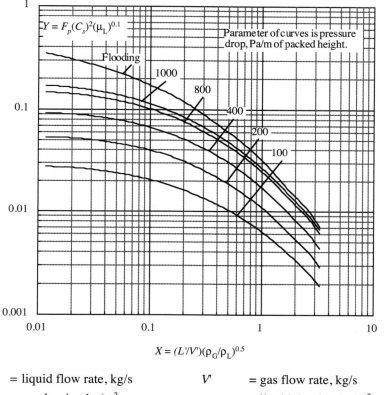

L' = liquid flow rate, kg/s V' = gas flow rate, kg/s

ρ_G = gas density, kg/m³ ρ_L = liquid density, kg/m³

C_S = $v_G[\rho_G/(\rho_L - \rho_G)]^{1/2}$ v_G = superficial gas velocity, m/s

F_p = packing factor, ft²/ft³ μ_L = liquid viscosity, Pa-s

Figure 4.5 Generalized pressure drop correlation for packed beds. (Generated using the *Packed Column Calculator,* version 1.1, Katmar Software, 1998.)

The flooding curve in Figure 4.5 can be accurately described by the polynomial regression

$$\ln Y_{flood} = -\left[3.5021 + 1.028 \ln X + 0.11093 \left(\ln X\right)^2\right]$$ (4-8)

> Notice that in the definition of Y used in Figure 4.5 and equation (4-8), the units of F_p are ft²/ft³. The unusual mixture of units that results is tolerated in an effort to avoid confusion since most of the packing factors reported by packing manufacturers, and most of those available in the literature, are in these units.

It has been found that the pressure drop at flooding is strongly dependent on the packing factor for both random and structured packings. Kister and Gill (1991) developed the empirical expression

$$\Delta P_{flood} = 93.9 F_p^{0.7}$$ (4-9)

where ΔP_{flood} has units of Pa per meter of packed height and F_p has units of ft²/ft³.

Usually, packed columns are designed based on either of two criteria: a fractional approach to flooding gas velocity, or a maximum allowable gas-pressure drop. For given fluid flow rates and properties, and a given packing material, equation (4.8) is used to compute the superficial gas velocity at flooding, v_{GF}. Then, according to the first sizing criterion, a fraction of flooding, f, is selected (usually from 0.5 to 0.7), followed by calculation of the tower diameter from

$$D = \left[\frac{4 Q_G}{f v_{GF} \pi}\right]^{0.5}$$ (4-10)

where Q_G is the volumetric flow rate of the gas.

According to the second sizing criterion, absorbers and strippers are usually designed for gas-pressure drops of 200 to 400 Pa/m of packed depth, atmospheric-pressure fractionators from 400 to 600 Pa/m, and vacuum stills for 8 to 40 Pa/m (Kister, 1992). The GPDC chart of Figure 4.5 can be used to obtain a first estimate of the column diameter that would result in the specified maximum pressure drop. As an alternative, theoretically based models to predict pressure drop in packed beds with countercurrent flows of gas and liquid have been presented by Stichlmair et al. (1989), who use a particle model, and Billet and Schultes (1991a), who use a channel model. Both models extend well-accepted equations for dry-bed pressure drop to account for the effect of liquid holdup. Based on extensive experimental studies using 54 different packing materials, including structured packings, Billet and Schultes (1991a) developed a correlation for dry-gas pressure drop, ΔP_o. Their dimensionally consistent correlating equation is

$$\frac{\Delta P_o}{Z} = \Psi_o \frac{a}{\varepsilon^3} \frac{\rho_G v_G^2}{2} \frac{1}{K_W} \qquad (4\text{-}11)$$

where Z = packing height, and K_W = wall factor. This wall factor can be important for columns with an inadequate ratio of effective particle diameter to inside column diameter, and is given by

$$\frac{1}{K_W} = 1 + \frac{2}{3}\left(\frac{1}{1-\varepsilon}\right)\frac{d_p}{D} \qquad (4\text{-}12)$$

where the effective particle diameter, d_p, is given by

$$d_p = 6\left(\frac{1-\varepsilon}{a}\right) \qquad (4\text{-}13)$$

The dry-packing resistance coefficient (a modified friction factor), Ψ_o, is given by the empirical expression

$$\Psi_o = C_p\left(\frac{64}{\mathrm{Re}_G} + \frac{1.8}{\mathrm{Re}_G^{0.08}}\right) \qquad (4\text{-}14)$$

where

$$\mathrm{Re}_G = \frac{v_G d_p \rho_G K_W}{(1-\varepsilon)\mu_G} \qquad (4\text{-}15)$$

and C_p is a packing constant determined from experimental data and tabulated for a number of packings in Table 4.1.

When the packed bed is irrigated, the liquid holdup causes the pressure drop to increase. The experimental data are reasonably well correlated by (Billet and Schultes, 1991a)

$$\frac{\Delta P}{\Delta P_o} = \left(\frac{\varepsilon}{\varepsilon - h_L}\right)^{1.5} \exp\left(\frac{\mathrm{Re}_L}{200}\right) \qquad (4\text{-}16)$$

where Re_L was defined under equation (4-3). For operation in the preloading region, as usual, the liquid holdup can be estimated from equations (4-3) to (4-5).

Example 4.3 Pressure Drop in Beds Packed with First-, and Third-Generation Random Packings

Air containing 5 mole% NH_3 at a total flow rate of 20 kmol/h enters a packed column operating at 293 K and 1 atm where 90% of the ammonia is scrubbed by a countercurrent flow of 1,500 kg/h of pure liquid water. Estimate the superficial gas velocity and pressure drop at flooding, and the column inside diameter and pressure drop for operation at 70% of flooding for two packing materials:

(a) 25-mm ceramic Raschig rings
(b) 25-mm metal Hiflow rings

Solution
Because the superficial gas velocity is highest at the bottom of the column, calculations are made for conditions there.

Inlet gas:
$$M_G = 0.95 \times 29 + 0.05 \times 17 = 28.4$$
$$V' = \frac{20 \times 28.4}{3,600} = 0.158 \text{ kg/s}$$
$$\rho_G = \frac{PM_G}{RT} = \frac{101.3 \times 28.4}{8.314 \times 293} = 1.181 \text{ kg/m}^3$$
$$Q_G = \frac{V'}{\rho_G} = 0.134 \text{ m}^3/s$$

Exiting liquid:
$$\text{Ammonia absorbed} = 20 \times 0.05 \times 0.9 \times 17$$
$$= 15.3 \text{ kg/h}$$
$$L' = \frac{1,500 + 15.3}{3,600} = 0.421 \text{ kg/s}$$
$$\rho_L = 1,000 \text{ kg/m}^3$$

Then,
$$X = \frac{L'}{V'}\left(\frac{\rho_G}{\rho_L}\right)^{0.5} = \frac{0.421}{0.158}\left(\frac{1.181}{1,000}\right)^{0.5} = 0.092$$

From equation (4-8), $Y_{flood} = 0.187$.

(a) For 25-mm ceramic Raschig rings ($F_p = 179 \text{ ft}^2/\text{ft}^3$):
From equation (4-6), with $\mu_L = 1 \text{ cP} = 0.001 \text{ Pa-s}$,
$$C_{s,flood} = \left[\frac{Y_{flood}}{F_p\mu_L^{0.1}}\right]^{0.5} = \left[\frac{0.187}{179 \times (0.001)^{0.1}}\right]^{0.5} = 0.0457 \text{ m/s}$$

From equation (4-7)
$$v_{GF} = \frac{C_{s,flood}}{\left[\dfrac{\rho_G}{\rho_L - \rho_G}\right]^{0.5}} = \frac{0.0457}{\left[\dfrac{1.181}{1,000 - 1.181}\right]^{0.5}} = 1.329 \text{ m/s}$$

From equation (4-9)

$$\Delta P_{flood} = 93.9 \times (179)^{0.7} = 3,545 \text{ Pa / m of packing}$$

For operation at 70% of the flooding velocity ($f = 0.7$), $v_G = 0.7 \times 1.329 = 0.9303$ m/s. From equation (4-10),

$$D = \left[\frac{4 \times 0.1336}{0.7 \times 1.329 \times \pi} \right]^{0.5} = 0.428 \text{ m}$$

Now, estimate the pressure drop for operation at 70% of flooding using the Billet and Schultes (1991a) correlation. From Table 4.1, for 25-mm ceramic Raschig rings:

$$\varepsilon = 0.68 \qquad a = 190 \text{ m}^2/\text{m}^3$$
$$C_h = 0.577 \qquad C_p = 1.329$$

From equation (4-13)

$$d_p = \frac{6 \times (1 - 0.68)}{190} = 0.0101 \text{ m}$$

From equation (4-12), $1/K_W = 1.049$. The viscosity of the gas phase is basically that of air at 293 K and 1 atm, $\mu_G = 1.84 \times 10^{-5}$ kg/m-s. Then, from equation (4-15)

$$\text{Re}_G = \frac{0.9303 \times 0.0101 \times 1.181}{(1 - 0.68) \times 1.84 \times 10^{-5} \times 1.049} = 1,796.3$$

From equation (4-14), $\Psi_o = 1.361$. From equation (4-11), the dry pressure drop is

$$\frac{\Delta P_o}{Z} = 1.361 \times \frac{190}{0.68^3} \times \frac{1.181 \times 0.9303^2}{2} \times 1.049 = 441 \text{ Pa / m}$$

Correct the pressure drop for liquid holdup with equation (4-16). Calculate first the liquid mass velocity, G_x

$$G_x = \frac{L'}{\dfrac{\pi D^2}{4}} = \frac{0.421 \times 4}{\pi \times (0.428)^2} = 2.926 \text{ kg / m}^2 - \text{s}$$

Then,

$$\mathrm{Re}_L = \frac{v_L \rho_L}{a \mu_L} = \frac{G_x}{a \mu_L} = \frac{2.926}{190 \times 0.001} = 15.4$$

$$\mathrm{Fr}_L = \frac{v_L^2 a}{g} = \frac{G_x^2 a}{\rho_L^2 g} = \frac{(2.926)^2 \times 190}{(1,000)^2 \times 9.8} = 1.659 \times 10^{-4}$$

From equation (4-5),

$$\frac{a_h}{a} = 0.85 \times 0.577 \times (15.4)^{0.25} \times \left(1.659 \times 10^{-4}\right)^{0.1} = 0.407$$

From equation (4-3),

$$h_L = \left[\frac{12 \times 1.659 \times 10^{-4}}{15.4}\right]^{1/3} \times (0.407)^{2/3} = 0.0278$$

From equation (4-16),

$$\frac{\Delta P}{\Delta P_o} = \left[\frac{0.680}{0.680 - 0.0278}\right]^{1.5} \times \exp\left[\frac{15.4}{200}\right] = 1.15$$

Then,

$$\frac{\Delta P}{Z} = 441 \times 1.15 = 507 \ \mathrm{Pa} \, / \, \mathrm{m}$$

(b) For 25-mm metal Hiflow rings ($F_p = 42$ ft^2/ft^3):
For this third-generation packing, the results are:

v_{GF}	= 2.742 m/s
ΔP_{flood}	= 1,286 Pa/m
v_G at 70% of flooding	= 1.920 m/s
D	= 0.298 m
$\Delta P/Z$ at 70% of flooding	= 445 Pa/m

Based on these results, the Hiflow rings have a much greater capacity than the Raschig rings, since the required column cross-sectional area is reduced by 50%.

4.2.5 Mass-Transfer Coefficients

In an extensive investigation, Billet and Schultes (1991b) measured and correlated mass-transfer coefficients for 31 different binary and ternary systems with 67 different types and sizes of packings in columns of diameter ranging from 6 cm to 1.4 m. The systems include some for which mass-transfer resistance resides mainly in the liquid phase and others for which resistance in the gas phase predominates. For the liquid phase resistance, the proposed correlation is

$$k_L = 0.757 C_L \left[\frac{D_L a v_L}{\varepsilon h_L} \right]^{0.5} \tag{4-17}$$

where C_L is an empirical constant characteristic of the packing, as shown in Table 4.2. For the gas phase, they proposed

$$k_y = 0.1304 \, C_V \frac{D_G P}{RT} \frac{a}{\left[\varepsilon (\varepsilon - h_L) \right]^{0.5}} \left[\frac{Re_G}{K_W} \right]^{3/4} Sc_G^{2/3} \tag{4-18}$$

where C_V is an empirical constant included in Table 4.2, and Re_G is as defined in equation (4-15).

Example 4.4 Design of a Packed Bed Ethanol Absorber

When molasses are fermented to produce a liquor containing ethanol, a CO_2-rich vapor containing a small amount of ethanol is evolved. The alcohol will be recovered by countercurrent absorption with water in a packed bed tower. The gas will enter the tower at a rate of 180 kmol/h, at 303 K and 110 kPa. The molar composition of the gas is 98% CO_2 and 2% ethanol. The required recovery of the alcohol is 97%. Pure liquid water at 303 K will enter the tower at the rate of 151.5 kmol/h, which is 50% above the minimum rate required for the specified recovery (Seader and Henley, 1998). The tower will be packed with 50-mm metal Hiflow rings, and will be designed for a maximum pressure drop of 300 Pa/m of packed height.

(a) Determine the column diameter for the design conditions.

(b) Estimate the fractional approach to flooding conditions

.
(c) Estimate the gas and liquid volumetric mass-transfer coefficients, $k_y a_h$ and $k_L a_h$.

Solution
(a) Because the superficial gas velocity and pressure drop are highest at the bottom of the column, calculations are made for conditions there.

Table 4.2 Mass-Transfer Parameters of Random Packings

Packing	Size	C_L	C_V
Berl saddle			
Ceramic	25 mm	1.246	0.387
Ceramic	13 mm	1.364	0.232
Bialecki ring			
Metal	50 mm	1.721	0.302
Metal	35 mm	1.412	0.390
Metal	25 mm	1.461	0.331
Envi Pac ring			
Plastic	80, no. 3	1.603	0.257
Plastic	60, no. 2	1.522	0.296
Plastic	32, no. 1	1.517	0.459
Hiflow ring			
Ceramic	50 mm	1.377	0.379
Ceramic	20 mm, 4stg.	1.744	0.465
Metal	50 mm	1.168	0.408
Metal	25 mm	1.641	0.402
Plastic	50 mm	1.487	0.345
Plastic	25 mm	1.577	0.390
NORPAC ring			
Plastic	50 mm	1.080	0.322
Plastic	35 mm	0.756	0.425
Plastic	25 mm, type B	0.883	0.366
Pall ring			
Ceramic	50 mm	1.227	0.415
Metal	50 mm	1.192	0.410
Metal	25 mm	1.440	0.336
Plastic	50 mm	1.239	0.368
Plastic	35 mm	0.856	0.380
Plastic	25 mm	0.905	0.446
Raschig ring			
Ceramic	25 mm	1.361	0.412
Ceramic	15 mm	1.276	0.401
Ceramic	10 mm	1.303	0.272
Top-Pak ring			
Aluminum	50 mm	1.326	0.389
VSP ring			
Metal	50 mm, no. 2	1.222	0.420
Metal	25 mm, no. 1	1.376	0.405

Source: Seader and Henley (1998).

Inlet gas:

$$M_G = 0.98 \times 44 + 0.02 \times 46 = 44.04$$

$$V' = \frac{180 \times 44.04}{3,600} = 2.202 \text{ kg / s}$$

$$\rho_G = \frac{PM_G}{RT} = \frac{110 \times 44.04}{8.314 \times 303} = 1.923 \text{ kg / m}^3$$

$$\mu_G = 1.45 \times 10^{-5} \text{ Pa - s}$$

$$Q_G = \frac{V'}{\rho_G} = 1.145 \text{ m}^3 / s$$

Exiting liquid:

$$\text{Ethanol absorbed} = 180 \times 0.02 \times 0.97 \times 46$$
$$= 160.6 \text{ kg / h}$$

$$L' = \frac{151.5 \times 18 + 160.6}{3,600} = 0.804 \text{ kg/s}$$

$$\rho_L = 986 \text{ kg / m}^3 \qquad \mu_L = 0.631 \text{ cP}$$

The tower diameter that will result in the specified pressure drop of 300 Pa/m must be determined by an iterative procedure. As a first estimate of the diameter, assume operation at 70% of flooding. Calculate the resulting pressure drop under those conditions by the Billet and Schultes (1991a) correlation and compare to the specified value. Continue iterating, changing the column diameter and calculating the corresponding pressure drop, until convergence to the specified value is achieved.. Then,

$$X = \frac{L'}{V'}\left(\frac{\rho_G}{\rho_L}\right)^{0.5} = \frac{0.804}{2.202}\left(\frac{1.923}{986}\right)^{0.5} = 0.016$$

From equation (4-8), $Y_{flood} = 0.317$.
For 50-mm metal Hiflow rings, from Table 4.1, $F_p = 16$ ft²/ft³.
From equation (4-6), with $\mu_L = 0.631$ cP $= 0.000631$ Pa-s,

$$C_{s,flood} = \left[\frac{Y_{flood}}{F_p\mu_L^{0.1}}\right]^{0.5} = \left[\frac{0.317}{16 \times (0.000631)^{0.1}}\right]^{0.5} = 0.203 \text{ m / s}$$

From equation (4-7),

$$v_{GF} = \frac{C_{S,flood}}{\left[\dfrac{\rho_G}{\rho_L - \rho_G}\right]^{0.5}} = \frac{0.202}{\left[\dfrac{1.923}{986 - 1.923}\right]^{0.5}} = 4.602 \text{ m/s}$$

For operation at 70% of the flooding velocity ($f = 0.7$), $v_G = 0.7 \times 4.602 = 3.22$ m/s. From equation (4-10),

$$D = \left[\frac{4 \times 1.145}{0.7 \times 4.602 \times \pi}\right]^{0.5} = 0.674 \text{ m}$$

Now, we estimate the pressure drop for operation at 70% of flooding using the Billet and Schultes (1991a) correlation. From Table 4.1, for 50-mm metal Hiflow rings:

$$\varepsilon = 0.977 \qquad a = 92.3 \text{ m}^2/\text{m}^3$$
$$C_h = 0.876 \qquad C_p = 0.421$$

From equation (4-13),

$$d_p = \frac{6 \times (1 - 0.977)}{92.3} = 0.0015 \text{ m}$$

From equation (4-12), $1/K_W = 1.066$, $K_W = 0.9381$. Then, from equation (4-15),

$$\text{Re}_G = \frac{3.22 \times 0.0015 \times 1.923 \times 0.9381}{(1 - 0.977) \times 1.45 \times 10^{-5}} = 26{,}130$$

From equation (4-14), $\Psi_o = 0.3365$. From equation (4-11), the dry pressure drop is

$$\frac{\Delta P_o}{Z} = 0.3365 \times \frac{92.3}{0.977^3} \times \frac{1.923 \times 3.22^2}{2} \times 1.066 = 354 \text{ Pa / m}$$

Correct the pressure drop for liquid holdup with equation (4-16). Calculate first the liquid velocity, v_L

$$v_L = \frac{L'}{\rho_L \dfrac{\pi D^2}{4}} = \frac{0.804 \times 4}{\pi \times (0.674)^2 \times 986} = 0.0023 \text{ m/s}$$

Then,

$$\mathrm{Re}_L = \frac{v_L \rho_L}{a \mu_L} = \frac{0.0023 \times 986}{92.3 \times 0.000631} = 39.0$$

$$\mathrm{Fr}_L = \frac{v_L^2 a}{g} = \frac{(0.0023)^2 \times 92.3}{9.8} = 4.98 \times 10^{-5}$$

From equation (4-5),

$$\frac{a_h}{a} = 0.85 \times 0.876 \times (39.0)^{0.25} \times \left(4.98 \times 10^{-5}\right)^{0.1} = 0.691$$

$$a_h = 0.691 \times 92.3 = 63.8 \ \mathrm{m^2/m^3}$$

From equation (4-3),

$$h_L = \left[\frac{12 \times 4.98 \times 10^{-5}}{39}\right]^{1/3} \times (0.691)^{2/3} = 0.0194$$

From equation (4-16),

$$\frac{\Delta P}{Z} = 354 \times \left[\frac{0.977}{0.977 - 0.0194}\right]^{1.5} \times \exp\left[\frac{39.0}{200}\right] = 443 \ \mathrm{Pa/m}$$

The resulting pressure drop is too high; therefore we must increase the tower diameter to reduce the pressure drop. Appendix D presents a Mathcad computer program designed to iterate automatically until the pressure drop criterion is satisfied. Convergence is achieved, as shown in Appendix D, at a tower diameter of $D = 0.738$ m, at which $\Delta P/Z = 300$ Pa/m of packed height.

(b) For the tower diameter of $D = 0.738$ m, the following intermediate results were obtained from the computer program in Appendix D:

$v_G = 2.68$ m/s,	$v_L = 0.00193$ m/s,	$h_L = 0.017$,
$a_h = 58.8$ m^2/m^3,	$\mathrm{Re}_G = 21{,}890$	$\mathrm{Re}_L = 32.6$

The fractional approach to flooding, f, is

$$f = \frac{v_G}{v_{GF}} = \frac{2.68}{4.56} = 0.58$$

(c) Next, we calculate the mass-transfer coefficients. Estimates of the diffusivities of ethanol in a dilute aqueous solution at 303 K (D_L) and in a gas mixture with CO_2 at 303K and 110 kPa (D_G) are needed. For ethanol, $V_c = 167.1$ cm^3/mole, $\sigma = 4.53$ Å, $\varepsilon/k = 362.6$ K; for CO_2, $\sigma = 3.94$ Å, $\varepsilon/k = 195.2$ K (Reid, et al., 1987).

From equation (1-48), $V_A = 60.9$ cm^3/mole.
From equation (1-53), $D_L = 1.91 \times 10^{-5}$ cm^2/s.
From equation (1-49), $D_G = 0.085$ cm^2/s.
From Table 4.2, for 50-mm metal Hiflow rings, $C_L = 1.168$, $C_V = 0.408$.
From equation (4.17), using the results obtained in (b), $k_L = 1.251 \times 10^{-4}$ m/s. Then, the volumetric liquid-phase mass-transfer coefficient is

$$k_L a_h = 1.251 \times 10^{-4} \times 58.84 = 7.33 \times 10^{-3} \ s^{\pm 1}$$

Calculate $Sc_G = \mu_G/\rho_G D_G = 0.887$. From equation (4-18), $k_y = 3.26$ mole/m^2-s. The volumetric gas-phase mass-transfer coefficient is

$$k_y a_h = 3.26 \times 10^{-3} \times 58.84 = 0.191 \ kmol / m^3\text{-}s$$

4.3 GAS–LIQUID OPERATIONS: GAS DISPERSED

Your objectives in studying this section are to be able to:

1. Design bubble columns.
2. Design sieve-tray towers for gas–liquid operations.
3. Estimate stage efficiencies for sieve-tray towers.

In this group are included those devices, such as *sparged* and agitated vessels and the various types of *tray towers*, in which the gas phase is dispersed into bubbles or foams. Tray towers are the most important of the group, since they produce countercurrent, multistage contact, but the simple vessel contactors have many applications (Treybal, 1980).

Gas and liquid can conveniently be contacted, with gas dispersed as bubbles, in agitated vessels whenever multistage, countercurrent effects are not required. This is particularly the case when a chemical reaction between the dissolved gas and a constituent of the liquid is required. The carbonation of a lime slurry, the chlorination of paper stock, the hydrogenation of vegetable oils, the aeration of fermentation broths, as in the production of penicillin, the production of citric acid from beet sugar by the action of microorganisms, and the aeration of activated sludge for biological oxidation are all examples (Treybal, 1980). The gas-liquid mixture can be mechanically agitated, or agitation can be accomplished by the gas itself in *sparged vessels*. The operation may be batch, semibatch with continuous flow of gas and a fixed quantity of liquid, or continuous with flow of both phases.

4.3.1 Sparged Vessels (Bubble Columns)

A *sparger* is a device for introducing a stream of gas in the form of small bubbles into a liquid. If the vessel diameter is small, the sparger, located at the bottom of the vessel, may simply be an open tube through which the gas issues into the liquid. For vessels of diameter greater than 0.3 m, it is better to use several orifices for introducing the gas to ensure good gas distribution. In that case, the orifices may be holes from 1.5 to 3.0 mm in diameter drilled in a pipe distributor placed horizontally at the bottom of the vessel. The purpose of the sparging may be contacting the sparged gas with the liquid, or it may simply be a device for agitation.

The size of gas bubbles depends upon the rate of flow through the orifices, the orifice diameter, the fluid properties, and the extent of turbulence prevailing in the liquid. What follows is for cases where turbulence in the liquid is solely that generated by the rising bubbles and when orifices are horizontal and sufficiently separated to prevent bubbles from adjacent orifices from interfering with each other (at least $3d_p$ apart). For air-water, the following correlations can be used to estimate the size of the bubbles, d_p, as they leave the orifices of the sparger (Leibson, et al., 1956):

$$d_p = 0.0287 d_o^{1/2} \text{Re}_o^{1/3} \qquad \text{for Re}_o \leq 2,100 \qquad (4\text{-}19)$$

$$d_p = 0.0071 \text{Re}_o^{-0.05} \qquad \text{for } 10,000 \leq \text{Re}_o < 50,000 \qquad (4\text{-}20)$$

where the bubble size, d_p, and the orifice diameter, d_o, are in meters and $\text{Re}_o = 4m_G/\pi d_o \mu_G$. For the transition range ($2,100 < \text{Re}_o < 10,000$), there is no correlation of data. It is suggested that d_p for air-water can be approximated by log-log interpolation between the points given by equation (4-19) at $\text{Re}_o = 2,100$ and by equation (4-20) at $\text{Re}_o = 10,000$ (Treybal, 1980).

The volume fraction of the gas–liquid mixture in the vessel which is occupied by the gas is called the *gas holdup*, φ_G. If the superficial gas velocity in the vessel is v_G, then v_G/φ_G is the true gas velocity relative to the vessel walls. If the liquid flows upward, cocurrently with the gas, at a velocity relative to the vessel walls $v_L/(1 - \varphi_G)$, the relative velocity of gas and liquid, or *slip velocity,* is

$$v_S = \frac{v_G}{\varphi_G} - \frac{v_L}{1 - \varphi_G} \qquad (4\text{-}21)$$

Equation (4-21) will also give the slip velocity for countercurrent flow of liquid if v_L for the downward liquid flow is assigned a negative size.

The holdup for sparged vessels can be correlated through the slip velocity by the following equation developed from data presented by Hughmark (1967):

$$\ln\left[\frac{V_G}{v_S}\right] = -0.847 + 0.0352 \ln V_G$$

$$-0.1835\left(\ln V_G\right)^2 - 0.01348\left(\ln V_G\right)^3$$

$$V_G = v_G\left[\frac{\rho_W}{\rho_L}\frac{\sigma_{AW}}{\sigma}\right]^{1/3} \qquad\qquad (4\text{-}22)$$

Knowing v_G, and v_L, equations (4-21) and (4-22) can be combined to estimate the gas holdup and slip velocity. The following restrictions apply when using equation (4-22), satisfactory for no liquid flow ($v_L = 0$), cocurrent liquid flow up to $v_L = 0.1$ m/s, and also for small countercurrent flow:

(1) Vessel diameter, D = above 0.1m

(2) Liquid density, ρ_L = 770 to 1,700 kg/m³

(3) Liquid viscosity, μ_L = 0.0009 to 0.152 Pa-s

(4) Density of water, ρ_W = 1,000 kg/m³

(5) Air-water surface tension, σ_{AW} = 0.072 N/m

(6) Surface tension, σ = 0.025 to 0.076 N/m

If a unit volume of a gas–liquid mixture contains a gas volume φ_G made up of n bubbles of diameter d_p, then $n = 6\varphi_G/\pi d_p^3$. If the interfacial area per unit volume is a, then $n = a/\pi d_p^2$. Equating the two expressions for n provides the specific area

$$a = \frac{6\varphi_G}{d_p} \qquad\qquad (4\text{-}23)$$

For low liquid velocities, the bubble size may be taken as that produced at the orifices of the sparger—according to equations (4-19) or (4-20)—corrected as necessary for pressure. For high liquid velocities, the bubble size may be altered by turbulent breakup and coalescence of bubbles. For example, for air-water in the range φ_G = 0.1 to 0.4 and $v_L/(1 - \varphi_G) = 0.15$ to 15 m/s, the bubble size is approximated by (Petrick, 1962)

$$d_p = \frac{0.002344}{\left[v_L / (1 - \varphi_G) \right]^{0.67}} \tag{4-24}$$

where d_p is in meters and v_L in m/s.

In practically all the gas-bubble liquid systems, the liquid-phase mass-transfer resistance is strongly controlling, and gas-phase coefficients are not needed. The liquid-phase coefficients are correlated by (Hughmark, 1967)

$$\text{Sh}_L = \frac{F_L d_p}{c D_L} = 2 + b' \, \text{Re}_G^{0.779} \, \text{Sc}_L^{0.546} \left(\frac{d_p g^{1/3}}{D_L^{2/3}} \right)^{0.116} \tag{4-25}$$

where

$$b' = \begin{cases} 0.061 & \text{single gas bubbles} \\ 0.0187 & \text{swarms of bubbles} \end{cases}$$

The gas-bubble Reynolds number must be calculated with the slip velocity: $\text{Re}_G = d_p v_s \rho_L / \mu_L$.

The power supplied to the vessel contents, which is responsible for the agitation and creation of large interfacial area, is derived from the gas flow. Bernoulli's equation, a mechanical energy balance written for the gas between location o (just above the sparger orifices) and location s (at the liquid surface), is

$$\frac{v_s^2 - v_o^2}{2 g_c} + (Z_s - Z_o) \frac{g}{g_c} + \int_o^s \frac{dP}{\rho_G} + W + h_f = 0 \tag{4-26}$$

The friction loss, h_f, and the velocity at the liquid surface, v_s, can be neglected, and the gas density can be described by the ideal gas law, whereupon

$$W = \frac{v_o^2}{2 g_c} + \frac{P_o}{\rho_{Go}} \ln \frac{P_o}{P_s} + (Z_o - Z_s) \frac{g}{g_c} \tag{4-27}$$

W is the work done by the gas on the vessel contents, per unit mass of gas. Work for the gas compressor will be larger to account for friction losses in the piping and orifices and compressor inefficiency.

Example 4.5 Stripping Chloroform from Water by Sparging with Air

Chlorinating drinking water kills microbes but produces trace amounts of a group of potentially harmful substances called trihalomethanes (THMs), the most prevalent of which is chloroform. Federal regulations have been proposed to limit THMs levels in drinking water to 40 µg/L (Davis and Cornwell, 1998). Sparging with air in a bubble column is an effective method to strip the water of THMs.

A vessel 1.0 m in diameter is to be used for stripping chloroform from water by sparging with air at 298 K. The water will flow continuously downward at the rate of 10.0 kg/s at 298 K. The water contains 240 µg/L of chloroform. It is desired to remove 90% of the chloroform in the water using an airflow that is 50% higher than the minimum required. At these low concentrations, chloroform-water solutions follow Henry's law ($y_i = mx_i$) with $m = 220$. The sparger is in the form of a ring located at the bottom of the vessel, 50 cm in diameter, containing 90 orifices, each 3 mm in diameter. Estimate the depth of the water column required to achieve the specified 90% removal efficiency. Estimate the power required to operate the air compressor, if the mechanical efficiency of the system is 60%.

Solution
Following the procedure discussed in Chapter 3 for countercurrent flow and dilute solutions, it is determined that the air flow rate corresponding to 50% above the minimum is 0.10 kg/s. The mass flow rate through each orifice is $m_G = 0.10/90 = 0.0011$ kg/s. The viscosity of air at 298 K is $\mu_G = 1.8 \times 10^{-5}$ kg/m-s. The orifice diameter is $d_o = 0.003$ m. Then, $Re_o = 25,940$. From equation (4-20), the bubble diameter at the orifice outlet is $d_p = 0.00427$ m.

To estimate the gas holdup, the density of the gas at the average pressure in the vessel must be estimated. Since the water column height is not known at this point, an iterative procedure must be implemented. Assume an initial value of the column height (say, $Z = 0.5$ m) to estimate the average pressure, as a first step in calculating an improved estimate of the column height. With the refined estimate of Z, calculate a new value of average pressure, and continue iterating until convergence is achieved. For $Z = 0.5$ m, $\rho_L = 1,000$ kg/m^3, and $P_s = 101.3$ kPa, $P_o = 101.3 + 1,000 \times 9.8 \times 0.5/1,000 = 106.2$ kPa. Then, the average pressure is $P_{av} = (P_s + P_o)/2 = 103.8$ kPa. From the ideal gas law, the gas density at this average pressure and 298 K is $\rho_G = 1.215$ kg/m^3. The cross-sectional area of the 1.0-m-diameter vessel is 0.785 m^2, therefore the gas superficial velocity is $v_G = 0.10/(1.215 \times 0.785) = 0.105$ m/s. In this case, $\rho_L = \rho_W$, $\sigma = \sigma_{AW}$. Then, from equation (4-22), $V_G = v_{G''}$, $v_G/v_S = 0.182$, $v_S = 0.577$ m/s. Taking into consideration that the water flows downward, the superficial liquid velocity is $v_L = -10/(1,000 \times 0.785) = -0.0127$ m/s. Substituting in equation (4-21):

$$0.577 = \frac{0.105}{\varphi_G} + \frac{0.0127}{1 - \varphi_G}$$

Solving, the gas holdup is $\varphi_G = 0.187$.

To estimate the interfacial area and the liquid-phase mass-transfer coefficient, the average bubble size along the water column must be estimated. In this case,

$$\frac{v_L}{1-\varphi_G} = \frac{0.0127}{1-0.187} << 0.15 \text{ m / s}$$

Then, the bubble diameter may be taken as that produced at the orifices of the sparger, corrected for pressure. Therefore, the bubble diameter at the average column pressure, calculated from d_p at the orifice, is

$$d_p = \left[(0.00427)^3 \frac{106.2}{103.8}\right]^{1/3} = 0.0043 \text{ m}$$

Then, from equation (4-23), $a = 261$ m^{-1}. Next, an estimate of the diffusivity of chloroform in dilute aqueous solution at 298 K, D_L, is needed. Use the corresponding Hayduk-Minhas correlation, equation (1-53). The molar volume for chloroform, estimated from its critical volume, is $V_A = 88.6$ cm^3/mole (Reid, et al., 1987). The viscosity of water at 298 K is approximately 0.9 cP. Then, $D_L = 1.08 \times 10^{-5}$ cm^2/s, and $Sc_L = 833$. Calculate

$$\frac{d_p g^{1/3}}{D_L^{2/3}} = \frac{0.0043 \times 9.8^{1/3}}{\left(1.08 \times 10^{-9}\right)^{2/3}} = 8,742$$

$$Re_G = \frac{d_p v_S \rho_L}{\mu_L} = \frac{0.0043 \times 0.577 \times 1,000}{0.9 \times 10^{-3}} = 2,760$$

From equation (4-25),

$$Sh_L = 2 + 0.0187 \times 2,760^{0.779} \times 833^{0.546} \times 8,742^{0.116} = 1012$$

For dilute aqueous solutions, $x_{B,M} \approx 1.0$, $c \approx 55.5$ kmol/m^3. Then, for $N_B = 0$

$$k_x \cong \frac{Sh_L c D_L}{d_p} = \frac{1012 \times 55.5 \times 1.08 \times 10^{-9}}{0.0043} = 0.0141 \text{ kmol / m}^2\text{-s}$$

The volumetric mass-transfer coefficient is

$$k_x a = 3.68 \frac{\text{kmol}}{\text{m}^3 - \text{s}}$$

Next, we estimate the height of the water column required for 90% removal of the chloroform. If we assume that all of the resistance to mass transfer in this case resides in the liquid phase, the mass-transfer circumstances are analogous to those of the dissolved-oxygen stripping of boiler feedwater described in Example 2.13. Therefore, equation (2-64) applies, modified here as

$$Z = \frac{G_{Mx}}{k_x a (1 - A)} \ln\left[\frac{x_{in}}{x_{out}} (1 - A) + A \right] \qquad (4\text{-}28)$$

where G_{Mx} is the liquid molar velocity, and A is the absorption factor, $A = L/(mV)$. For this case, $L = 10/18 = 0.556$ kmol/s; $G_{Mx} = 0.556/0.785 = 0.708$ kmol/m²-s; $V = 0.10/29 = 0.00344$ kmol/s; $A = 0.732$. Substituting in equation (4-28), for $x_{in}/x_{out} = 10$, $Z = 0.90$ m.

With this new estimate of Z, calculate again the average pressure in the column of water: $P_o = 110.1$ kPa, $P_{av} = 105.7$ kPa. The gas density at this pressure is $\rho_G = 1.237$ kg/m³, and the average gas velocity is $v_G = 0.103$ m/s. This is very close to the value used in the first iteration ($v_G = 0.105$ m/s); therefore there will be only a small change in the estimate of the column height. Convergence is achieved in three iterations at the value of $Z = 0.904$ m.

Use equation (4-27) to calculate the work done by the gas on the vessel contents per unit mass of gas, W. Calculate:

$$\rho_{Go} = \frac{P_o M}{RT} = 1.289 \frac{\text{kg}}{\text{m}^3}$$

$$v_o = \frac{4 m_G}{\pi d_o^2 \rho_{Go}} = \frac{4 \times 0.0011}{\pi \times (0.003)^2 \times 1.289} = 121.9 \; \frac{\text{m}}{\text{s}}$$

$$\frac{v_o^2}{2 g_c} = \frac{(121.9)^2}{2 \times 1} = 7{,}431 \; \text{J} / \text{kg}$$

$$\frac{P_o}{\rho_{Go}} \ln \frac{P_o}{P_s} = \frac{RT}{M} \ln \frac{P_o}{P_s} = \frac{8{,}314 \times 298}{29} \ln \frac{110.1}{101.3} = 7{,}165 \; \text{J} / \text{kg}$$

$$\left(Z_o - Z_s \right) \frac{g}{g_c} = \frac{0.904 \times 9.8}{1.0} = 9.0$$

Adding all the contributions, $W = 14{,}605$ J/kg of air. For a total air flowrate of 0.1 kg/s and 60% mechanical efficiency, the air-compressor power is

$$\overline{W} = \frac{14{,}605 \times 0.1}{0.6} = 2{,}434 \text{ W} = 2.43 \text{ kW}$$

4.3.2 Tray Towers

Tray towers are vertical cylinders in which the liquid and gas are contacted in stepwise fashion on trays or plates. The liquid enters at the top and flows downward by gravity. On the way, it flows across each tray and through a downspout to the tray below. The gas passes upward through openings of one sort or another in the tray, then bubbles through the liquid to form a froth, disengages from the froth, and passes on to the next tray above. The overall effect is a multiple countercurrent contact of gas and liquid, although each tray is characterized by a cross flow of the two. Each tray of the tower is a stage, since on the tray the fluids are brought into intimate contact, interphase diffusion occurs, and the fluids are then separated.

The number of equilibrium stages (theoretical trays) in a tower is dependent only upon the difficulty of the separation to be carried out and is determined solely from material balances and equilibrium considerations. The stage efficiency, and therefore the number of real trays, is determined by the mechanical design used and the conditions of operation. The diameter of the tower, on the other hand, depends upon the quantities of gas and liquid flowing through the tower per unit time. Once the number of theoretical plates, or equilibrium stages, required has been determined, the principal problem in the design of the tower is to choose dimensions and arrangements which will represent the best compromise between several opposing tendencies, since it is generally found that conditions leading to high tray efficiencies will ultimately lead to operational problems (Treybal, 1980).

High efficiencies require deep pools of liquid on the tray (long contact time) and relatively high gas velocities (large interfacial contact areas and mass-transfer coefficients). These conditions, however, lead to a number of difficulties. One is the mechanical entrainment of liquid droplets in the rising gas stream. At high gas velocities, when the gas disengages from the froth, small droplets of liquid are carried by the gas to the tray above. Liquid carried up the tower in this manner reduces mass transfer and consequently adversely affects the tray efficiency.

Furthermore, great liquid depths on the tray and high gas velocities both result in high pressure drop for the gas in flowing through the tray. For absorbers and humidifiers, high pressure drop results in high fan power to force the gas through the tower, and consequently high operating costs. In the case of distillation, high pressure at the bottom of the tower results in high boiling temperatures, which in turn may lead to heating difficulties and possibly damage to heat-sensitive compounds.

Ultimately, purely mechanical difficulties arise. High pressure drop may lead directly to flooding. With a large pressure difference in the space between trays, the level of liquid leaving a tray at relatively low pressure and entering one of high pressure must necessarily assume an elevated position in the downspouts, as shown in Figure 4.6. As the pressure difference increases due to an increased rate of flow of the gas, the level in the downspout will rise further to permit the liquid to enter the lower tray. Ultimately, the liquid level in the downcomer may reach that on the tray above and the liquid will fill the entire space between the trays. The tower is then flooded, the tray efficiency falls to a very low value, the flow of gas is erratic, and liquid may be forced out of the gas exit pipe at the top of the tower.

For liquid–gas combinations which tend to foam excessively, high gas velocities may lead to a condition of *priming*, which is also an inoperative situation. Here, the foam persists throughout the space between trays, and a great deal of liquid is carried by the gas from one tray to the tray above. This is an exaggerated condition of entrainment. The liquid so carried recirculates between trays, and the added liquid-handling load increases the gas pressure drop sufficiently to lead to flooding.

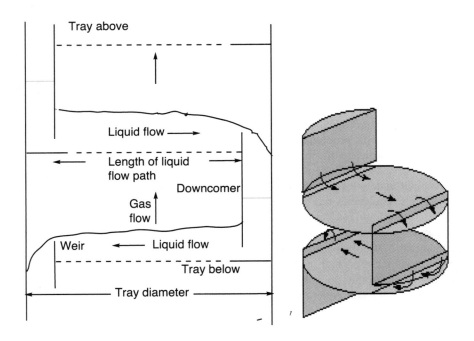

Figure 4.6 Cross-flow tray tower.

If liquid rates are too low, the gas rising through the openings of the trays may push the liquid away (*coning*), and contact of the gas and liquid is poor. If the gas rate is too low, much of the liquid may rain down through the openings of the tray (*weeping*), thus failing to obtain the benefit of complete flow over the tray; and at very low gas rates, none of the liquid reaches the downspouts (*dumping*). The relations between these conditions are shown schematically in Figure 4.7, and all types of trays are subject to these difficulties in some form. The various arrangements, dimensions, and operating conditions chosen for design are those which experience have proved to be reasonably good compromises. The general design procedure involves a somewhat empirical application of them, following by computational check to ensure that pressure drop and flexibility to handle varying flow quantities are satisfactory.

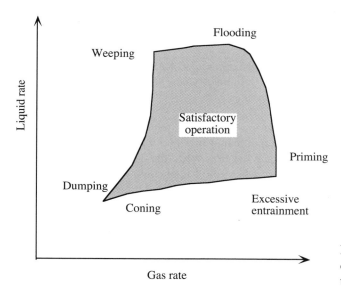

Figure 4.7 Operating characteristics of sieve trays.

A great variety of tray designs have been and are being used. The simplest is a sieve tray—it has a perforated tray deck with a uniform hole diameter of from less than a millimeter to about 25 mm. Trays with valves, which can be fixed or floating, are also very common. During the first half of the twentieth century, practically all towers were fitted with bubble-cap trays, but new installations now use either sieve trays or one of the proprietary designs which have proliferated since 1950. Figure 4.8 shows samples of each family of tray-deck design.

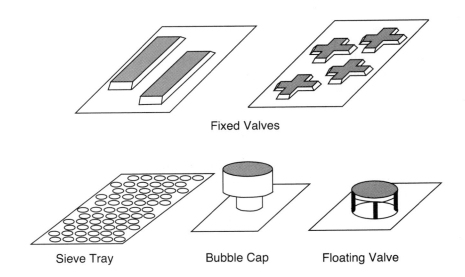

Fixed Valves

Sieve Tray   Bubble Cap   Floating Valve

Figure 4.8 Types of tray-deck mass-transfer devices (Taken from Bennet, D. L. and K. W. Kovak, "Optimize Distillation Columns," *Chem. Eng. Progress,* **96** (5), 20, May 2000).

Because of their simplicity and low cost, sieve (perforated) trays are now the most important of tray devices. In the design of sieve trays, the diameter of the tower must be chosen to accommodate the flow rates, the details of the tray layout must be selected, estimates must be made of the gas-pressure drop and approach to flooding, and assurance against excessive weeping and entrainment must be established.

4.3.3 Tray Diameter

The tower diameter and consequently its cross-sectional area must be sufficiently large to handle the gas and liquid rates within the satisfactory region of Figure 4.7. For a given type of tray at flooding, the superficial velocity of the gas v_{GF} (volumetric rate of gas flow, Q_G per net cross-section area for gas flow in the space between trays, A_n) is related to the fluid densities by

$$v_{GF} = C \left[\frac{\rho_L - \rho_G}{\rho_G} \right]^{1/2} \qquad (4\text{-}29)$$

The net cross-section area A_n is the tower cross-section area A_t minus the area taken up by the downspout A_d, in the case of a cross-flow tray as in Figure 4-6. The value of the empirical constant C depends on the tray design, tray spacing, flow rates, liquid surface tension, and foaming tendency. It is estimated according to the empirical relationship (Seader and Henley, 1998; Fair, 1961)

$$C = F_{ST} F_F F_{HA} C_F \qquad (4\text{-}30)$$

where $\quad F_{ST}\quad$ = surface tension factor = $(\sigma/20)^{0.2}$

$\qquad\quad \sigma \quad$ = liquid surface tension, dyne/cm

$\qquad\quad F_F \quad$ = foaming factor = 1.0 for nonfoaming systems; for many absorbers may be 0.75 or even less (Kister, 1992)

$\qquad\quad F_{HA} \quad$ = 1.0 for $A_h/A_a \geq 0.10$, and $5(\,A_h/A_a) + 0.5$ for $A_h/A_a < 0.1$

$\qquad\quad A_h/A_a \quad$ = ratio of vapor hole area to tray active area

$$C_F = \alpha \log \frac{1}{X} + \beta \qquad (4\text{-}31)$$

$$\alpha = 0.0744t + 0.01173$$
$$\beta = 0.0304t + 0.015 \qquad (4\text{-}32)$$

$\quad X \qquad$ = flow parameter = $(L'/G')(\rho_G/\rho_L)^{0.5}$

$\quad t \qquad$ = tray spacing, in m

If the value of X is in the range 0.01 *to* 0.10, *use X* = 0.10 *in equation* (4-31).

Typically, the column diameter D is based on a specified fractional approach to flooding f. Then,

$$D = \left[\frac{4Q_G}{f v_{GF}(1 - A_d/A_t)\,\pi} \right]^{0.5} \qquad (4\text{-}33)$$

Oliver (1966) suggests that A_d/A_t must be chosen, based on the value of X, as

$$\frac{A_d}{A_t} = \begin{cases} 0.1, & X \leq 0.1 \\ 0.1 + \dfrac{X - 0.1}{9}, & 0.1 \leq X \leq 1.0 \\ 0.2, & X \geq 1.0 \end{cases} \qquad (4\text{-}34)$$

In most sieve trays, the holes are placed in the corners of equilateral triangles at distances between centers (pitch, p') of from 2.5 to 5 hole diameters d_o, as shown in Figure 4.9 (a). For such an arrangement, it is easily shown that

$$\frac{A_h}{A_a} = \frac{\pi}{4\sin\left(60^\circ\right)}\left[\frac{d_o}{p'}\right]^2 = 0.907\left[\frac{d_o}{p'}\right]^2 \qquad (4\text{-}35)$$

The downcomer geometry is shown in Figure 4.9 (b). From this and geometric relationships, the weir length, L_w, and the distance of the weir from the center of the tower, r_w, can be estimated for a specified value of A_d/A_t:

$$\frac{A_d}{A_t} = \frac{\theta - \sin\theta}{2\pi}, \qquad \theta \text{ in radians}$$

$$\frac{L_w}{D} = \sin\left[\frac{\theta}{2}\right], \qquad \frac{r_w}{D} = \frac{1}{2}\cos\left[\frac{\theta}{2}\right] \qquad (4\text{-}36)$$

Because of the need for internal access to columns with trays, a packed column is generally used if the diameter calculated from equation (4-33) is less than 60 cm. Tray spacing must be specified to compute column diameter, as shown by equation (4-32). On the other hand, recommended values of tray spacing depend on column diameter, as summarized in Table 4.3. Therefore, calculation of the tower diameter involves an iterative procedure: (a) an initial value of tray spacing is chosen (usually, $t = 0.6$ m); (b) the tower diameter is calculated based on this value of tray spacing; (c) the value of t is modified as needed according to the recommendations of Table 4.3 and the current estimate of D; (d) the procedure is repeated until convergence is achieved.

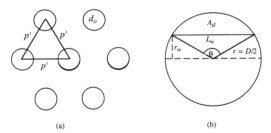

Figure 4.9 Tray geometry: (a) equilateral-triangular arrangement of holes in a sieve tray; (b) downcomer area geometry.

Table 4.3 Recommended Tray Spacing

Tower diameter, D, m	Tray spacing, t, m
1 or less	0.50
1 – 3	0.60
3 – 4	0.75
4 – 8	0.90

Source: Treybal (1980).

Example 4.6 Design of a Sieve-Tray Column for Ethanol Absorption

Design a sieve-tray column for the ethanol absorber of Example 4.4. For alcohol absorbers, Kister (1992) recommends a foaming factor $F_F = 0.9$. The liquid surface tension is estimated as $\sigma = 70$ dyne/cm. Take $d_o = 5$ mm on an equilateral-triangular pitch 15 mm between hole centers, punched in stainless steel sheet metal 2 mm thick. Design for an 80% approach to the flood velocity.

Solution

From Example 4.4, $X = 0.016$. From equation 4-35, $A_h/A_a = 0.907(5/15)^2 = 0.101$. Assume, $t = 0.5$ m; then, from equation (4-32), $\alpha = 0.0489$, $\beta = 0.0302$. Since $X < 0.1$, using $X = 0.1$ in equation (4-31), $C_F = 0.0791$ m/s. Since $A_h/A_a > 0.1$, $F_{HA} = 1$.

$$F_{ST} = \left[\frac{\sigma}{20}\right]^{0.2} = \left[\frac{70}{20}\right]^{0.2} = 1.285 \qquad F_F = 0.90$$

From equation (4-30), $C = 0.091$ m/s. From Example 4.4, $\rho_G = 1.923$ kg/m^3, $\rho_L = 986$ kg/m^3, $Q_G = 1.145$ m^3/s; then, from equation (4-29), $v_{GF} = 2.07$ m/s. Since $X < 0.1$m, equation (4-34) recommends $A_d/A_t = 0.1$. For an 80% approach to flooding, equation (4-33) yields

$$D = \left[\frac{4 \times 1.145}{0.8 \times 2.07 \times (1-0.1) \times \pi}\right]^{0.5} = 0.989 \text{ m}$$

> At this point, the assumed value of tray spacing ($t = 0.5$ m) must be checked against the recommended values of Table 4.3. Since the calculated value of $D < 1.0$ m, $t = 0.5$ m is the recommended tray spacing and no further iteration is needed.

Next, we calculate some further details of the tray design. From equation (4-36), for $A_d/A_t = 0.1$ and $D = 0.989$ m, $\theta = 1.627$ rads; $L_w = 0.727 \times 0.972 = 0.707$ m; $r_w = 0.3435 \times 0.972 = 0.334$ m.

Summarizing, the details of the sieve tray design are as follows:

Diameter $D = 0.989$ m
Tray spacing $t = 0.5$ m
Total cross-sectional area $A_t = 0.768$ m^2
Downcomer area $A_d = 0.077$ m^2
Active area over the tray $A_a = A_t - 2A_d = 0.615$ m^2
Weir length $L_w = 0.719$ m

Distance from tray center to weir $r_w = 0.34$ m
Total hole area $A_h = 0.101$ $A_a = 0.062$ m^2
Hole arrangement: 5-mm diameter on an equilateral-triangular pitch 15 mm between hole centers, punched in stainless steel sheet metal 2 mm thick

4.3.4 Tray Gas-Pressure Drop

Typical tray pressure drop for flow of vapor in a tower is from 0.3 to 1.0 kPa/tray. Pressure drop (expressed as head loss) for a sieve tray is due to friction for vapor flow through the tray perforations, holdup of the liquid on the tray, and a loss due to surface tension:

$$h_t = h_d + h_l + h_\sigma \qquad (4\text{-}37)$$

where h_t = total head loss/tray, cm of liquid

h_d = dry tray head loss, cm of liquid

h_l = equivalent head of clear liquid on tray, cm of liquid

h_σ = head loss due to surface tension, cm of liquid

The dry sieve tray pressure drop is given by a modified orifice equation (Ludwig, 1979),

$$h_d = 0.0051 \left[\frac{v_o}{C_o}\right]^2 \rho_G \left[\frac{\rho_W}{\rho_L}\right]\left[1 - \left(A_h / A_a\right)^2\right] \qquad (4\text{-}38)$$

where v_o is the hole velocity, in m/s, ρ_W is the density of liquid water at the liquid temperature, and the orifice coefficient C_o can be determined by the following equation (Wankat, 1988):

$$C_o = 0.85032 - 0.04231\frac{d_o}{l} + 0.0017954\left(\frac{d_o}{l}\right)^2$$

$$\frac{d_o}{l} \geq 1.0 \qquad (4\text{-}39)$$

where l is the tray thickness.

The equivalent height of clear liquid holdup on a tray depends on weir height, h_W, liquid and vapor densities and flow rates, and downcomer weir length, as given by the following empirical expression developed from experimental data (Bennett, et al., 1983):

$$h_l = \phi_e \left[h_w + C_l \left(\frac{q_L}{L_w \phi_e} \right)^{2/3} \right] \qquad (4\text{-}40)$$

where h_w = weir height, cm (typical values are from 2.5 to 7.5 cm)
 ϕ_e = effective relative froth density (height clear liquid/froth height)
      = $\exp(-12.55 K_s^{0.91})$
 K_s = capacity parameter, m/s

$$K_s = v_a \left(\frac{\rho_G}{\rho_L - \rho_G} \right)^{1/2} \qquad (4\text{-}41)$$

 v_a = superficial gas velocity based on tray active area, m/s
 q_L = liquid flow rate across tray, m³/s
 C_l = $50.12 + 43.89 \exp(-1.378 h_w)$

As the gas emerges from the tray perforations, the bubbles must overcome surface tension. The pressure drop due to surface tension is given by the difference between the pressure inside the bubble and that of the liquid according to the theoretical relation

$$h_\sigma = \frac{6\sigma}{g \rho_L d_o} \qquad (4\text{-}42)$$

where it is assumed that the maximum bubble size may be taken as the perforation diameter d_o.

Methods for estimating gas-pressure drop for bubble-cap trays and valve trays are discussed by Kister (1992).

Example 4.7 Gas-Pressure Drop in a Sieve-Tray Ethanol Absorber.

Estimate the tray gas-pressure drop for the ethanol absorber of Examples 4.4 and 4.6. Use a weir height h_w of 50 mm.

Solution
Calculate the dry tray pressure drop: From Example 4.4, $Q_G = 1.145$ m³/s; from Example 4.6, $A_h = 0.062$ m². Then, $v_o = 1.145/0.062 = 18.48$ m/s. From Example 4.6, $d_o/l = 5/2 = 2.5$; $A_h/A_a = 0.101$; $\rho_G = 1.923$ kg/m³; $\rho_L = 986$ kg/m³. The density of liquid water at 303 K is $\rho_W = 995$ kg/m³. From equation (4-39), $C_o = 0.756$. Then, from equation (4-38), $h_d = 5.85$ cm.

Calculate the equivalent head of clear liquid on the tray: From Example 4.6, A_a = 0.615 m², then v_a = 1.145/0.615 = 1.863 m/s. From equation (4-41), K_s = 0.082 m/s; ϕ_e = 0.274. From Example 4.4, q_L = 0.804/986 = 0.000815 m³/s; From Example 4.6, L_w = 0.719 m. For h_w = 5.0 cm, C_l = 50.16. Substituting in equation (4-40), h_l = 1.73 cm.

Calculate head loss due to surface tension: From Example 4.6, the surface tension is 70 dyne/cm = 0.07 N/m. Substituting in equation (4-42), h_σ = 0.0087 m = 0.87 cm. Calculate the total head loss/tray: Substituting in equation (4-37), h_t = 8.45 cm of clear liquid/tray. For a liquid density of 986 kg/m³, this is equivalent to a gas-pressure drop of ΔP_G = 0.0845 × 986 × 9.8 = 817 Pa/tray.

4.3.5 Weeping and Entrainment

For a tray to operate at high efficiency, weeping of liquid through the tray perforations must be small compared to flow over the outlet weir and into the downcomer, and entrainment of liquid by the gas must not be excessive. Weeping occurs at low vapor velocities and/or high liquid rates. Bennett, et al. (1997) report that weeping did not appear to substantially degrade tray performance as long as the orifice Froude number:

$$\text{Fr}_o = \left(\frac{\rho_G}{\rho_L} \frac{v_o^2}{g h_l} \right)^{0.5} \geq 0.5 \tag{4-43}$$

At high vapor rates, entrainment becomes significant and decreases tray performance. To take this into account, Bennett et al. (1997) defined the fractional entrainment E as

$$E = \frac{\text{Liquid entrainment mass flow rate}}{\text{Upward gas mass flow rate}} \tag{4-44}$$

and developed the correlation

$$E = 0.00335 \left(\frac{h_{2\phi}}{t} \right)^{1.1} \left(\frac{\rho_L}{\rho_G} \right)^{0.5} \left(\frac{h_l}{h_{2\phi}} \right)^{\kappa} \tag{4-45}$$

where t = tray spacing
       $h_{2\phi}$ = height of two-phase region on the tray, given by

$$h_{2\phi} = \frac{h_l}{\phi_e} + 7.79\left[1 + 6.9\left(\frac{d_o}{h_l}\right)^{1.85}\right]\frac{K_s^2}{\phi_e g\left(\frac{A_h}{A_a}\right)} \qquad (4\text{-}46)$$

κ = constant defined by (Bennett, et al., 1995)

$$\kappa = 0.5\left[1 - \tanh\left(1.3\ln\frac{h_l}{d_o} - 0.15\right)\right] \qquad (4\text{-}47)$$

Example 4.8 Weeping and Entrainment in a Sieve-Tray Ethanol Absorber

Estimate the entrainment flow rate for the ethanol absorber of Examples 4.4, 4.6, and 4.7. Determine whether significant weeping occurs.

Solution

First, check whether weeping is significant. From equation (4-43), using information from the previous examples,

$$Fr_o = \left[\frac{1.923}{986} \times \frac{18.48^2}{9.8 \times 0.0173}\right]^{0.5} = 1.98 > 0.5$$

It can be concluded from this result that weeping is not a problem under these circumstances. Next, estimate the entrainment flow rate. From equation (4-47),

$$\kappa = 0.5\left[1 - \tanh\left(1.3 \times \ln\frac{17.3}{5} - 0.15\right)\right] = 0.0511$$

From equation (4-46),

$$h_{2\phi} = \frac{0.0173}{0.274} + 7.79\left[1 + 6.9 \times \left(\frac{5}{17.3}\right)^{1.85}\right] \times \frac{0.082^2}{0.274 \times 9.8 \times 0.101} = 0.394 \text{ m}$$

From equation (4-45),

$$E = 0.00335\left(\frac{0.394}{0.500}\right)^{1.1}\left(\frac{986}{1.923}\right)^{0.5}\left(\frac{0.0173}{0.394}\right)^{0.0546} = 0.050$$

From Example 4.4, the gas mass flow rate is $V' = 2.202$ kg/s. From equation (4-44), the entrainment mass flow rate, L_e is

$$L_e = E \times V' = 0.050 \times 2.202 = 0.110 \text{ kg / s}$$

4.3.6 Tray Efficiency

The graphical and algebraic methods for determining stage requirements for absorption and stripping discussed in Chapter 3 assume that the streams leaving each stage are in equilibrium with respect to both heat and mass transfer. The assumption of thermal equilibrium is reasonable, but the assumption of equilibrium with respect to mass transfer is seldom justified. To determine the actual number of stages required for a given separation, the number of equilibrium stages must be adjusted with a *stage efficiency* (or *tray efficiency*).

> Stage efficiency concepts are applicable only to devices in which the phases are contacted and then separated, that is, when discrete stages can be identified. This is not the case for packed columns or other continuous-contact devices. For these, the efficiency is already imbedded into the design equation.

Tray efficiency is the fractional approach to an equilibrium stage which is attained by a real tray. Ultimately, we require a measure of approach to equilibrium of all the vapor and liquid from the tray, but since the conditions at various locations on the tray may differ, we begin by considering the local or *point efficiency* of mass transfer at a particular place on the tray surface.

Figure 4.10 is a schematic representation of one tray of a multitray tower. The tray n is fed from tray $(n - 1)$ above by liquid of average composition x_{n-1}, and it delivers liquid of average composition x_n to the tray below. At the place under consideration, a pencil of gas of composition $y_{n+1, local}$ rises from below and, as a result of mass transfer, leaves with a concentration $y_{n, local}$. At the place in question, it is assumed that the local liquid concentration x_{local} is constant in the vertical direction. The point efficiency is then defined by

$$\mathbf{E}_{OG} = \frac{y_{n,local} - y_{n+1, local}}{y^*_{local} - y_{n+1, local}} \tag{4-48}$$

Here, y^*_{local} is the concentration in equilibrium with x_{local}, and equation (4-48) then represents the change in gas concentration that actually occurs as a fraction of that which would occur if equilibrium were established. The subscript G signifies that gas concentrations are used, and the O emphasizes that $\mathbf{E}_{OG}$ is a measure of the overall resistance to mass transfer for both phases.

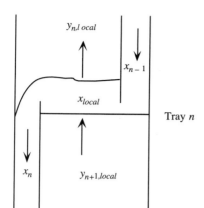

Figure 4.10 Point efficiency.

Consider that the gas rises at a rate G_M mole/(area)(time). Let the interfacial surface between gas and liquid be a area/volume of liquid-gas froth. As the gas rises a differential height dh_l, the area of contact is $a\,dh_l$ per unit active area of the tray. If, while of concentration y, it undergoes a concentration change dy in this height, and if the total quantity of gas is assumed to remain essentially constant, the rate of solute transfer is $G_M\,dy$:

$$G_M\,dy = K_y a\,dh_l\left(y^*_{local} - y\right) \tag{4-49}$$

Then,

$$\int_{y_{n+1,local}}^{y_{n,local}} \frac{dy}{y^*_{local} - y} = \frac{K_y a\,h_l}{G_M} \tag{4-50}$$

Since y^*_{local} is constant for constant x_{local}

$$-\ln\frac{y^*_{local} - y_{n,local}}{y^*_{local} - y_{n+1,local}} = -\ln\left(1 - \frac{y_{n,local} - y_{n+1,local}}{y^*_{local} - y_{n+1,local}}\right)$$
$$= -\ln(1 - \mathbf{E}_{OG}) = \frac{K_y a h_l}{G_M} \tag{4-51}$$

Therefore,

$$\mathbf{E}_{OG} = 1 - \exp\left(-\frac{K_y a h_l}{G_M}\right) \tag{4-52}$$

A method to estimate the overall volumetric mass-transfer coefficient at any point over the tray, $K_y\, a$, is needed in order to use equation (4-52). Bennett et al. (1997) proposed the following correlation for estimating sieve-tray point efficiency:

$$\mathbf{E}_{OG} = 1 - \exp\left(-\frac{0.0029}{1+m\dfrac{c_G}{c_L}\sqrt{\dfrac{D_G(1-\phi_e)}{D_L\left(\dfrac{A_h}{A_a}\right)}}}\,\mathrm{Re}_{Fe}^{a1}\left(\frac{h_l}{d_o}\right)^{a2}\left(\frac{A_h}{A_a}\right)^{a3}\right) \tag{4-53}$$

where

$$\mathrm{Re}_{Fe} = \frac{\rho_G v_o h_l}{\mu_G \phi_e} \tag{4-54}$$

$a1$	=	0.4136
$a2$	=	0.6074
$a3$	=	-0.3195
m	=	local slope of the equilibrium curve
c_G, c_L	=	molar concentration of gas and liquid, respectively
D_G, D_L	=	diffusivities of gas and liquid, respectively

The bulk-average concentration of all the local pencils of gas of Figure 4.10 are y_{n+1} and y_n. The efficiency of the entire tray, also known as the Murphree efficiency, is then

$$\mathbf{E}_{MG} = \frac{y_n - y_{n+1}}{y^* - y_{n+1}} \tag{4-55}$$

where y^* is in equilibrium with the liquid leaving the tray of concentration x_n. The relationship between $\mathbf{E}_{MG}$ and $\mathbf{E}_{OG}$ can then be derived by integrating the local $\mathbf{E}_{OG}$ over the surface of the tray. Four cases will be considered, depending upon the degree of mixing of the vapor and liquid phases:

(1) If the liquid on the tray is completely back-mixed, everywhere of uniform concentration x_n, and the vapor entering the tray is perfectly mixed, then $\mathbf{E}_{MG} = \mathbf{E}_{OG}$.

(2) If the movement of the liquid on the tray is in plug flow, with no mixing, while the vapor entering the tray is perfectly mixed (Kister, 1992)

$$\mathbf{E}_{MG} = \frac{1}{\lambda}\left[\exp(\lambda E_{OG}) - 1\right] \tag{4-56}$$

where $\lambda = mV/L$.

(3) If the entering vapor is well mixed, but the liquid is only partially mixed (Gerster, et al., 1958):

$$\frac{E_{MG}}{E_{OG}} = \frac{1 - \exp\left[-(\eta + Pe_L)\right]}{(\eta + Pe_L)\left[1 + \dfrac{(\eta + Pe_L)}{\eta}\right]} + \frac{\exp(\eta) - 1}{\eta\left[1 + \dfrac{\eta}{\eta + Pe_L}\right]}$$

$$\eta = \frac{Pe_L}{2}\left[\left(1 + \frac{4\lambda E_{OG}}{Pe_L}\right)^{1/2} - 1\right] \tag{4-57}$$

For $\lambda < 3.0$, equation (4-57) can be approximated to within an average error of about 0.4 % by (Bennett, et al., 1997):

$$\mathbf{E}_{MG} = \frac{\left[1 + \dfrac{\lambda E_{OG}}{N}\right]^N - 1}{\lambda} \tag{4-58}$$

where

$$N = \frac{Pe_L + 2}{2}$$

$$Pe_L = \frac{4q_L r_w^2}{A_a h_l D_{E,L}} \tag{4-59}$$

$D_{E,L}$ = liquid eddy diffusivity, estimated from (Bennett, et al., 1997)

$$D_{E,L} = 0.1\left[gh_{2\phi}^3\right]^{0.5} \tag{4-60}$$

(4) If both the vapor and liquid are only partially mixed, for cross-flow trays and $\lambda < 3.0$ (Bennett, et al., 1997):

$$E_{MG} = \frac{\left[1 + \dfrac{\lambda E_{OG}}{N}\right]^N - 1}{\lambda} \times \left[1 - 0.0335\lambda^{1.073} E_{OG}^{2.518} Pe_L^{0.175}\right] \quad (4\text{-}61)$$

The degree of vapor mixing can be estimated from Katayama and Imoto (1972), who report that the vapor can be considered unmixed if the vapor-Peclet number, $Pe_G > 50$, or if $h_{2\phi}/t \geq 1$, where

$$Pe_G = \frac{4Q_G r_w^2}{A_a\left(t - h_{2\phi}\right) D_{E,G}} \qquad \text{for} \quad \frac{h_{2\phi}}{t} < 1 \qquad (4\text{-}62)$$

Lockett (1986) recommends a value of gas eddy diffusivity $D_{E,G} = 0.01$ m^2/s.

A further correction of the tray efficiency is required for the damage done by entrainment, a form of back-mixing which acts to destroy the concentration changes produced by the trays. Bennett et al. (1997) suggest

$$E_{MGE} = E_{MG}\left[1 - \frac{0.8 E_{OG}\lambda^{1.543} E}{m}\right] \qquad (4\text{-}63)$$

where E_{MGE} is the Murphree tray efficiency corrected for the effect of entrainment.

Example 4.9 Murphree Efficiency of a Sieve-Tray Ethanol Absorber

Estimate the entrainment-corrected Murphree tray efficiency for the ethanol absorber of Examples 4.4, 4.6, 4.7, and 4.8.

Solution

From equation (4-54), $Re_{Fe} = 1.543 \times 10^5$. The molar densities are $c_G = \rho_G/M_G = 0.044$ kmol/m^3, and $c_L = \rho_L/M_L = 52.93$ kmol/m^3. For the low concentrations prevailing in the liquid phase, the ethanol-water solution at 303 K obeys Henry's law, and the slope of the equilibrium curve is $m = 0.57$ (Seader and Henley, 1998). Equation (4-53) yields a point efficiency of $E_{OG} = 0.809$. Next, check the degree of mixing of the vapor phase. Equation (4-62) yields $Pe_G = 810 > 50$. Therefore, the vapor is unmixed. From equation (4-60), $D_{E,L} = 0.077$ m^2/s; from equation (4-59), $Pe_L = 0.459$, $N = 1.229$. Next, calculate $\lambda = mG/L = 0.57 \times (2.202/44.04)/(0.804/18.63) = 0.66$. Equation (4-58) yields a Murphree tray efficiency of $E_{MG} = 0.836$. From Example 4.8, the fractional entrainment is $E = 0.05$. Substituting in equation (4-63), the tray efficiency corrected for entrainment is $E_{MGE} = 0.811$.

Example 4.10 Sieve-Tray Distillation Column for Ethanol Purification

In the development of a new process, it will be necessary to fractionate 910 kg/h of an ethanol-water solution containing 0.3 mole fraction ethanol. It is desired to produce a distillate containing 0.80 mole fraction ethanol, with negligible loss of ethanol in the residue, using a sieve-tray distillation column at atmospheric pressure. Preliminary calculations show that the maximum vapor-volumetric flow rate occurs at a point in the column, just above the feed tray, where the liquid and the vapor contain 0.5 and 0.58 mole fraction of ethanol, respectively. The temperature at that point in the column is about 353 K; the local slope of the equilibrium curve is $m = 0.42$. The liquid and vapor molar flow rates are 39 and 52 kmol/hr, respectively.

(a) Specify a suitable sieve-tray design for this application. Check your design for vapor-pressure drop, weeping, and excessive entrainment. Estimate the corresponding Murphree tray efficiency, corrected for entrainment.

(b) Check your design for the conditions prevailing at the bottom of the tower where the vapor molar flow rate remains the same, but the liquid molar flow rate is 73 kmol/hr. The temperature is 373 K, and both the liquid and vapor phases are virtually pure water.

Solution

(a) For the given concentrations of the liquid and vapor phases, the average molecular weights are $M_G = 34.2$, $M_L = 32.0$. The corresponding mass flow rates are $V' = 0.494$ kg/s and $L' = 0.347$ kg/s. The density of the vapor is from the ideal-gas law, $\rho_G = 1.18$ kg/m^3. The viscosity of the vapor, as estimated from the Lucas' method, is $\mu_G = 105$ μP; the diffusivity is from the Wilke-Lee equation, $D_G = 0.158$ cm^2/s. Properties of the liquid phase are (Treybal, 1980): $\rho_L = 791$ kg/m^3, $D_L = 2.07 \times 10^{-5}$ cm^2/s, $\sigma = 21$ dyne/cm, $m = 0.42$. The density of liquid water at 353 K is $\rho_W = 970$ kg/m^3. The foaming factor for alcohol distillation is $F_F = 0.9$ (Kister, 1992). Use stainless steel, 2-mm-thick sheet metal with 4.5-mm perforations on an equilateral-triangular pitch 12 mm between hole centers. Set the weir height at 5 cm, and design for an 80% fractional approach to flooding.

Appendix E presents a Mathcad sieve-tray design computer program. For a given set of data on properties of the gas and liquid streams, tray geometry, and fractional approach to flooding, the program calculates the tray diameter, weir length and location, and recommended tray spacing; estimates the vapor-pressure drop per tray; checks the design for weeping; calculates fractional liquid entrainment; and estimates the tray Murphree efficiency, corrected for entrainment.

For the conditions specified at a point in the column, just above the feed tray of the ethanol fractionator, the results are:

Tray diameter = 0.631 m

Recommended tray spacing = 0.5 m
Weir length = 0.458 m
Weir location = 0.217 m from center of tray
Gas pressure drop = 358 Pa/tray
Orifice Froude number = 1.11 > 0.5; no excessive weeping
Fractional liquid entrainment = 0.024
Point efficiency = 0.760
Murphree tray efficiency = 0.795
Murphree tray efficiency, corrected for entrainment = 0.783

(b) At the bottom of the tower the liquid and vapor phases are very dilute solutions of ethanol in water, therefore $M_G = M_L = 18$. The corresponding mass flow rates are $V' = 0.494$ kg/s and $L' = 0.347$ kg/s. The density of the vapor and of the liquid are from the Steam Tables (Smith et al., 1996) for saturated water at 373 K, $\rho_L = \rho_W = 958$ kg/m^3, $\rho_G = 0.600$ kg/m^3. The viscosity of the gas is estimated by Lucas' method, $\mu_G = 1.21 \times 10^{-5}$ kg/m-s; the surface tension of the liquid is estimated as $\sigma = 63$ dyne/cm (Reid, et al., 1987). The gas-phase diffusivity is from the Wilke-Lee correlation, $D_G = 0.177$ cm^2/s; the liquid-phase diffusivity is from the Hayduk-Minhas correlation for aqueous solutions, $D_L = 5.54 \times 10^{-5}$ cm^2/s. The slope of the equilibrium curve for very dilute solutions can be derived from the modified Raoult's law, equation (3-2), written here as

$$ m = \lim_{x_A \to 0} \frac{\gamma_A P_A}{P} = \frac{\gamma_A^\infty P_A(@373K)}{P} \tag{4-63} $$

The activity coefficient at infinite dilution for ethanol in water at 373 K is 5.875 (Reid, et al., 1987); the vapor pressure of ethanol at 373 K is 226 kPa (Smith, et al., 1996). For a total pressure of 101.3 kPa, equation (4-63) yields $m = 13.1$.

Next, enter the data presented above into the Mathcad sieve-tray design computer program of Appendix E. Since the dimensions of the tray are known, the fractional approach to flooding is adjusted until the tray design coincides with the tray dimensions determined in part (a) of this example. Convergence is achieved at a value of $f = 0.431$. This means that at the bottom of the distillation column the gas velocity is only 43.1% of the flooding velocity. Other important results obtained from the program are as follows:

Gas pressure drop = 434 Pa/tray
Orifice Froude number = 0.65 > 0.5; no excessive weeping
Fractional liquid entrainment = 0.019
Point efficiency = 0.500
Murphree tray efficiency = 0.807
Murphree tray efficiency, corrected for entrainment = 0.792

Notice that, even though the point efficiency for the conditions prevailing at the bottom of the column is significantly smaller than the point efficiency just above the feed tray, the entrainment-corrected Murphree tray efficiencies at both points are very similar.

PROBLEMS

The problems at the end of each chapter have been grouped into four classes (designated by a superscript after the problem number).

Class a: Illustrates direct numerical application of the formulas in the text.
Class b: Requires elementary analysis of physical situations, based on the subject material in the chapter.
Class c: Requires somewhat more mature analysis.
Class d: Requires computer solution.

4.1[a]. Void fraction near the walls of packed beds.

Consider a cylindrical vessel with a diameter of 305 mm packed with solid spheres with a diameter of 50 mm.

(a) From equation (4-1), calculate the asymptotic porosity of the bed.

Answer: 40.1%

(b) Estimate the void fraction at a distance of 100 mm from the wall.

Answer: 47.7%

4.2[b]. Void fraction near the walls of packed beds.

Because of the oscillatory nature of the void-fraction radial variation of packed beds, there are a number of locations close to the wall where the local void fraction is exactly equal to the asymptotic value (see Figure 4.4). For the bed described in Example 4.1, calculate the distance from the wall to the first five such locations.

Answer: 7.2, 16.4, 25.8, 35.1, 44.5 mm

4.3^{c,d}. Void fraction near the walls of packed beds.

(a) Show that the radial location of the maxima and minima of the function described by Equation (4-1) are the roots of the equation

$$J_0(\alpha r^*) + \frac{\alpha}{\beta} J_1(\alpha r^*) = 0 \qquad (4\text{-}64)$$

(b) For the packed bed of Example 4.1, calculate the radial location of the first five maxima, and of the first five minima; calculate the amplitude of the void fraction oscillations at those points.

(c) Calculate the distance from the wall at which the absolute value of the porosity fluctuations has been dampened to less than 10% of the asymptotic bed porosity.

Answer: 69 mm

(d) What fraction of the cross-sectional area of the packed bed is characterized by porosity fluctuations which are within ±10% of the asymptotic bed porosity?

Answer: 30%

(e) Show that the average void fraction of a bed packed with equal-sized spheres ε_{av} is given by

$$\varepsilon_{av} = \frac{2}{(R^*)^2} \int_0^{R^*} (R^* - r^*)\varepsilon(r^*)\,dr^* \qquad \text{where } R^* = \frac{D}{2d_p} \qquad (4\text{-}65)$$

(f) For the packed bed of Example 4.1, estimate the average void fraction by numerical integration of equation (4-65) and estimate the ratio $\varepsilon_{av}/\varepsilon_b$.

Answer: $\varepsilon_{av}/\varepsilon_b = 1.064$

4.4^c. Void fraction near the walls of annular packed beds.

Annular packed beds (APBs) involving the flow of fluids are used in many technical and engineering applications, such as in chemical reactors, heat exchangers, and fusion reactor blankets. It is well known that the wall in a packed bed affects the radial void fraction distribution. Since APBs have two walls that can simultaneously affect the radial void fraction distribution, it is essential to include this variation in transport models. A correlation for this purpose was recently formulated (Mueller, 1999). The correlation is restricted to randomly packed beds in annular cylindrical containers of outside diameter D_o, inside diameter D_i, equivalent diameter $D_e = D_o - D_i$, consisting of equal-sized spheres of diameter d_p, with diameter aspect ratios of $4 \leq D_e/d_p \leq 20$. The correlation is

$$\varepsilon = \varepsilon_b + (1-\varepsilon_b)\{J_0(\alpha r^*)\exp(\beta r^*) + J_0(\alpha[R^*-r^*])\exp(\beta[R^*-r^*])\} \quad (4\text{-}66)$$

where

$$r^* = \frac{r}{d_p} \qquad R^* = \frac{D_e}{2d_p} \qquad 0 \leq r^* \leq R^* \tag{4-67}$$

$$r = \text{radial position measured from the outer wall}$$

$$\alpha = 6.64 - 6.1e^{-R^*}$$
$$\beta = -(0.69 + 0.015R^*)$$
$$\varepsilon_b = 0.456 - 0.3e^{-R^*} \tag{4-68}$$
$$J_0 = \text{Bessel function of the first kind and order zero}$$

Consider an APB with outside diameter of 140 mm, inside diameter of 40 mm, packed with identical 10-mm diameter spheres.

(a) Estimate the void fraction at a distance from the outer wall of 25 mm.

Answer: 42.2%

(b) Plot the void fraction, as predicted by equation. (4-66), for r^* from 0 to R^*.

(c) Show that the average porosity for an APB is given by

$$\varepsilon_{av} = \frac{2}{(R_o^*)^2 - (R_i^*)^2} \int_0^{R^*} (R_o^* - r^*)\varepsilon(r^*)dr^*$$

where

$$R_o^* = \frac{D_o}{2d_p}, \qquad R_i^* = \frac{D_i}{2d_p} \tag{4-69}$$

(d) Estimate the average porosity for the APB described above.

Answer: 48.7%

4.5[a]. Minimum liquid mass velocity for proper wetting of packing.

A 1.0-m-diameter bed used for absorption of ammonia with pure water at 298 K is packed with 25-mm plastic Intalox saddles. Calculate the minimum water flow rate, in kg/s, neded to ensure proper wetting of the packing surface.

Answer: 0.94 kg/s

4.6[a]. Minimum liquid mass velocity for proper wetting of packing.

Repeat Problem 4.5 using ceramic instead of plastic Intalox saddles.

Answer: 0.12 kg/s

4.7[b]. Specific liquid holdup and void fraction in first-generation random packing.

Repeat Example 4.2 using 25-mm ceramic Berl saddles as packing material.

Answer: $\varepsilon = 0.589$

4.8[b]. Specific liquid holdup and void fraction in structured packing.

Repeat Example 4.2, but using Montz metal B1-200 structured packing (very similar to the one shown in Figure 4.3). For this packing, $a = 200$ m^{-1}, $\varepsilon = 0.979$, $C_h = 0.547$ (Seader and Henley, 1998).

Answer: $\varepsilon = 0.907$

4.9[b]. Specific liquid holdup and void fraction in first-generation random packing.

A tower packed with 25-mm ceramic Raschig rings is to be used for absorbing benzene vapor from a dilute mixture with an inert gas using a wash oil at 300 K. The viscosity of the oil is 2.0 cP and its density is 840 kg/m^3. The liquid mass velocity is $G_x = 2.71$ kg/m^2-s. Estimate the liquid holdup, the void fraction, and the hydraulic specific area of the packing.

Answer: $\varepsilon = 0.646$, $a_h = 65$ m^{-1}

4.10[b, d]. Pressure drop in beds packed with first-generation random packings.

Repeat Example 4.3 using 15-mm ceramic Raschig rings as packing material. Assume that, for this packing, $C_p = 1.783$.

Answer: $\Delta P = 525$ Pa/m, $D = 0.517$ m

4.11[b, d]. Pressure drop and approach to flooding in structured packing.

Repeat Example 4.3 using Montz metal B1-200 structured packing (very similar to the one shown in Figure 4.3). For this packing, $F_p = 22$ ft^2/ft^3, $a = 200$ m^{-1}, $\varepsilon = 0.979$, $C_h = 0.547$, $C_p = 0.355$ (Seader and Henley, 1998).

Answer: $\Delta P = 421$ Pa/m, $D = 0.254$ m

4.12[b, d]. Pressure drop in beds packed with second-generation random packings.

A packed tower is to be designed for the countercurrent contact of a benzene-nitrogen gas mixture with kerosene to wash out the benzene from the gas. The gas enters the tower at the rate of 1.5 m^3/s, measured at 110 kPa and 298 K, containing 5 mole% benzene. Essentially, all the benzene is absorbed by the kerosene. The liquid enters the tower at the rate of 4.0 kg/s; the liquid density is 800 kg/m^3, viscosity is 2.3 cP. The packing will be 50-mm metal Pall rings, and the tower diameter will be chosen to produce a gas-pressure drop of 400 Pa/m of irrigated packing.

(a) Calculate the tower diameter to be used, and the resulting fractional approach to flooding.

Answer: $f = 0.825$, $D = 0.913$ m

(b) Assume that, for the diameter chosen, the irrigated packed height will be 5 m and that 1 m of unirrigated packing will be placed over the liquid inlet to act as entrainment separator. The blower-motor combination to be used at the gas inlet will have an overall mechanical efficiency of 60%. Calculate the power required to blow the gas through the packing.

Answer: 5.68 kW

(c) Estimate the volumetric mass-transfer coefficients for the gas and liquid phases. Assume that $D_L = 5.0 \times 10^{-10}$ m^2/s.

Answer: $k_L a_h = 0.00675$ s^{-1}

4.13[b, d]. Pressure drop in beds packed with structured packings.

Redesign the packed bed of Problem 4.12 using Montz metal B1-200 structured packing (very similar to the one shown in Figure 4.3). Estimate the corresponding mass-transfer coefficients. For this packing, $F_p = 22$ ft^2/ft^3, $a = 200$ m^{-1}, $\varepsilon = 0.979$, $C_h = 0.547$, $C_p = 0.355$, $C_L = 0.971$, $C_V = 0.390$ (Seader and Henley, 1998).

Answer: $D = 0.85$ m; $k_y a_h = 0.376$ kmol/m^3-s

4.14[c, d]. Air stripping of wastewater in a packed column.

A wastewater stream of 0.038 m^3/s, containing 10 ppm (by weight) of benzene, is to be stripped with air in a packed column operating at 298 K and 2 atm to reduce the benzene concentration to 0.005 ppm. The packing specified is 50-mm plastic Pall rings. The air flow rate to be used is 5 times the minimum. Henry's law constant for benzene in water at this temperature is 0.6 kPa-m^3/mole (Davis and Cornwell, 1998). Calculate the tower diameter if the gas-pressure drop is not to exceed 500 Pa/m of packed height. Estimate the corresponding mass-transfer coefficients. The diffusivity of benzene vapor in air at 298 K and 1 atm is 0.096 cm^2/s; the diffusivity of liquid benzene in water at infinite dilution at 298 K is 1.02×10^{-5} cm^2/s (Cussler, 1997).

Answer: $D = 1.145$ m; $k_L a_h = 0.032$ s^{-1}

4.15[b]. Stripping chloroform from water by sparging with air.

Repeat Example 4.5 using an air flow rate that is twice the minimum required.

Answer: $W = 4.09$ kW

4.16[b]. Stripping chloroform from water by sparging with air.

Repeat Example 4.5 using the same air flow rate used in Problem 4.15, and specifying a chloroform removal efficiency of 99%.

Answer: $Z = 1.30$ m

4.17[b]. Stripping chlorine from water by sparging with air.

A vessel 2.0 m in diameter and 2.0 m deep (measured from the gas sparger at the bottom to liquid overflow at the top) is to be used for stripping chlorine from water by sparging with air. The water will flow continuously downward at the rate of 7.5 kg/s with an initial chlorine concentration of 5 mg/L. Airflow will be 0.22 kg/s at 298 K. The sparger is in the form of a ring, 25 cm in diameter, containing 200 orifices, each 3.0 mm in diameter. Henry's law constant for chlorine in water at this temperature is 0.11 kPa-m^3/mole (Perry and Chilton, 1973). The diffusivity of chlorine at infinite dilution in water at 298 K is 1.25×10^{-5} cm^2/s (Cussler, 1997).

(a) Assuming that all the resistance to mass transfer resides in the liquid phase, estimate the chlorine removal efficiency achieved.

Answer: 99.34%

(b) Estimate the power required to operate the air compressor if the mechanical efficiency of the system is 65%.

Answer: 7.17 kW

4.18[c,d]. Batch wastewater aeration using spargers.

In the treatment of wastewater, undesirable gases are frequently stripped or desorbed from the water, and oxygen is adsorbed into the water when bubbles of air are dispersed near the bottom of aeration tanks or ponds. As the bubbles rise, solute can be transferred from the gas to the liquid or from the liquid to the gas depending upon the concentration driving force.

For batch aeration in a constant volume tank, an oxygen mass-balance can be written as

$$\frac{dc_A}{dt} = K_L a\left(c_A^* - c_A\right) \tag{4-70}$$

where c_A^* is the oxygen saturation concentration. Integrating between the time limits zero and t and the corresponding dissolved oxygen concentration limits $c_{A,0}$ and $c_{A,t}$; assuming that c_A^* remains essentially constant, and that all the resistance to mass transfer resides in the liquid phase:

$$\ln\left[\frac{c_A^* - c_{A,0}}{c_A^* - c_{A,t}}\right] = k_L a t \tag{4-71}$$

In aeration tanks, where air is released at an increased liquid depth, the solubility of oxygen is influenced both by the increasing pressure of the air entering the aeration tank and by the decreasing oxygen partial pressure in the air bubble as oxygen is absorbed. For these cases, the use of a mean saturation value corresponding to the aeration tank middepth is suggested (Eckenfelder, 2000):

$$c_A^* = c_s \times \frac{1}{2}\left(\frac{P_o}{P_s} + \frac{O_t}{20.9}\right) \tag{4-72}$$

where

$c_s =$ saturation dissolved oxygen concentration in fresh water exposed to atmospheric air at 101.3 kPa containing 20.9% oxygen

$P_o =$ absolute pressure at the depth of air release

$P_s^o =$ atmospheric pressure

$O_t =$ molar oxygen percent in the air leaving the aeration tank

The molar oxygen percent in the air leaving the aeration tank is related to the oxygen transfer efficiency, O_{eff}, through

$$O_t = \frac{21(1 - O_{eff})}{79 + 21(1 - O_{eff})} \times 100 \tag{4-73}$$

where

$$O_{eff} = \frac{\text{Mass of oxygen absorbed by the water}}{\text{Total mass of oxygen supplied}} \tag{4-74}$$

Consider a 567 m³ aeration pond aerated with 15 spargers, each using compressed air at a rate of 0.01 kg/s. Each sparger is in the form of a ring, 100 cm in diameter, containing 20 orifices, each 3.0 mm in diameter. The spargers will be located 5 m below the surface of the pond. The water temperature is 298 K; atmospheric conditions are 298 K and 101.3 kPa. Under these conditions, $c_s = 8.38$ mg/L (Davis and Cornwell, 1998).

(a) Estimate the volumetric mass-transfer coefficient for these conditions from equations (4-23) and (4-25).

Answer: $k_L a = 2.83$ hr^{-1}

(b) Estimate the time required to raise the dissolved oxygen concentration from 0.5 mg/L to 6.0 mg/L, and calculate the resulting oxygen transfer efficiency.

Answer: 17.9 min

(c) Estimate the power required to operate the 15 spargers, if the mechanical efficiency of the compressor is 60%.

Answer: 8.65 kW

4.19[c,d]. Batch wastewater aeration using spargers; effect of liquid depth.

Consider the situation described in Problem 4.18. According to Eckenfelder (2000), for most types of bubble-diffusion aeration systems the volumetric mass-transfer coefficient will vary with liquid depth Z according to the relationship

$$\frac{k_L a(Z_1)}{k_L a(Z_2)} = \left[\frac{Z_1}{Z_2}\right]^n \tag{4-75}$$

where the exponent n has a value near 0.7 for most systems. For the aeration pond of Problem 4.18, calculate $k_L a$ at values of $Z = 3$ m, 4 m, 6 m, and 7 m. Estimate the corresponding value of n from regression analysis of the results. *Hint:* Remember that the total volume of the pond must remain constant, therefore the cross-sectional area of the pond must change as the water depth changes.

4.20[c]. Flooding conditions in a packed cooling tower.

A cooling tower, 2 m in diameter, packed with 75-mm ceramic Hiflow rings, is fed with water at 316 K at a rate of 25 kg/m²-s. The water is contacted with air, at 300 K and 101.3 kPa essentially dry, drawn upward countercurrently to the water flow. Neglecting evaporation of the water and changes in the air temperature, estimate the volumetric rate of airflow, in m³/s, which would flood the tower.

Answer: 11.80 m³/s

4.21[c,d]. Design of a sieve-tray column for ethanol absorption.

Repeat the calculations of Examples 4.6, 4.7, 4.8, and 4.9 for a column diameter corresponding to 50% of flooding.

Answer: $D = 1.176$ m

4.22[c,d]. Design of a sieve-tray column for aniline stripping.

A sieve-tray tower is to be designed for stripping an aniline (C_6H_7N)-water solution with steam. The circumstances at the top of the tower, which are to be used to establish the design, are:

Temperature = 371.5 K   Pressure = 100 kPa

Liquid:

     Rate = 10.0 kg/s   Composition = 7.00 mass % aniline

     Density = 961 kg/m^3 Viscosity = 0.3 cP

     Surface tension = 58 dyne/cm

     Diffusivity = 4.27×10^{-5} cm^2/s (est.)   Foaming factor = 0.90

Vapor:

     Rate = 5.0 kg/s   Composition = 3.6 mole % aniline

     Density = 0.670 kg/m^3 Viscosity = 118 µP (est.)

     Diffusivity = 0.116 cm^2/s (est.)

The equilibrium data at this concentration indicate that $m = 0.0636$ (Treybal, 1980).

(a) Design a suitable cross-flow sieve-tray for such a tower. Take $d_o = 5.5$ mm on an equilateral-triangular pitch 12 mm between hole centers, punched in stainless-steel sheet metal 2 mm thick. Use a weir height of 40 mm. Design for a 75% approach to the flood velocity. Report details respecting tower diameter, tray spacing, weir length, gas-pressure drop, and entrainment in the gas. Check for excessive weeping.

Answer: $D = 1.93$ m

(b) Estimate the tray efficiency corrected for entrainment for the design reported in part a.

Answer: $E_{MGE} = 0.72$

4.23[c,d]. Design of a sieve-tray column for aniline stripping.

Repeat Problem 4.22, but for a 45% approach to flooding. Everything else remains the same as in Problem 4.22.

Answer: $D = 2.49$ m; $E_{MGE} = 0.67$; excessive weeping

4.24[c,d]. Design of a sieve-tray column for methanol stripping.

A dilute aqueous solution of methanol is to be stripped with steam in a sieve-tray tower. The conditions chosen for design are:

Temperature = 368 K   Pressure = 101.3 kPa
Liquid:
    Rate = 0.25 kmol/s   Composition = 15.0 mass % methanol
    Density = 961 kg/m^3 Viscosity = 0.3 cP
    Surface tension = 40 dyn/cm
    Diffusivity = 5.70 × 10^{-5} cm^2/s (est.)   Foaming factor = 1.0
Vapor:
    Rate = 0.1 kmol/s   Composition = 18 mole % methanol
    Viscosity = 125 μP (est.)   Diffusivity = 0.213 cm^2/s (est.)

The equilibrium data at this concentration indicate that m = 2.5 (Perry and Chilton, 1973).

(a) Design a suitable cross-flow sieve-tray for such a tower. Take d_o = 6.0 mm on an equilateral-triangular pitch 12 mm between hole centers, punched in stainless-steel sheet metal 2 mm thick. Use a weir height of 50 mm. Design for 80% approach to the flood velocity. Report details respecting tower diameter, tray spacing, weir length, gas-pressure drop, and entrainment in the gas. Check for excessive weeping.

Answer: D = 1.174 m; ΔP = 386 Pa/tray

(b) Estimate the tray efficiency corrected for entrainment for the design reported in part a.

Answer: E_{OG} = 0.60

4.25[c,d]. Sieve-tray column for methanol stripping; effect of hole size.

Repeat Problem 4.24, but changing the perforation size to 4.5 mm, keeping everything else constant.

Answer: ΔP = 686 Pa/tray; E_{OG} = 0.793

4.26^{c,d}. Design of sieve-tray column for butane absorption.

A gas containing methane, propane, and n-butane is to be scrubbed countercurrently in a sieve-tray tower with a hydrocarbon oil to absorb principally the butane. It is agreed to design a tray for the circumstances existing at the bottom of the tower, where the conditions are:

Temperature = 310 K   Pressure = 350 kPa

Liquid:

   Rate = 0.50 kmol/s   Average molecular weight = 150

   Density = 850 kg/m^3 Viscosity = 1.6 cP

   Surface tension = 25 dyn/cm

   Diffusivity = 1.14×10^{-5} cm^2/s (est.)   Foaming factor = 0.9

Vapor:

   Rate = 0.3 kmol/s   Composition = 86% CH_4, 12% C_3H_8, 2% C_4H_{10}

   Viscosity = 113 µP (est.)   Diffusivity = 0.035 cm^2/s (est.)

The system obeys Raoult's law; the vapor pressure of n-butane at 310 K is 3.472 bar (Reid et al., 1987).

(a) Design a suitable cross-flow sieve-tray for such a tower. According to Bennett and Kovak (2000), the optimal value of the ratio A_h/A_a is that which yields an orifice Froude number, $Fr_o = 0.5$. Design for the optimal value of d_o on an equilateral-triangular pitch 12 mm between hole centers, punched in stainless-steel sheet metal 2 mm thick. Use a weir height of 50 mm. Design for a 75% approach to the flood velocity. Report details respecting tower diameter, tray spacing, weir length, gas-pressure drop, and entrainment in the gas.

Answer: $D = 2.73$ m

(b) Estimate the tray efficiency corrected for entrainment for the design reported in part a.

Answer: $E_{MG\,E} = 1.038$

4.27^{c,d}. Design of sieve-tray column for ammonia absorption.

A process for making small amounts of hydrogen by cracking ammonia is being considered, and residual uncracked ammonia is to be removed from the resulting gas. The gas will consist of H_2 and N_2 in the molar ratio 3:1, containing 3% NH_3 by volume. The ammonia will be removed by scrubbing the gas countercurrently with pure liquid water in a sieve-tray tower. Conditions at the bottom of the tower are:

Temperature = 303 K   Pressure = 200 kPa

Liquid:

  Rate = 6.0 kg/s   Average molecular weight = 18

  Density = 996 kg/m^3 Viscosity = 0.9 cP

  Surface tension = 68 dyn/cm

  Diffusivity = 2.42×10^{-5} cm^2/s (est.)   Foaming factor = 1.0

Vapor:

  Rate = 0.7 kg/s

  Viscosity = 113 μP (est.)   Diffusivity = 0.230 cm^2/s (est.)

For dilute solutions, NH_3-H_2O follows Henry's law, and at 303 K the slope of the equilibrium curve is $m = 0.85$ (Treybal, 1980).

(a) Design a suitable cross-flow sieve-tray for such a tower. Take $d_o = 4.75$ mm on an equilateral-triangular pitch 12.5 mm between hole centers, punched in stainless-steel sheet metal 2 mm thick. Use a weir height of 40 mm. Design for an 80% approach to the flood velocity. Report details respecting tower diameter, tray spacing, weir length, gas-pressure drop, and entrainment in the gas. Check for excessive weeping.

Answer: $D = 0.775$ m

(b) Estimate the tray efficiency corrected for entrainment for the design reported in part a.

Answer: $E_{MGE} = 0.847$

4.28[c,d]. Design of sieve-tray column for toluene-methylcyclohexane distillation.

A sieve-tray tower is to be designed for distillation of a mixture of toluene and methylcyclohexane. The circumstances which are to be used to establish the design, are:

Temperature = 380 K   Pressure = 98.8 kPa

Liquid:

    Rate = 4.8 mole/s   Composition =48.0 mol % toluene

    Density = 726 kg/m^3 Viscosity = 0.22 cP

    Surface tension = 16.9 dyne/cm

    Diffusivity = 7.08×10^{-5} cm^2/s (est.)   Foaming factor = 0.80

Vapor:

    Rate = 4.54 mole/s   Composition =44.6 mole % toluene

    Density = 2.986 kg/m^3 Viscosity = 337 μP (est.)

    Diffusivity = 0.0386 cm^2/s (est.)

The equilibrium data at this concentration indicate that $m = 1.152$.

(a) Design a suitable cross-flow sieve-tray for such a tower. Take $d_o = 4.8$ mm on an equilateral-triangular pitch 12.7 mm between hole centers, punched in stainless-steel sheet metal 2 mm thick. Use a weir height of 50 mm. Design for a 60% approach to the flood velocity. Report details respecting tower diameter, tray spacing, weir length, gas-pressure drop, and entrainment in the gas. Check for excessive weeping.

Answer: $D = 0.6$ m

(b) Estimate the tray efficiency corrected for entrainment for the design reported in part a.

Answer: $E_{MGE} = 0.71$

REFERENCES

Bennett, D. L., R. Agrawal, and P. J. Cook, *AIChE J.*, **29**, 434–442 (1983).

Bennett D. L., A. S. Kao, and L. W. Wong, *AIChE J.*, **41**, 2067–2082 (Sept. 1995).

Bennett D. L., D. N. Watson, and M. A. Wiescinski, *AIChE J.*, **43**, 1611–1626 (June 1997).

Bennet, D. L., and K. W. Kovak, *Chem. Eng. Progress*, **96** (5), 20 (May 2000).

Billet, R. *Packed Column Analysis and Design*, Ruhr-University Bochum, (1989).

Billet, R., and M. Schultes, *Chem. Eng. Technol.*, **14**, 89–95 (1991a).

Billet, R., and M. Schultes, *Beitrage zur Verfahrens–und Umwelt–technik*, Ruhr-Universitat-Bochum 88–106, (1991b).

Billet, R., and M. Schultes, *Packed Towers in Processing and Environmental Technology*, VCH, New York, (1995).

Cussler E. L., *Diffusion*, 2nd ed., Cambridge University Press, Cambridge, UK (1997).

Davis, M. L., and D. A. Cornwell, *Introduction to Environmental Engineering*, 3rd ed., McGraw-Hill, New York (1998).

Eckenfelder, W. W. Jr., *Industrial Water Pollution Control*, 3rd ed., McGraw-Hill, Boston (2000).

Fair, J. R., *Petro/Chem. Eng.*, **33**, 211–218 (Sept. 1961).

Gerster, J. A., A. B. Hill, N. H. Hochgraf, and D. G. Robinson, "Tray Efficiencies in Distillation Columns," Final Report from the University of Delaware, AIChE, New York (1958).

Govindarao, V. M. H., and G. F. Froment, *Chem. Eng. Sci.*, **41**, 533 (1986).

Hughmark, G. A., *Ind. Eng. Chem. Process Des. Dev.*, **6**, 218 (1967).

Hulswitt, C., and J. A. Mraz, *Chem. Eng.*, **79**, 80 (May 15, 1972).

Katayama, H., and T. Imoto, *J. Chem. Soc. Japan*, **9**, 1745 (1972).

Katmar Software, Packed Column Calculator, version 1.1, South Africa (1998).

Kister, H. Z., and D. R. Gill, *Chem. Eng. Prog.*, **87** (2), 32–42 (1991).

Kister, H. Z., *Distillation Design*, McGraw-Hill, New York (1992).

Leibson, I., E. G. Holcomb, A. G. Cocoso, and J. J. Jacmie, *AIChE J.* **2**, 296 (1956).

Leva, M., *Chem. Eng. Progr. Symp. Ser*, **50** (10), 51 (1954).

Lewis, W. K., *Ind. Eng. Chem.*, **28**, 399 (1936).

Lockett, M. J., *Distillation Tray Fundamentals*, Cambridge Univ. Press, New York (1986).

Ludwig, E. E., *Applied Process Design for Chemical and Petrochemical Plants*, vol. 2, 2nd ed., Gulf Pub. Company, Houston, TX (1979).

Mueller, G. E., *Powder Technol.*, **72**, 269 (1992).

Mueller, G. E., *AIChE J.*, **45**, 2458–60 (Nov. 1999).

Oliver, E. D., *Diffusional Separation Processes. Theory, Design, and Evaluation*, Wiley, New York (1966).

Perry, R. H., and C. H. Chilton (eds.), *Chemical Engineers' Handbook*, 5th ed., McGraw-Hill, New York (1973).

Petrick, M., *U. S. Atom. Energy Comm.*, **ANL-658** (1962).

Reid, R. C., J. M. Prausnitz, and B. E. Poling, *The Properties of Gases and Liquids*, 4th ed., McGraw-Hill, Boston (1987).

Seader, J. D., and E. J. Henley, *Separation Process Principles*, Wiley, New York, (1998).

Sherwood, T. K., G. H. Shipley, and F. A. L. Holloway, *Ind. Eng. Chem.*, **30**, 765–769 (1938).

Smith, J. M., H. C. Van Ness, and M. M. Abbott, *Introduction to Chemical Engineering Thermodynamics*, 5th ed., McGraw-Hill, New York (1996).

Stichlmair, J., J. L. Bravo, and J. R. Fair, *Gas Separation and Purification*, **3**, 19–28 (1989).

Treybal, R. E., *Mass-Transfer Operations*, 3rd ed., McGraw-Hill, New York, (1980).

Wankat, P. C., *Equilibrium Staged Separations*, Elsevier, New York (1988).

5

Absorption and Stripping

5.1 INTRODUCTION

Gas absorption is an operation in which a gas mixture is contacted with a liquid for the purposes of preferentially dissolving one or more components of the gas into the liquid. For example, the gas from byproduct coke ovens is washed with water to remove ammonia, and again with an oil to remove benzene and toluene vapors. Objectionable hydrogen sulfide is removed from such a gas, or from naturally occurring hydrocarbon gases by washing with various alkaline solutions in which it is absorbed. Acid-rain precursor sulfur dioxide is removed from stack gases by scrubbing the gaseous effluent with an alkaline aqueous solution.

The operations described above require mass transfer of a substance from the gas stream to the liquid. When mass transfer occurs in the opposite direction (i.e., from the liquid to the gas) the operation is called *desorption*, or *stripping*. For example, the benzene and toluene are removed from the absorption oil mentioned above by contacting the liquid solution with steam, whereupon the aromatic vapors enter the gas stream and are carried away, and the absorption oil can be used again. Since the principles of both absorption and desorption are the same, we can study both operations simultaneously.

Absorption and stripping are usually conducted in packed columns or in trayed towers. Packed columns are preferred when (1) the required column diameter is less than 60 cm; (2) the pressure drop must be low, as for a vacuum service; (3) corrosion considerations favor the use of ceramic or polymeric materials; and/or (4) low liquid holdup is desirable. Trayed towers are preferred when (1) the liquid/gas ratio is very low; and (2) frequent cleaning is required. If there is no overriding consideration, cost is the major factor to be taken into account when choosing between packed columns and trayed towers for absorption or stripping.

Absorption and stripping are technically mature separation operations. Design procedures are well developed for both packed columns and tray towers, and commercial processes are common. In most applications, the solutes are contained in gaseous effluents from chemical reactors. Passage of strict environmental standards with respect to air pollution by emission of noxious gases from industrial sources has greatly increased the use of gas absorbers (also known as *scrubbers*) in the past decade.

As was shown in Chapter 3, the fraction of a component absorbed in a countercurrent cascade depends on the number of equilibrium stages and the absorption factor, $A = L_S/mV_S$, for that component. If the value of A is greater than 1.0, any degree of absorption can be achieved: the larger the value of A, the fewer the number of stages required to absorb a desired fraction of the solute. However, very large values of A can correspond to absorbent flow rates that are larger than necessary. From an economic standpoint, the value of A, for the main species to be absorbed, should be in the range of 1.25 to 2.0, with 1.4 being a frequently recommended value. For a stripper, the stripping factor, $S = 1/A$, is crucial. The optimum value of the stripping factor is also in the vicinity of 1.4 (Seader and Henley, 1998).

5.2 COUNTERCURRENT MULTISTAGE EQUIPMENT

Your objectives in studying this section are to be able to:

1. Estimate the number of real trays required for a sieve-tray absorber or stripper by graphical stepwise construction on an *xy* diagram, incorporating in the analysis the Murphree tray efficiency.
2. Estimate the number of real trays required for a sieve-tray absorber or stripper by algebraic methods, for the case of dilute solutions and equilibrium described by Henry's law.

5.2.1 Graphical Determination of the Number of Ideal Trays

Tray towers and similar devices bring about stepwise contact of the liquid and gas and are, therefore, countercurrent multistage cascades. On each tray of a sieve-tray tower, for example, the gas and liquid are brought into intimate contact and separated—somewhat in the manner of Figure 3.23—and the tray thus constitutes a stage. Few of the tray devices described in Chapter 4 actually provide the parallel flow on each tray as shown in Figure 3.23. Nevertheless, it is convenient to use the latter as an arbitrary standard for design and for assessment of performance of actual trays regardless of their method of operation. For this purpose, a *theoretical,* or *ideal*, tray is defined as one where the average composition of all the gas leaving the tray is in equilibrium with the average composition of all the liquid leaving the tray.

   The number of ideal trays required to bring about a given change in composition of the liquid or the gas, for either absorbers or strippers, can be determined graphically on an *XY* diagram in the manner of Figure 3.24, with each step representing an ideal stage. The nearer the operating line is to the equilibrium curve, the more steps will be required. Should the two curves touch at any time corresponding to a minimum L_S/V_S ratio, the number of steps will be infinite. The construction for strippers is the same, except, of course, that the operating line lies below the equilibrium curve. The steps can be equally constructed on diagrams plotted in terms of any concentration units, such as mole fractions or partial pressures.

5.2.2 Tray Efficiencies and Real Trays by Graphical Methods

   Methods for estimating the Murphree tray efficiency corrected for entrainment E_{MGE} for sieve trays were discussed in detail in Chapter 4. For a given stripper or absorber, these allow estimation of the tray efficiency as a function of fluid compositions and flow rates as they vary from one end of the tower to the other. Usually, it is sufficient to make such computations at only three or four locations and then proceed as in Figure 5.1. The broken line is drawn on an *xy* diagram between the operating line and the equilibrium curve at a fractional vertical distance from the operating line equal to the prevailing Murphree gas efficiency, E_{MGE}.Thus, the value of E_{MGE} for the bottom tray is the ratio of the line segments *AB/AC*. Since the broken line then represents the real effluent compositions from the trays, it is used instead of the equilibrium curve to complete the tray construction, which now provides the number of real trays. Notice that on the *xy* diagram, the operating line is usually curved.

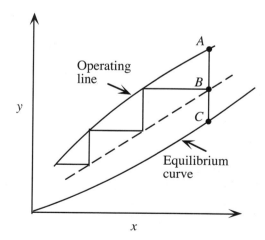

Figure 5.1 Use of Murphree efficiencies for an absorber.

5.2.3 Dilute Mixtures

Where both operating line and equilibrium curve are straight, the number of ideal trays can be determined without recourse to graphical methods. This will frequently be the case for relatively dilute gas and liquid mixtures. In this case, Henry's law usually applies; therefore the equilibrium relation in terms of mole fractions is a straight line. If the quantity of gas absorbed is small, the total flow of liquid entering and leaving the absorber remains substantially constant; therefore $L_0 \approx L_N \approx L$, and similarly the total flow of gas remains relatively constant at V. An operating line plotted in terms of mole fractions then will be virtually straight. For such cases, the absorption factor, $A = L_S/mV_S \approx L/mV$. Small variations in A from one end of the tower to the other due to changing L/V as a result of absorption or stripping can be roughly allowed for by using the geometric average of the values of A at top and bottom (Hachmuth, 1951). The following modified version of the Kremser equations can be used to estimate the number of equilibrium stages required for a given separation.

For tansfer from L to V (stripping of L)

$A = L/mV \neq 1$:

$$N = \frac{\ln\left[\dfrac{x_0 - y_{N+1}/m}{x_N - y_{N+1}/m}(1 - A) + A\right]}{\ln(1/A)} \tag{5-1}$$

$A = 1$

$$N = \frac{x_0 - x_N}{x_N - y_{N+1}/m} \tag{5-2}$$

For tansfer from V to L (absorption into L)

$A \neq 1$:

$$N = \frac{\ln\left[\dfrac{y_{N+1} - mx_0}{y_1 - mx_0}\left(1 - \dfrac{1}{A}\right) + \dfrac{1}{A}\right]}{\ln A} \tag{5-3}$$

$A = 1$

$$N = \frac{y_{N+1} - y_1}{y_1 - mx_0} \tag{5-4}$$

When the Murphree efficiency is constant for all trays, and under conditions such that the operating line and equilibrium curve are straight, the overall tray efficiency can be computed and the number of real trays can be determined analytically from:

$$E_O = \frac{\text{equilibrium trays}}{\text{real trays}} = \frac{\ln\left[1 + E_{MGE}(1/A - 1)\right]}{\ln(1/A)} \qquad (5\text{-}5)$$

Example 5.1 Number of Real Sieve-Trays in an Absorber

A sieve-tray tower is being designed for a gas absorption process. The entering gas contains 1.8% (molar) of A, the component to be absorbed. The gas should leave the tower containing no more than 0.1% (molar) of A. The liquid to be used as absorbent initially contains 0.01% (molar) of A. The system obeys Henry's law with $m = y_i/x_i = 1.41$. At the bottom of the tower, the molar liquid-to-gas ratio is $L/V = 2.115$, while at the other extreme is $L/V = 2.326$. For these conditions, it has been found that the Murphree efficiency is constant at $E_{MGE} = 0.65$.
(a) Estimate the number of trays required.

(b) Based on the criteria presented in Chapter 4, the tower diameter must be 1.5 m. Estimate the tower height.

Solution
(a) Calculate the absorption factor for the molar liquid-to-gas ratio given at the extremes of the column and the slope of the equilibrium curve: $A_1 = 2.115/1.41 = 1.50$; $A_2 = 2.326/1.41 = 1.65$. As recommended in the text, a geometric average is calculated: $A = (A_1 A_2)^{0.5} = 1.573$. From the concentration data given: $y_{N+1} = 0.018$, $y_1 = 0.001$, $x_0 = 0.0001$. Substituting in equation (5-3), $N = 4.647$ ideal stages. From equation (5-5), $E_O = 0.596$. Therefore, the number of real trays required = 4.647/0.596 = 7.79. Use 8 trays, since it is not possible to specify a fractional number of trays. We can back-calculate the actual concentration of the gas leaving the tower when 8 trays are used. For an overall efficiency of 0.596 and an actual number of trays of 8, equation (5-5) yields $N = 4.784$ ideal trays. Solving equation (5-3), $y_1 = 0.094\%$, which satisfies the requirement that the gas exit concentration should not exceed 0.1%.

(b) For a tower diameter of 1.5 m, Table 4.3 recommends a plate spacing of 0.6 m. Therefore, the tower height will be $Z = (8)(0.6) = 4.8$ m.

Example 5.2 Sieve-Tray Absorber for Recovery of Benzene Vapors

Consider the absorber of Example 3.7. Assume that the wash oil is n-tetradecane $(C_{14}H_{30})$. The absorber will be a cross-flow sieve-tray tower with $d_o = 4.5$ mm on an equilateral-triangular pitch 12 mm between hole centers, punched in stainless steel sheet metal 2 mm thick, with a weir height of 50 mm. Estimate the number of real trays required, the dimensions of the absorber, and the power required to pump the gas and the liquid through the tower. Design for a 65% approach to the flooding velocity.

Solution

To determine the tower diameter, consider the conditions at the bottom of the absorber where the maximum gas and liquid flow rates are found. From Examples 3.7 and 3.10:

Gas in:

Rate = 1.0 m³/s = 0.0406 kmol/s at 300 K and 1 atm

Benzene content: $y_{N+1} = 0.074$ mole fraction, $Y_{N+1} = 0.080$ kmol/kmol

Average molecular weight, $M_G = 32.63$ kg/kmole

Density, $\rho_G = 1.325$ kg/m³ (from the ideal gas law)

Mass flow rate, $V' = 1.325$ kg/s

Viscosity, $\mu_G = 166$ μP (estimated by the Lucas method)

Diffusivity, $D_G = 0.096$ cm²/s (Appendix A)

Slope of equilibrium curve = 0.136

Liquid out:

Rate = 0.00888 kmol/s (ideal solution of benzene in tetradecane)

Benzene content: $x_N = 0.324$ mole fraction, $X_N = 0.48$ kmol/kmol

Average molecular weight, $M_L = 159.12$ kg/kmole

Mass flow rate, $L' = 1.413$ kg/s

Density, $\rho_L = 780$ kg/m³ (estimated from the pure-component densities)

Viscosity, $\mu_L = 1.41$ cP (estimated from the pure-component viscosities)

Diffusivity, $D_L = 0.93 \times 10^{-5}$ cm²/s (estimated from equation 1-56 applied to an ideal solution, $\Gamma = 1.0$)

Surface tension, $\sigma = 24$ dyne/cm (estimated)

Foaming factor, $F_F = 0.9$

Run the Sieve-Tray Design Program of Appendix E using these data. The program converges to a tower diameter $D = 1.036$ m, and a tray spacing $t = 0.6$ m. When convergence is achieved, the results at the bottom of the tower are:

$f = 65\%$	$\Delta P_G = 375$ Pa/tray
$Fr_0 = 0.966$	$E = 0.0169$
$E_{OG} = 0.705$	$E_{MGE} = 0.77$

Consider now the conditions at the top of the 1.036-m diameter tower:

Gas out:

Rate = 0.0381 kmol/s at 300 K and 1 atm
Benzene content; $y_1 = 0.0119$ mole fraction, $Y_1 = 0.012$ kmol/kmol
Average molecular weight, $M_G = 29.58$ kg/kmole

Density, $\rho_G = 1.201$ kg/m^3 (from the ideal gas law)
Mass flow rate, $V' = 1.126$ kg/s
Viscosity, $\mu_G = 180$ μP (estimated by the Lucas method)

Diffusivity, $D_G = 0.096$ cm^2/s (Appendix A)

Liquid in:

Rate = 0.00630 kmol/s (ideal solution of benzene in tetradecane)
Benzene content; $x_0 = 0.0476$ mole fraction, $X_0 = 0.05$ kmol/kmol
Average molecular weight, $M_L = 192.7$ kg/kmole
Mass flow rate, $L' = 1.214$ kg/s
Density, $\rho_L = 765$ kg/m^3 (estimated from the pure-component densities)
Viscosity, $\mu_L = 1.98$ cP (estimated from the pure-component viscosities)

Diffusivity, $D_L = 0.83 \times 10^{-5}$ cm^2/s (estimated from equation (1-56) applied to an ideal solution, $\Gamma = 1.0$)
Surface tension, $\sigma = 22$ dyne/cm (estimated)
Foaming factor, $F_F = 0.9$

Run the Sieve-Tray Design Program of Appendix E using these data. Try different values of the fractional approach to flooding, f, until the program converges to the specified tower diameter $D = 1.036$ m. When convergence is achieved, the results at the top of the tower are:

$f = 59.6\%$	$\Delta P_G = 348$ Pa/tray

$$Fr_o = 0.847 \qquad\qquad E = 0.0146$$
$$E_{OG} = 0.656 \qquad\qquad E_{MGE} = 0.730$$

From these results, we conclude that the plate design is appropriate from the pont of view of pressure drop, entrainment, weeping, and plate efficiency. The next step in the design is to determine the number of real stages required. Since the plate efficiency does not change much from one end of the tower to the other, we will assume a constant value taken as the arithmetic average of the efficiency values at the tower ends: $E_{MGE} = (0.77 + 0.73)/2 = 0.75$.

To determine the number of real stages, we must generate the operating line and the equilibrium distribution curve on an xy diagram. For this system, Raoult's law applies; therefore the equilibrium relation is straight on the xy diagram ($y_i = 0.136\ x_i$). However, the operating line will not be straight since we are dealing with fairly concentrated solutions. From Example 3.7, the operating line can be written in terms of mole ratios as

$$\frac{Y_{N+1} - Y}{X_N - X} = \frac{L_S}{V_S}$$
$$\frac{0.08 - Y}{0.48 - X} = \frac{0.006}{0.038} = 0.158$$

After some algebraic manipulation, the last equation can be written in terms of mole fractions as

$$y = \frac{0.004 + 0.154x}{1.004 - 0.846x} \qquad\qquad \text{(operating line)}$$

Figure 5.2 shows the xy diagram for this example. Notice how the operating line curves upwards. The broken line is a pseudoequilibrium curve located between the operating and the equilibrium lines, at 75% (the average value of E_{MGE}) of the distance between the lines, measured from the operating line. The pseudoequilibrium curve is used to complete the tray construction, which now yields the number of real trays. As Figure 5.2 shows, in this case 5 real trays are not enough to achieve the desired separation, while 6 real trays accomplish more than the specified 85% benzene recovery. We choose 6 trays. For a tray spacing of 0.6 m, the total height of the mass-transfer region of the absorber is $Z = 3.6$ m.

A conservative estimate of the total gas-pressure drop can be obtained assuming that the pressure drop through each tray remains constant at the value estimated for the bottom of the absorber where the gas and liquid velocities are the highest. Therefore, the total gas-pressure drop is approximately $\Delta P_G = 375 \times 6 = 2{,}250$ Pa.

Assuming a mechanical efficiency of the motor-fan system $E_m = 60\%$, the power required to force the gas through the tower is

$$W_G = \frac{Q_G \Delta P_G}{E_m} = \frac{1.0 \times 2,250}{0.60} = 3,750 \text{ watt}$$

The liquid must be pumped to the top of the absorber, and gravity will force it through the tower. The power required to pump the liquid to the top of the absorber is

$$W_L = \frac{L_0 \overline{M}_{L} g Z}{E_m} = \frac{1.214 \times 9.8 \times 3.6}{0.60} = 71.4 \text{ watt}$$

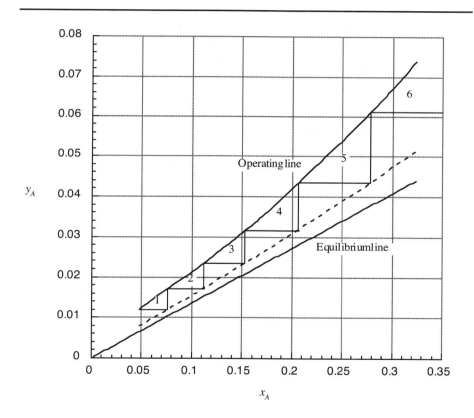

Figure 5.2 Example 5.2: absorber real stages on xy diagram.

5.3 COUNTERCURRENT CONTINUOUS-CONTACT EQUIPMENT

Your objectives in studying this section are to be able to:

1. Define the concepts: number of mass-transfer units (NTU), and height of a mass-transfer unit (HTU).
2. Estimate the required packed height for a continuous-contact absorber or stripper in terms of the number and height of mass-transfer units.

Countercurrent packed towers operate in a different manner from plate towers in that the fluids are in contact continuously in their path through the tower, rather than intermittently. Thus, in a packed bed the liquid and gas compositions change continuously with height of packing. Every point on an operating line therefore represents conditions found somewhere in the tower; whereas for tray towers, only the isolated points on the operating line corresponding to the trays have real significance.

The development of the design equation for a countercurrent packed tower absorber or stripper begins with a differential mass balance of component A in the gas phase, in a manner similar to that of Example 2.11. However, this time we do not restrict the analysis to dilute solutions nor to constant molar velocity. If only component A is transferred ($\Psi_{A,G} = \Psi_{A,L} = 1.0$), considering the fact that the gas-phase molar velocity will change along the column, and that F-type mass-transfer coefficients are required for concentrated solutions, the mass balance is

$$\frac{d(Vy)}{Sdz} = F_G a_h \ln\left[\frac{1-y_i}{1-y}\right] \tag{5-6}$$

where S = cross-sectional area of the tower

   a_h = effective specific area of the packing

Both V and y vary from one end of the tower to the other, but V_S does not. Therefore,

$$d(Vy) = d\left(\frac{V_S y}{1-y}\right) = \frac{V_S dy}{(1-y)^2} = \frac{V dy}{1-y} \tag{5-7}$$

Substituting in equation (5-6), rearranging, and integrating,

$$Z = \int_{y_2}^{y_1} \frac{G_{My} \, dy}{F_G a_h (1-y) \ln\left[\dfrac{1-y_i}{1-y}\right]} \tag{5-8}$$

where Z is the total packed height and G_{My} is the gas-phase molar velocity. The value of y_i can be found by the methods of Chapter 3. For any value of (x, y) on the operating curve plotted in terms of mole fractions a value of (x_i, y_i) on the equilibrium curve can be found using

$$\frac{1-y_i}{1-y} = \left(\frac{1-x}{1-x_i}\right)^{F_L/F_G} \tag{5-9}$$

This provides the local y and y_i for use in equation (5-8). Equation (5-8) can then be integrated numerically or graphically after plotting the integrand as ordinate vs. y as abscissa.

However, it is more customary to proceed as follows. Since

$$y - y_i = (1-y_i) - (1-y) \tag{5-10}$$

the numerator and denominator of the integral of equation (5-8) can be multiplied respectively by the right- and left-hand sides of equation (5-10) to provide

$$Z = \int_{y_2}^{y_1} \frac{G_{My} (1-y)_{i,M} \, dy}{F_G a_h (1-y)(y-y_i)} \tag{5-11}$$

where $(1-y)_{i,M}$ is the logarithmic mean of $(1-y)$ and $(1-y_i)$.

It has been observed that the ratio $G_{My}/F_G a_h$ is very much more constant along the tower than either G_{My} or $F_G a_h$, and in many cases may be considered constant within the accuracy of the available data. Define the *height of a gas-phase transfer unit* H_{tG} (HTU) as

$$H_{tG} = \frac{G_{My}}{F_G a_h} \tag{5-12}$$

(Since there will be some variation of the molar gas velocity, the mass-transfer coefficients, and the effective specific area along the tower, estimate these values at both ends of the tower and use arithmetic averages to estimate the HTU and the local values of y_i.) Equation (5-11) becomes

$$Z = H_{tG} \int_{y_2}^{y_1} \frac{(1-y)_{i,M}\, dy}{(1-y)(y-y_i)} = H_{tG} N_{tG} \qquad (5\text{-}13)$$

where the integral is called the *number of gas transfer units*, N_{tG}. (NTU). Equation (5-13) can be further simplified by substituting the arithmetic average for the logarithmic average

$$(1-y)_{i,M} \approx \frac{(1-y_i)+(1-y)}{2} \qquad (5\text{-}14)$$

which involves little error. Then

$$N_{tG} = \int_{y_2}^{y_1} \frac{dy}{y-y_i} + \frac{1}{2}\ln\frac{1-y_2}{1-y_1} \qquad (5\text{-}15)$$

which makes for simpler numerical or graphical integration.

The above relationships all have their counterparts in terms of liquid concentrations, derived in exactly the same way

$$Z = H_{tL} N_{tL} \qquad (5\text{-}16)$$

$$H_{tL} = \frac{G_{Mx}}{F_L\, a_h} \qquad (5\text{-}17)$$

$$N_{tL} = \int_{x_2}^{x_1} \frac{dx}{x_i-x} + \frac{1}{2}\ln\frac{1-x_2}{1-x_1} \qquad (5\text{-}18)$$

where G_{Mx} = the liquid-phase molar velocity
  H_{tL} = the *height of a liquid-phase transfer unit* (HTU)
  N_{tL} = the *number of liquid-phase transfer units* (NTU)

Example 5.3 Packed-Tower Absorber for Recovery of Benzene Vapors

Consider the absorber of Example 3.7 and Example 5.2. The adsorber will be packed with 50-mm metal Hiflow rings. Estimate the dimensions of the absorber, and the power required to pump the gas and the liquid through the tower. Design for a gas-phase pressure drop of 300 Pa/m of packed height.

Solution

To determine the tower diameter, consider the conditions at the bottom of the absorber where the maximum gas and liquid flow rates are found. Run the Packed-Tower Design Program of Appendix D using the data from Example 5.2 and packing parameters from Chapter 4. From Table 4.1 for 50-mm metal Hiflow rings:

$$F_p = 16 \text{ ft}^2/\text{ft}^3 \qquad a = 92.3 \text{ m}^2/\text{m}^3 \quad \varepsilon = 0.977 \qquad C_h = 0.876 \qquad C_p = 0.421$$

From Table 4.2: $\qquad C_L = 1.168 \qquad C_V = 0.408$

For a pressure drop of 300 Pa/m, the program converges to a tower diameter $D = 0.641$ m. When convergence is achieved, the results at the bottom of the tower are:

$$f = 73.3\% \qquad a_h = 73.52 \text{ m}^{-1} \quad G_{My} = 126 \text{ mole/m}^2\text{-s}$$

$$k_y = 3.417 \text{ mole/m}^2\text{-s} \qquad k_L = 9.74 \times 10^{-5} \text{ m/s}$$

From equations (2-6) and (2-11),

$$F_G = k_y y_{B,M} = k_y (1 - y)_{i,M}$$
$$F_L = k_x x_{B,M} = k_x (1 - x)_{i,M} = k_L c_L (1 - x)_{i,M} \qquad (5\text{-}19)$$

We do not know yet the interfacial concentrations along the column, therefore we canot evaluate the terms $(1 - y)_{i,M}$ and $(1 - x)_{i,M}$. However, a good estimate of these terms is

$$(1 - y)_{i,M} \approx (1 - y_1) = 0.926$$
$$(1 - x)_{i,M} \approx (1 - x_1) = 0.676 \qquad (5\text{-}20)$$

The liquid molar concentration is $c_L = \rho_L/M_L = 780/159.12 = 4.902 \text{ kmole/m}^3$.

Substituting in equation (5-19), $F_G = 3.164$ mole/m²-s, $F_L = 0.323$ mole/m²-s:

$$H_{tG} = \frac{G_{My}}{F_G \, a_h} = \frac{126}{3.164 \times 73.52} = 0.542 \text{ m}$$

Now, we consider the conditions at the top of the absorber. Run the Packed-Tower Design Program of Appendix D using the data for that part of the column from Example 5.2. Try different values of the pressure drop per unit packed height until the program converges to the tower diameter of 0.641 m already selected. Convergence is achieved at a gas-pressure drop of 228 Pa/m. The results at the top of the tower are:

$f = 66.8\%$ $a_h = 63.31$ m⁻¹ $G_{My} = 118$ mole/m²-s

$k_y = 3.204$ mole/m²-s $k_L = 8.72 \times 10^{-5}$ m/s

The liquid molar concentration is $c_L = \rho_L / M_L = 765/192.7 = 3.970$ kmole/m³.

Substituting in equation (5-19), $F_G = 3.172$ mole/m²-s, $F_L = 0.330$ mole/m²-s:

$$H_{tG} = \frac{G_{My}}{F_G \, a_h} = \frac{118}{3.172 \times 63.31} = 0.588 \text{ m}$$

The average HTU value is $H_{tG} = 0.565$ m. The average mass-transfer coefficients are $F_G = 3.168$ mole/m²-s, $F_L = 0.327$ mole/m²-s.

The next step is to estimate the number of gas transfer units from equation (5-15). The operating curve equation for this system in terms of mole fractions was developed in Example 5.2 and is

$$y = \frac{0.004 + 0.154x}{1.004 - 0.846x} \qquad \text{(operating line)}$$

Figure 5.3 presents a Mathcad computer program to estimate N_{tG}. A vector of 12 x-values from $x_2 = 0.0476$ to $x_1 = 0.324$ is created. The operating curve equation is used to generate the corresponding y vector. The average values of F_G and F_L are supplied to the program. For each point (x, y) on the operating curve, equation (5-9)

$$x := \begin{bmatrix} 0.0476 \\ .075 \\ .1 \\ .125 \\ .15 \\ .175 \\ .2 \\ .225 \\ .25 \\ .275 \\ .3 \\ .324 \end{bmatrix}$$

$j := 0 .. 11$

$$y := \overrightarrow{\frac{0.004 + .154 \cdot x}{1.004 - 0.846 \cdot x}}$$

$$FL := 0.330 \cdot \frac{mole}{m^2 \cdot sec} \qquad FG := 3.172 \cdot \frac{mole}{m^2 \cdot sec}$$

$yi := 0.03$ Initial estimate

Given

$$\frac{1 - yi}{1 - y} = \left(\frac{1 - x}{1 - 7.353 \cdot yi} \right)^{\frac{FL}{FG}}$$

$yi(x, y) := Find(yi)$

$$y_{i_j} := yi(x_j, y_j)$$

$$f := \overrightarrow{(y - y_i)^{-1}}$$

$y^T =$

	0	1	2	3	4	5	6	7	8	9	10	11
0	0.012	0.017	0.021	0.026	0.031	0.036	0.042	0.048	0.054	0.06	0.067	0.074

$y_i{}^T =$

	0	1	2	3	4	5	6	7	8	9	10	11
0	$9.402 \cdot 10^{-3}$	0.014	0.018	0.022	0.026	0.03	0.035	0.039	0.044	0.048	0.053	0.0

$f^T =$

	0	1	2	3	4	5	6	7	8	9	10	11
0	424.7	349.1	290.6	241.8	201.6	168.6	141.5	119.3	100.9	85.7	73.1	62.9

$vs := cspline(y, f)$

$$N_{tG} := \int_{y_0}^{y_{11}} interp(vs, y, f, \theta) \, d\theta + 0.5 \cdot \ln\left(\frac{1 - y_0}{1 - y_{11}} \right)$$

$N_{tG} = 10.569$

Figure 5.3 Example 5.3: calculation of N_{tG}.

is solved for the interfacial gas-phase composition y_i (notice that $x_i = y_i/m = 7.353$ y_i). The y- and y_i-vectors are combined to generate the integrand in equation (5-15), $(y - y_i)^{-1}$. An interpolation formula relating the integrand to y is generated in the form of a cubic spline, which is then used to evaluate the integral in equation (5-15) numerically. The result is $N_{tG} = 10.57$ units. The total packed height is from equation (5-13)

$$Z = H_{tG}N_{tG} = 0.588 \times 10.57 = 6.22 \text{ m}$$

A conservative estimate of the total gas-pressure drop can be obtained assuming that the pressure drop through the tower remains constant at the value estimated for the bottom of the absorber where the gas and liquid velocities are the highest. Therefore, the total gas-pressure drop is approximately $\Delta P_G = 300 \times 6.22 = 1{,}866$ Pa. Assuming a mechanical efficiency of the motor-fan system $E_m = 60\%$, the power required to force the gas through the tower is

$$W_G = \frac{Q_G \Delta P_G}{E_m} = \frac{1.0 \times 1{,}866}{0.60} = 3{,}110 \text{ watt}$$

The liquid must be pumped to the top of the absorber, and gravity will force it through the tower. The power required to pump the liquid to the top of the absorber is

$$W_L = \frac{L_2' g Z}{E_m} = \frac{1.214 \times 9.8 \times 6.22}{0.60} = 123 \text{ watt}$$

5.3.1 Dilute Solutions; Henry's Law

For dilute solutions and when Henry's law applies, the equilibrium curve and the operating lines are straight lines. Overall mass-transfer coefficients are convenient in this case. The expression for the height of packing can be written as

$$Z = H_{tOG}N_{tOG} \qquad\qquad (5\text{-}21)$$

where Z = the total packed height
           H_{tOG} = the *overall height of a gas-phase transfer unit* (HTU)
           N_{tOG} = the *overall number of gas-phase transfer units* (NTU)

$$H_{tOG} = \frac{G_{My}}{K_y a_h} = H_{tG} + \frac{H_{tL}}{A} \qquad (5\text{-}22)$$

$$N_{tOG} = \int_{y_2}^{y_1} \frac{dy}{y - y*} + \frac{1}{2}\ln\frac{1-y_2}{1-y_1} \qquad (5\text{-}23)$$

In equation (5.22), K_y is the gas-phase overall mass-transfer coefficient as defined in equation (3-12). For dilute solutions in both phases and equilibrium described by Henry's law, equation (5-23) becomes (Treybal, 1980)

$$N_{tOG} = \frac{\ln\left[\frac{y_1 - mx_2}{y_2 - mx_2}\left(1-\frac{1}{A}\right)+\frac{1}{A}\right]}{1-1/A} \qquad (5\text{-}24)$$

The corresponding expressions for strippers are:

$$Z = H_{tOL}N_{tOL} \qquad (5\text{-}25)$$

$$H_{tOL} = \frac{G_{Mx}}{K_x a_h} \qquad (5\text{-}26)$$

$$N_{tOL} = \frac{\ln\left[\frac{x_2 - y_1/m}{x_1 - y_1/m}(1-A)+A\right]}{1-A} \qquad (5\text{-}27)$$

where G_{Mx} is the liquid-phase molar velocity.

Example 5.4 Packed Height of an Ethanol Absorber

Consider the ethanol absorber of Example 4.4, packed with 50-mm metal Hiflow rings. Estimate the packed height required to recover 97% of the alcohol, using pure water at a rate 50% above the minimum, and with a gas-pressure drop of 300 Pa/m. The system ethanol-CO_2-water at the given temperature and pressure obeys Henry's law with $m = 0.57$.

Solution

The conditions specified correspond to those of Example 4.4 where the tower diameter ($D = 0.738$ m) and the volumetric mass-transfer coefficients for both phases were evaluated. The amount of ethanol absorbed is $(180)(0.02)(0.97) = 3.5$ kmole/hr. The inlet gas molar velocity is

$$G_{My1} = 180(4)/[(3600)(\pi)(0.738)^2] = 0.117 \text{ kmole/m}^2\text{-s}$$

the outlet gas velocity is

$$G_{My2} = (180 - 3.5)(4)/[(3600)(\pi)(0.738)^2] = 0.115 \text{ kmole/m}^2\text{-s}$$

The average molar gas velocity is

$$G_{My} = (G_{My1} + G_{My2})/2 = 0.116 \text{ kmole/m}^2\text{-s}$$

The inlet liquid molar velocity is

$$G_{Mx2} = 151.5(4)/[(3600)(\pi)(0.738)^2] = 0.098 \text{ kmole/m}^2\text{-s}$$

the outlet liquid molar velocity is

$$G_{Mx1} = (151.5 + 3.5)(4)/[(3600)(\pi)(0.738)^2] = 0.101 \text{ kmole/m}^2\text{-s}$$

Calculate the absorption factor at both ends of the column:

$$A_1 = 0.101/[(0.57)(0.117)] = 1.511$$
$$A_2 = 0.098/[(0.57)(0.115)] = 1.506$$

Calculate the geometric average,

$$A = (A_1 \times A_2)^{1/2} = 1.508$$

For 97% removal of the ethanol, $y_2 \approx (0.03)(0.02) = 0.0006$. Since pure water is used, $x_2 = 0$. Substituting in equation (5-24),

$$N_{tOG} = 7.35.$$

From Example 4.4, $k_y a_h = 0.191$ kmole/m^3-s; $k_L a_h = 0.00733$ s^{-1}. The total molar concentration in the liquid phase is

$$c \approx \rho_L/M_L = 986/18 = 54.78 \text{ kmole/m}^3.$$

Therefore,

$$k_x a_h = k_L a_h c = 0.402 \text{ kmole/m}^3\text{-s}$$

The overall volumetric mass-transfer coefficient is given by

$$K_y a_h = [k_y a_h^{-1} + m/k_x a_h]^{-1} = 0.150 \text{ kmole/m}^3\text{-s}$$

From equation (5-22),

$$H_{tOG} = 0.116/0.150 = 0.771 \text{ m}$$

The packed height is given by equation (5-21),

$$Z = (7.35)(0.771) = 5.67 \text{ m}$$

5.4 THERMAL EFFECTS DURING ABSORPTION AND STRIPPING

Your objectives in studying this section are to be able to:

1. Combine material and energy balances with equilibrium considerations to determine the number of equilibrium stages required by a tray absorber or stripper when heat effects are important.
2. Combine material and energy balances with mass-transfer considerations to determine the height required by a packed-bed absorber or stripper when heat effects are important.

Many absorbers and strippers deal with dilute gas mixtures and liquid solutions, and it is satisfactory in these cases to assume that the operation is isothermal. But actually absorption operations are usually exothermic, and when large quantities of solute gas are absorbed to form concentrated solutions, the thermal effects cannot be ignored. If by absorption the temperature of the liquid is raised to a considerable extent, the equilibrium solubility of the solute will be appreciably reduced and the capacity of the absorber decreased (or else much larger flow rates of liquid will be required). For stripping, an endothermic process, the temperature of the liquid tends to fall. To take into account thermal effects during absorption and stripping, energy balances must be combined with the material balances presented in Chapter 3.

5.4.1 Adiabatic Operation of a Tray Absorber

Consider *adiabatic operation* of a countercurrent tray absorber represented by a cascade of ideal stages such as illustrated in Figure 3.23. The temperature of the streams leaving the absorber will generally be higher than the entering temperatures owing to the heat of solution. The design of such absorbers may be done numerically, calculating tray by tray from the bottom (gas entrance) to the top. The principle of an ideal tray, that the effluent streams from the tray are in equilibrium both with respect to temperature and composition, is utilized for each tray. Thus, total and solute balances around an envelope from the bottom of the absorber up to tray n are

$$L_n + V_{N+1} = L_N + V_{n+1} \qquad (5\text{-}28)$$

$$L_n x_n + V_{N+1} y_{N+1} = L_N x_N + V_{n+1} y_{n+1} \qquad (5\text{-}29)$$

from which L_n and x_n are computed. An energy balance around the same envelope is

$$L_n H_{L,n} + V_{N+1} H_{V,N+1} = L_N H_{L,N} + V_{n+1} H_{V,n+1} \qquad (5\text{-}30)$$

where H represents in each case the molal enthalpy of the stream at its particular thermodynamic state.

The temperature of stream L_n can be obtained from equation (5-30). Stream V_n is then at the same temperature as L_n and in composition equilibrium with it. Equations (5-28) to (5-30) are then applied to tray $n - 1$, and so forth. To get started, since usually only the temperatures of the entering streams L_0 and V_{N+1} are known, it is necessary to make an initial estimate of the temperature T_1 of the gas leaving V_1. An energy balance over the entire absorber,

$$L_0 H_{L,0} + V_{N+1} H_{V,N+1} = L_N H_{L,N} + V_1 H_{V,1} \qquad (5\text{-}31)$$

is used to compute T_N, the temperature of the liquid leaving at the bottom of the tower. The estimate of T_1 is checked when the calculations reach the top tray, and if necessary the entire computation is repeated. The method is best illustrated by the following example.

Example 5.5 Tray Tower for Adiabatic Pentane Absorption

One kmole/s of a gas consisting of 75% methane and 25% n-pentane at 300 K and 1 atm is to be scrubbed with 2 kmole/s of a nonvolatile paraffin oil entering the absorber free of pentane at 308 K. Estimate the number of ideal trays for adiabatic absorption of 98.6% of the pentane. Neglect the solubility of methane in the oil, and assume operation to be at constant pressure. The pentane forms ideal solutions with the paraffin oil. The average molecular weight of the oil is 200, heat capacity is 1.884 kJ/kg-K. The heat capacity of methane over the range of temperatures to be encountered is 35.6 kJ/kmole-K; for liquid pentane is 177.5 kJ/kmole-K; for pentane vapor is 119.8 kJ/kmole-K. The latent heat of vaporization of n-pentane at 273 K is 27.82 MJ/kmole (Treybal, 1980).

Solution
It is convenient to refer all enthalpies to the condition of pure liquid solvent, pure diluent gas, and pure solute at some base temperature T_0, with each substance assigned zero enthalpy for its normal state of aggregation at T_0 and 1 atm pressure. In this case we use a base temperature $T_0 = 273$ K. Enthalpies referred to 273 K, liq-

uid pentane, liquid paraffin oil, and gaseous methane, are then

$$H_L = (1 - x)(1.884)(200)(T_L - 273) + x(177.5)(T_L - 273) + \Delta H_S$$

where ΔH_S is the heat of solution. Since pentane forms ideal solutions with the paraffin oil, $\Delta H_S = 0$ and

$$H_L = (T_L - 273)(376.8 - 199.3x) \text{ kJ/kmole liquid solution}$$

$$H_V = (1 - y)(35.6)(T_G - 273) + y[(119.8)(T_G - 273) + 27,820]$$
$$= (T_G - 273)(35.6 + 84.2y) + 27,820y \text{ kJ/kmole gas-vapor mixture}$$

In this case, equilibrium is described by Raoult's law, $y_i = P_A x_i / P = m x_i$. The vapor pressure for n-pentane as a function of temperature can be estimated from the Antoine equation, in this case (Smith, et al., 1996)

$$P_A = \exp\left[13.8183 - \frac{2,477.07}{T - 40}\right]$$

where P_A is in kPa. Then, the slope of the equilibrium curve, m, as a function of temperature is given by

$$m(T) = \frac{\exp\left[13.8183 - \frac{2,477.07}{T - 40}\right]}{101.3}$$

Calculate the flow rate and composition of the two streams leaving the tower. From material balances for the required 98.6% pentane recovery:

Pentane entering with the incoming gas = (1 kmole/s)(0.25) = 0.25 kmole/s
Pentane absorbed = (0.986)(0.25) = 0.2465 kmole/s
Methane in the outgoing gas = methane in the incoming gas = 0.75 kmole/s

$V_1 = 0.75 + (0.014)(0.25) = 0.7535 \text{ kmole/s}$ $y_1 = (0.014)(0.25)/0.7535 = 0.0046$

$L_N = 2.0 + 0.2465 = 2.2465 \text{ kmole/s}$ $x_N = 0.2465/2.2465 = 0.1097$

Calculate the enthalpy of the two streams entering the tower:

$H_{L,0} = 13,190 \text{ kJ/kmole}$ $H_{V,N+1} = 8,480 \text{ kJ/kmole}$

Assume $T_1 = 308.5$ K (to be checked later), and calculate $H_{V,1} = 1,405$ kJ/kmole.

Substitute into the energy balance for the entire absorber, equation (5-31):

$$2 \times 13,190 + 1 \times 8,480 = 2.2465 \times H_{L,N} + 0.7535 \times 1,405$$

Then,

$$H_{L,N} = 15,050 = (T_N - 273)[376.8 - (199.3 \times 0.1097)]$$

$$T_N = 315.4K$$

$$m(T_N) = 1.228 \qquad\qquad y_N = m(T_N)x_N = (1.228)(0.1097) = 0.135$$

$$V_N = V_S/(1 - y_N) = 0.75/(1 - 0.135) = 0.867 \text{ kmole/s}$$

From equation (5-28) with $n = N - 1$,

$$L_{N-1} = L_N + V_N - V_{N+1} = 2.2465 + 0.867 - 1.000 = 2.114 \text{ kmole / s}$$

From equation (5-29) with $n = N - 1, x_{N-1} = 0.0521$.
From equation (5-30) with $n = N - 1, H_{L,N-1} = 14,300$ kJ/kmole, then $T_{N-1} = 312$ K.

The computations are continued upward through the tower in this manner until the gas composition falls at least to $y = 0.0046$. The results are:

n = tray number	T_n, K	x_n	y_n
$N = 4$	315.3	0.1091	0.1340
$N - 1 = 3$	312.0	0.0521	0.0568
$N - 2 = 2$	309.8	0.0184	0.0187
$N - 3 = 1$	308.5	0.0046	0.0045

The required $y_1 = 0.0046$ occurs at about 4 ideal trays, and the temperature on the top tray is esentially that assumed. Had this not been so, a new assumed value of T_1 and a new stage-by-stage calculation would have been required.

5.4.2 Adiabatic Operation of a Packed-Bed Absorber

During absorption in a packed bed, release of energy at the interface due to latent heat and heat of solution raises the interface temperature above that of the bulk liquid. This changes physical properties, mass-transfer coefficients, and equilibrium concentrations. Because of the possibility of substantial temperature effects, there is the possibility that the solvent evaporates in the warm parts of the tower and recondenses in the cooler parts. This has been confirmed experimentally (Raal and Khurani, 1973).

For this problem, therefore, the components are defined as follows:

A = principal transferred solute, present in both gas and liquid phases
B = carrier gas, not dissolving in the liquid, present only in the gas
C = principal liquid solvent; can evaporate and condense; present in both phases

Consider material and energy balances over a differential section of the packed tower. The mass and energy fluxes are considered positive in the direction gas to liquid, negative in the opposite direction. Looking at the gas phase: solute A and solvent vapor C may transfer, but carrier gas B will not ($N_B = 0$):

$$N_A a_h\, dz = \Psi_A F_{G,A} \ln\left[\frac{\Psi_A - y_{A,i}}{\Psi_A - y_A}\right] a_h\, dz = -G_{M,B}\, dY_A \tag{5-32}$$

$$N_C a_h\, dz = \Psi_C F_{G,C} \ln\left[\frac{\Psi_C - y_{C,i}}{\Psi_C - y_C}\right] a_h\, dz = -G_{M,B}\, dY_C \tag{5-33}$$

where Y_A = moles A/mole B, Y_C = moles C/mole B. Since the fluxes N_A and N_C may have either sign, the Ψ's can be grater or less than 1.0, positive or negative. However, $\Psi_A + \Psi_C = 1.0$. Neglecting the liquid heat-transfer resistance ($T_L = T_i$), the heat-transfer rate for the gas is

$$q_G a_h\, dz = h_{G,c} a_h (T_G - T_L)\, dz \tag{5-34}$$

where $h_{G,c}$ is the gas-phase convection heat-transfer coefficient corrected for mass transfer (Treybal, 1980). Similarly, for the liquid,

$$N_A a_h\, dz = \Psi_A F_{L,A} \ln\left[\frac{\Psi_A - x_A}{\Psi_A - x_{Ai}}\right] a_h\, dz \tag{5-35}$$

$$N_C a_h\, dz = \Psi_C F_{L,C} \ln\left[\frac{\Psi_C - x_C}{\Psi_C - x_{C,i}}\right] a_h\, dz \tag{5-36}$$

At steady-state, with no chemical reaction, a mass balance on the differential section yields (for countercurrent flow)

$$dL = dV \qquad (5\text{-}37)$$

Since

$$V = V_B\left(1 + Y_A + Y_C\right) \qquad (5\text{-}38)$$

Then

$$dL = V_B\left(dY_A + dY_C\right) \qquad (5\text{-}39)$$

The concentration gradients along the column are provided by equations (5-32) and (5-33) written here as

$$\frac{dY_A}{dz} = -\frac{\Psi_A F_{G,A}\, a_h}{G_{M,B}} \ln\left[\frac{\Psi_A - y_{A,i}}{\Psi_A - y_A}\right] \qquad (5\text{-}40)$$

$$\frac{dY_C}{dz} = -\frac{\Psi_C F_{G,C}\, a_h}{G_{M,B}} \ln\left[\frac{\Psi_C - y_{C,i}}{\Psi_C - y_C}\right] \qquad (5\text{-}41)$$

The gas-phase temperature gradient is from an energy balance over the gas phase in the differential section:

$$\frac{dT_G}{dz} = -\frac{h_{G,c}\, a_h\left(T_G - T_L\right)}{G_{M,B}\left(C_B + Y_A C_A + Y_C C_C\right)} \qquad (5\text{-}42)$$

The interface concentration of A is obtained combining equations (5-32) and (5-35)

$$y_{A,i} = \Psi_A - \left(\Psi_A - y_A\right)\left[\frac{\Psi_A - x_A}{\Psi_A - x_{A,i}}\right]^{F_{L,A}/F_{G,A}} \qquad (5\text{-}43)$$

Similarly, for component C

$$y_{C,i} = \Psi_C - \left(\Psi_C - y_C\right)\left[\frac{\Psi_C - x_C}{\Psi_C - x_{C,i}}\right]^{F_{L,C}/F_{G,C}} \qquad (5\text{-}44)$$

These equations are solved simultaneously with their respective equilibrium-distribution curves, as described in Chapter 3. The calculation is by trial-and -error: Ψ_A is assumed, whence $\Psi_C = 1 - \Psi_A$. The correct Ψ's are those for which $x_{A,i} + x_{B,i} = 1.0$.

Heat-transfer coefficients can be estimated, if not available otherwise, through the heat-mass-transfer analogies discussed in Chapter 2. For the gas phase, the heat-transfer coefficient must be corrected for mass transfer (Treybal, 1980):

$$h_{G,c}a_h = \frac{-G_{M,B}(C_A dY_A / dz + C_C dY_C / dz)}{1 - \exp\left[\dfrac{G_{M,B}}{h_G a_h}(C_A dY_A / dz + C_C dY_C / dz)\right]} \qquad (5\text{-}45)$$

Effective diffusivities for the three-component gas mixtures are estimated through equation (1-58), which reduces in this case to

$$D_{A,eff} = \frac{\Psi_A - y_A}{\Psi_A\left(\dfrac{y_B}{\mathcal{D}_{AB}} + \dfrac{y_A + y_C}{\mathcal{D}_{AC}}\right) - \dfrac{y_A}{\mathcal{D}_{AC}}} \qquad (5\text{-}46)$$

$$D_{C,eff} = \frac{\Psi_C - y_C}{\Psi_C\left(\dfrac{y_B}{\mathcal{D}_{CB}} + \dfrac{y_A + y_C}{\mathcal{D}_{AC}}\right) - \dfrac{y_C}{\mathcal{D}_{AC}}} \qquad (5\text{-}47)$$

In the design of an absorption tower, the cross-sectional area and hence the mass velocities of gas and liquid are established through pressure drop considerations as described in Chapter 4. Assuming that entering flow rates, compositions, and temperatures, pressure of absorption, and percentage absorption (or stripping) of one component are specified, the packed height is then fixed. The problem is therefore to estimate the packed height and the conditions (temperature and composition) of the outlet streams. Fairly extensive trial-and-error is required, for which the relations outlined above can best be solved by a digital computer (Taylor and Krishna, 1993; Seader and Henley, 1998). The procedure for the three-component case is outlined in the following example.

Example 5.6 Packed Tower for Adiabatic Ammonia Absorption

A gas consisting of 41.6% ammonia (A), 58.4% air (B) at 293 K and 1 atm, flowing at the rate of 34 moles/s is to be scrubbed countercurrently with water (C) entering at 293 K at the rate of 271 moles/s to remove 99% of the ammonia. The adiabatic absorber is to be packed with 25-mm ceramic Raschig rings. Estimate the packed height if the cross-sectional area is 1.0 m^2.

Solution

Basis for enthalpy calculations: NH_3 gas; H_2O liquid; air at 1 atm; $T_0 = 293$ K. Since ammonia is a gas at T_0, its latent heat of vaporization, $\lambda_{A0} = 0$; on the other hand, $\lambda_{C0} = 44.24$ kJ/mole (Perry and Chilton, 1973). From the same reference, the gas-phase heat capacities at 293 K are $C_A = 36.4$, $C_B = 29.1$, $C_C = 33.96$ kJ/kmol-K.

Gas in:

$$G_{M,B} = (34)(0.584) = 19.8 \text{ mole air/m}^2\text{-s}$$
$$y_A = 0.416 \quad Y_A = 0.416/(1\text{-}0.416) = 0.7123 \text{ moles } NH_3/\text{mole air}$$
$$y_C = Y_C = 0 \quad H_G = 0$$

Liquid in:

$$G_{M,x} = 271 \text{ mole/m}^2\text{-s} \quad x_A = 0 \quad x_C = 1.0 \quad H_L = 0$$

Gas out:

$$Y_A = (0.7123)(1 - 0.99) = 0.0071 \text{ moles } NH_3/\text{mole air}$$

Assume outlet gas $T_G = 296$ K (to be checked later)

Assume $y_C = 0.0293$ (saturated with water vapor, to be checked)

$$y_C = Y_C/(Y_C + 0.0071 + 1.0) \quad Y_C = 0.0304 \text{ moles } H_2O/\text{mole air}$$

$$H_G = C_B(T_G - T_0) + Y_A\left[C_A(T_G - T_0) + \lambda_{A0}\right] + Y_C\left[C_C(T_G - T_0) + \lambda_{C0}\right]$$
$$= 1{,}436 \text{ kJ / kmole air}$$

Liquid out:

$$H_2O \text{ content} = 271 - (19.8)(0.0304) = 270.5 \text{ moles/m}^2\text{-s}$$

$$NH_3 \text{ content} = (19.8)(0.7123\text{-}0.0071) = 13.96 \text{ moles/m}^2\text{-s}$$

$$G_{M,x} = (270.5 + 13.96) = 284.5 \text{ mole/m}^2\text{-s}$$
$$x_A = 13.96/284.5 = 0.0491 \quad x_C = 0.9509$$

$$H_L = C_L(T_L - T_0) + \Delta H_S$$

At $x_A = 0.0491$, $T_0 = 293$ K, $\Delta H_S = -1701$ kJ/kmole of solution

("International Critical Tables," 1929)

$$H_L = [C_L(T_L - 293) - 1{,}710] \text{ kJ/kmole solution}$$

Energy balance around the entire tower (adiabatic):

Enthalpy gas in + enthalpy liquid in = enthalpy gas out + enthalpy liquid out

$$0 + 0 = (0.0198)(1,436) + (0.2845)\,[C_L(T_L - 293) - 1,710]$$

The heat capacity of liquid mixtures of ammonia and water is related to the temperature and composition through (Smith, et al., 1996):

$$\frac{C_L}{R} = 8.712 + 13.914 x_A + \left(1.25 - 101.25 x_A\right) \times 10^{-3} T_L$$

$$+ \,(192.9 - 0.18) \times 10^{-6} T_L^2$$

The value of C_L, obtained by trial at $(T_L + 293)/2$, equals 75.8 kJ/kmol-K, whence $T_L = 314.2$ K.

Estimate the properties of the bulk gas at the bottom of the absorber (41.6% NH_3, 58.4% air, 293 K, 1 atm):

$$M_{G,av} = (0.416)(17) + (0.584)(29) = 24.02$$

$\rho_G = 999$ kg/m³ (ideal gas law)

The viscosity of the mixture is estimated from the viscosity of the pure components at the temperature of the mixture using Wilke's method (Reid, et al., 1987):

$$\mu_G = \sum_{i=1}^{n} \frac{y_i \mu_i}{\displaystyle\sum_{j=1}^{n} y_j \phi_{ij}} \qquad\qquad \phi_{ij} = \frac{\left[1 + \left(\mu_i / \mu_j\right)^{1/2} \left(M_j / M_i\right)^{1/4}\right]^2}{\left[8\left(1 + M_j / M_i\right)\right]^{1/2}}$$

$\mu_G = 1.52 \times 10^{-5}$ kg/m-s

The thermal conductivity is also estimated using Wilke's method:

k_G (thermal conductivity) = 0.0261 W/m-K

The heat capacity of the mixture is estimated from

$$C_p = \frac{y_A C_A + y_B C_B}{M_{G,av}} = 1.336 \text{ kJ / kg - K}$$

Gas Prandtl number, $Pr_G = C_p \mu_G / k_G = 0.725$

The binary diffusivities are from the Wilke-Lee equation:

$$\mathfrak{D}_{AB} = 0.2194 \text{ cm}^2/\text{s}; \ \mathfrak{D}_{AC} = 0.1944 \text{ cm}^2/\text{s}; \ \mathfrak{D}_{CB} = 0.2150 \text{ cm}^2/\text{s}$$

Estimate the properties of the liquid at the bottom of the absorber (4.9% NH_3, 95.1% water, 314.2 K):

$$M_{L,av} = (0.049)(17) + (0.951)(18) = 17.97$$

$\rho_L = 953 \text{ kg/m}^3$ ("International Critical Tables," 1929)

$\mu_L = 0.64$ cP (approximately equal to the viscosity of pure water)

k_L (thermal conductivity) = 0.64 W/m-K (pure water; Holman, 1990)

Liquid Prandtl number, $Pr_L = C_L \mu_L / k_L M_{L,av} = 4.20$

$D_{AL} = 2.71 \times 10^{-5} \text{ cm}^2/\text{s}$ (estimated from Hayduk and Minhas)

$Sc_L = \mu_L / \rho_L D_{AL} = 248$

Calculate the mass velocities at the bottom of the absorber:

$$G_y = G_{M,y} M_{G,av} = (0.034)(24) = 0.814 \text{ kg/m}^2\text{-s}$$

$$G_x = G_{M,x} M_{L,av} = (0.2845)(17.97) = 5.110 \text{ kg/m}^2\text{-s}$$

Calculate the mass-transfer coefficients and interfacial conditions at the bottom. To obtain the gas-phase mass-transfer coefficients, the effective diffusivities of A and C, and hence Ψ_A and Ψ_C must be known. Assume $\Psi_A = 1.32$, therefore $\Psi_C = 1 - 132 = -0.32$. With $y_A = 0.416$, $y_B = 0.584$, and $y_C = 0.0$, equations (5-46) and (5-47) yield $D_{A,eff} = 0.2156 \text{ cm}^2/\text{s}$, $D_{C,eff} = 0.2187 \text{ cm}^2/\text{s}$. From the data of Chapter 4, for 25-mm ceramic Raschig rings: $F_p = 179 \text{ ft}^{-1}$, $a = 190 \text{ m}^{-1}$, $\varepsilon = 0.680$, $C_h = 0.577$, $C_p = 1.329$, $C_L = 1.361$, and $C_V = 0.412$. Calculate the mass-trandfer coefficients using the Mathcad program in Appendix D. For the conditions at the bottom of the absorber, the gas-pressure drop is 380 Pa/m at 72% approach to flooding velocity. The liquid-phase mass-transfer coefficient is $k_L = 3.36 \times 10^{-4}$ m/s. The gas-phase mass-transfer coefficient for ammonia is $k_y = 3.00 \text{ mole/m}^2\text{-s}$; for water vapor is $k_y = 3.13 \text{ mole/m}^2\text{-s}$. The effective specific area is $a_h = 112.3 \text{ m}^{-1}$. Convert these to F-type coefficients:

$$F_L = k_L c_L (1-x)_{i,M} = \frac{k_L \rho_L (1-x)}{M_{L,av}} = 0.0169 \text{ kmole / m}^2\text{-s}$$

$$F_G = k_y (1-y)_{i,M}$$

$$= 1.752 \text{ mole / m}^2\text{-s for ammonia}$$

$$= 1.828 \text{ mole / m}^2\text{-s for water}$$

To estimate the gas-phase heat-transfer coefficient, the heat-transfer analogous of equation (4-18) is written as

$$\frac{h_G}{a k_G (\text{conductivity})} = \frac{0.1304 C_V}{[\varepsilon(\varepsilon - h_L)]^{0.5}} \left[\frac{\text{Re}_G}{K_W} \right]^{3/4} \text{Pr}_G^{2/3} \qquad (5\text{-}48)$$

which yields $h_G = 86.0$ W/m^2-s.

To calculate the interfacial concentrations, equilibrium data are needed for the system NH$_3$-air-H$_2$O at a total pressure of 1 atm, and different temperatures in the operational range of the absorber. Tables 5.1 and 5.2 summarize the equilibrium data (Perry and Chilton, 1973).

The Mathcad multivariate cubic spline function is used to interpolate in these tables for different values of temperature and ammonia liquid mole fraction. Equations (5-43) and (5-44) become

$$y_{A,i} = 1.32 - (1.32 - 0.416) \left[\frac{1.32 - 0.0491}{1.32 - x_{A,i}} \right]^{0.0169/0.001752} \qquad (5\text{-}49)$$

Table 5.1 Equilibrium Gas-Phase NH$_3$ Molar Fractions Over Aqueous Ammonia Solutions (1 atm Total Pressure)

Temperature	NH$_3$ mole fraction in liquid solution, x_A, percentages			
(K)	5	10	15	20
289	0.042	0.081	0.136	0.218
294	0.056	0.103	0.177	0.291
306	0.093	0.171	0.289	0.468
317	0.146	0.272	0.452	0.724
322	0.182	0.337	0.559	0.890

Source: Perry and Chilton, 1973.

Table 5.2 Equilibrium Gas-Phase H₂O Molar Fractions Over Aqueous Ammonia Solutions (1 atm Total Pressure)

Temperature	H₂O mole fraction in liquid solution, x_A, percentages			
(K)	100	95	90	85
289	0.0177	0.0163	0.0156	0.0143
294	0.0245	0.0231	0.0218	0.0204
306	0.0476	0.0449	0.0429	0.0395
317	0.0864	00816	0.0776	0.0728
322	0.1150	0.1088	0.1027	0.0966

Source: Perry and Chilton, 1973.

$$y_{C,i} = -0.32 - (-0.32 - 0.0)\left[\frac{-0.32 - 0.9509}{-0.32 - x_{C,i}}\right]^{0.0169/0.001828} \qquad (5\text{-}50)$$

Equations (5-49) and (5-50) are solved simultaneously with the equilibrium data of Tables 5.1 and 5.2 to obtain the interfacial ammonia and water concentrations at the bottom of the absorber, as shown in Figure 5.4 for the case of ammonia. The results are: $x_{A,i} = 0.0779$, $y_{A,i} = 0.1926$, $x_{C,i} = 0.9241$, and $y_{C,i} = 0.0696$. If the assumed values of the Ψ's are correct, the interfacial liquid mole fractions must add up to 1.0. In this case, $x_{A,i} + x_{C,i} = 1.0020$. By trial-and-error, it was found convergence is achieved at $\Psi_A = 1.29$, $\Psi_C = -0.29$; interfacial concentrations are: $x_{A,i} = 0.0780$, $y_{A,i} = 0.1929$, $x_{C,i} = 0.9220$, and $y_{C,i} = 0.0695$.

Move up the tower in small intervals. From equation (5-40), $dY_A/dz = -2.914$ kmole NH₃/kmole air-m. From equation (5-41), $dY_C/dz = 0.647$ kmole H₂O/kmole air-m. From equation (5-45), $h_{G,c}a_h = 10.514$ kW/m³-K. From equation (5-42), $dT_G/dz = 204.6$ K/m. Take a small increment ΔY_A. For illustration purposes, $\Delta Y_A = -0.1$ kmole NH₃/kmole air is suitable (for higher accuracy, using a digital computer, -0.01 would be more appropriate). Then, $\Delta z = \Delta Y/(dY_A/dz) = -0.1/(-2.914) = 0.0343$ m. At this next level in the column,

$$Y_{A,next} = Y_A + \Delta Y_A = 0.7123 - 0.1 = 0.6123 \text{ kmole NH}_3/\text{kmole air}$$

$$Y_{C,next} = Y_C + (dY_C/dz)\Delta z = 0 + (0.647)(0.0343) = 0.0222 \text{ kmole H}_2\text{O/kmole air}$$

$$T_{G,next} = T_G + (dT_G/dz)\Delta z = 293 + (204.6)(0.0343) = 300.02 \text{ K}$$

$$G_{M,x\,next} = G_{M,x} + G_{M,B}(\Delta Y_A + \Delta Y_C)$$

$$G_{M,x \text{ next}} = 0.2845 + (0.0198)(-0.1 + 0.0222) = 0.2830 \text{ kmole/m}^2\text{-s}$$

$$x_{A,next} = \frac{G_{M,x}x_A + G_{M,B}\Delta Y_A}{G_{M,x}} = 0.0421$$

$$H_{G,next} = 1{,}347 \text{ kJ/kmole air}$$

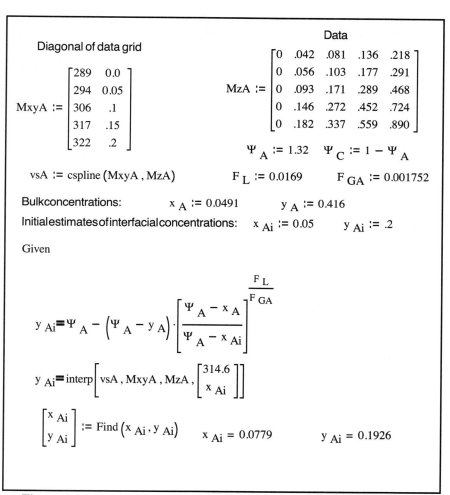

Figure 5.4 Example 5.6: calculation of interfacial ammonia concentrations

$$H_{L,next} = \frac{\left(G_{M,x}H_L\right)_{top} + G_{M,B}\left(H_{G,next} - H_{G,top}\right)}{G_{Mx,next}} = -6.23 \text{ kJ / kmole}$$

$$T_{L,next} = 312.2 \text{ K}$$

The previous computations can now be repeated at this level, leading ultimately to a new ΔY_A. The calculations are continued until the specified gas outlet
composition is reached, whereupon the assumed outlet gas temperature and water
concentration can be checked. The latter are adjusted, as necessary, and the entire
computation is repeated until convergence is achieved. The packed depth required is
then the sum of the final Δz's. The computations started above lead to $Z = 0.513$ m.
The gas leaves at the top at a temperature of 295 K and a water vapor content of
2.90 molar percent. These results agree reasonably well with the assumed values of
296 K and 2.93 molar percent. Figures 5.5 and 5.6 show the temperature and water
vapor concentration profiles along the tower. Figure 5.6 shows that water is stripped
in the lower part of the tower and reabsorbed in the upper part.

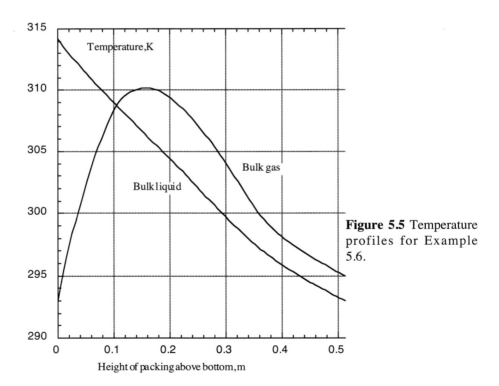

Figure 5.5 Temperature profiles for Example 5.6.

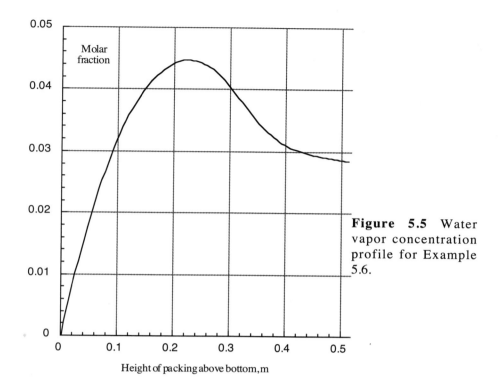

Figure 5.5 Water vapor concentration profile for Example 5.6.

PROBLEMS

The problems at the end of each chapter have been grouped into four classes (designated by a superscript after the problem number).

Class a: Illustrates direct numerical application of the formulas in the text.
Class b: Requires elementary analysis of physical situations, based on the subject material in the chapter.
Class c: Requires somewhat more mature analysis.
Class d: Requires computer solution.

5.1ª. Overall tray efficiency

Consider a tray absorber with a constant Murphree efficiency E_{MGE} = 0.75, and an average absorption factor A = 1.25.

(a) Estimate the overall tray efficiency.

Answer: 72.8%

(b) If the absorber requires 5.34 equilibrium stages, calculate the number of real trays.

Answer: 8 trays

5.2ᶜ. Relation between Murphree tray efficiency and overall efficiency.

Derive equation (5-5). _Hint:_ Start with the definition of E_{MGE} and locate the pseudoequilibrium line which could be used together with the operating line for graphically constructing steps representing real trays. Then use the Kremser equation (5-3) with the pseudoequilibrium line by moving the origin of the xy diagram to the intercept of the pseudoequilibrium line with the x axis.

5.3ª. Ammonia stripping from a wastewater in a tray tower.

A tray tower providing six equilibrium stages is used for stripping ammonia from a wastewater stream by means of countercurrent air at 1 atm and 300 K. Calculate the concentration of ammonia in the exit water if the inlet liquid concentration is 0.1 mole% ammonia, the inlet air is free of ammonia, and 1.873 standard cubic meter of air are fed to the tower per kilogram of wastewater. The equilibrium data for this system, in this range of concentrations and 300 K, can be represented by $y_{A,i} = 1.414\, x_{A,i}$ (King 1971).

Answer: 5.74×10^{-4} mole%

5.4^a. Ammonia stripping from a wastewater in a tray tower.

The Murphree plate efficiency for the ammonia stripper of Problem 5.3 is constant at 58.1%. Estimate the number of real trays required.

Answer: 9 trays

5.5^b. Ammonia stripping from a wastewater in a tray tower.

If the air flow rate to the absorber of Problems 5.3 and 5.4 is reduced to 1.0 standard m^3/kg of water, calculate concentration of ammonia in the exit water if 9 real trays are used and the Murphree efficiency remains constant at 58.1%.

Answer: 1.1×10^{-2} mole%

5.6^b. Absorption of an air pollutant in a tray tower.

A heavy-oil stream at 320 K is used in an absorber to remove dilute quantities of pollutant A from an air stream. The heavy oil is then recycled back to the process where A is stripped. The process is being run on a pilot plan basis, and information for scale-up is desired. The current absorber is a 16-plate sieve-tray column. Pilot plant data are as follows:

Liquid flow rate = 5.0 moles/hr

Gas flow rate = 2.5 moles/hr

$y_{A,in} = 0.04$ $\qquad$ $y_{A,out} = 0.0001$ $\qquad$ $x_{A,in} = 0.0001$

Equilibrium for A is given as $y_{A,i} = 0.7 x_{A,i}$. Find the overall column efficiency, and the Murphree plate efficiency. assume that the liquid and gas flow rates are roughly constant.

Answer: $E_{MGE} = 0.53$

5.7[b]. Absorption of ammonia in a laboratory-scale tray tower.

An absorption column for laboratory use has been carefully constructed so that it has exactly 4 equilibrium stages, and is being used to measure equilibrium data. Water is used as the solvent to absorb ammonia from air. The system operates isothermally at 300 K and 1 atm. The inlet water is pure distilled water. The ratio of $L/V = 1.2$, inlet gas concentration is 0.01 mole fraction ammonia; the measured outlet gas concentration is 0.0027 mole fraction ammonia. Assuming that Henry's law applies, calculate the slope of the equilibrium line.

Answer: $m = 1.413$

5.8[c,d]. Absorption of ammonia in a seive-tray tower.

A process for making small amounts of hydrogen by cracking ammonia is being considered and residual, uncracked ammonia is to be removed from the resulting gas. The gas will consist of H_2 and N_2 in the molar ratio 3:1, containing 3% NH_3 by volume, at a pressure of 2 bars and a temperature of 303 K.

There is available a sieve-tray tower, 0.75 m diameter, containing 6 cross-flow trays at 0.5 m tray spacing. The perforations are 4.75 mm in diameter, arranged in triangular pitch on 12.5 mm centers, punched in sheet metal 2 mm thick. The weir height is 40 mm. Assume isothermal scrubbing with pure water at 303 K. The water flow rate to be used should not exceed 50% of the maximum recommended for cross-flow sieve trays, which is 0.015 m^3/s-m of tower diameter (Treybal, 1980). The gas flow rate should not exceed 80% of the flooding value.

(a) Estimate the gas flow rate that can be processed in the column under the circumstances described above.

(b) Calculate the concentration of the gas leaving the absorber in part a.

Answer: 3.16×10^{-5} mole fraction

(c) Estimate the total gas-pressure drop, and check the operation for excessive weeping and entrainment.

Answer: $\Delta P = 3.7$ kPa

Data:

Liquid density = 996 kg/m^3 Surface tension = 0.068 N/m

Foaming factor = 1.0   Liquid diffusivity = 2.42×10^{-9} m^2/s

Gas viscosity = 1.122×10^{-5} kg/m-s   Gas diffusivity = 0.23 cm²/s

Slope of the equilibrium line, $m = 0.85$

5.9[b]. Absorption of carbon dioxide in a bubble-cap tray tower.

A plant manufacturing dry ice will burn coke in air to produce a flue gas which, when cleaned and cooled, will contain 15% CO_2, 6% O_2, and 79% N_2. The gas will be blown into a bubble-cap tower scrubber at 1.2 atm and 298 K, to be scrubbed countercurrently with a 30 wt% monoethanolamine (C_2H_7ON) aqueous solution entering at 298 K. The scrubbing liquid, which is recycled from a stripper, will contain 0.058 mole CO_2/mol solution. The gas leaving the scrubber is to contain 2% CO_2. A liquid-to-gas ratio of 1.2 times the minimum is specified. Assume isothermal operation. At 298 K and 1.2 atm, the equilibrium mole fraction of carbon dioxide over aqueous solutions of monoethanolamine (30 wt%) is given by

$$y_{A,i} = 7.5503 - 272.01 x_{A,i} + 2452.4 x_{A,i}^2 \qquad 0.055 \le x_{A,i} \le 0.065$$

where $x_{A,i}$ is the mole fraction of CO_2 in the liquid solution.

(a) Calculate the kilograms of solution entering the tower per cubic meter of entering gas.

Answer: 21.0 kg

(b) Determine the number of theoretical trays required for part a.

Answer: 2.5

(c) The monoethanolamine solution has a viscosity of 6.0 cP and a density of 1,012 kg/m³. Estimate the overall tray efficiency for the absorber, and the number of real trays required. Seader and Henley (1998) proposed the following empirical correlation to estimate the overall efficiency of absorbers and strippers using bubble-cap trays (it has also been used to obtain rough estimates for sieve-tray towers):

$$\log E_O = -0.773 - 0.415 \log\left(\frac{m M_L \mu_L}{\rho_L}\right) - 0.0896 \left[\log\left(\frac{m M_L \mu_L}{\rho_L}\right)\right]^2$$

where: E_O = overall fractional efficiency

m = slope of equilibrium curve

μ_L = liquid viscosity, in cP

ρ_L = liquid density, in kg/m^3

Hint: In this problem, the equilibrium-distribution curve is not a straight line, therefore m is not constant. Estimate the average value of m in the range of liquid concentrations along the operating line and use the average value in the correlation given above.

Answer: 20 trays

5.10^c. Minimum liquid rate for absorption: Henry's law.

Consider an absorber in which the equilibrium is described by Henry's law: $y_i = mx_i$.

(a) Show that the equilibrium-distribution curve in terms of mole ratios is given by

$$Y_i = \frac{mX_i}{1+(1-m)X_i} \tag{5-51}$$

(b) Show that, for $m < 1.0$, the equilibrium-distribution curve described by equation (5-51) is always concave downward; therefore the operating line that gives the minimum liquid rate will be tangent to the equilibrium curve at some point $T(X_T, Y_T)$ in the interval $X_{in} < X_T < X_{out}$.

(c) Show that, given m, V_S, X_{in}, Y_{out}, the coordinates of the tangent point, and the minimum flow rate, L_{Smin}, are given by simultaneously solving the following three equations:

$$\frac{L_{S\,min}}{V_S} = \frac{Y_T - Y_{out}}{X_T - X_{in}} \tag{5-52}$$

$$\frac{L_{S\,min}}{V_S} = \frac{m}{\left[1+(1-m)X_T\right]^2} \tag{5-53}$$

$$Y_T = \frac{mX_T}{\left[1+(1-m)X_T\right]} \tag{5-54}$$

(d) Write a Mathcad program to solve equations (5-52) to (5-54) and test it with data from Example 3.7.

(e) Show that a similar analysis is possible for strippers when $m > 1$.

5.11^c. Absorption of carbon disulfide in a sieve-tray tower.

Carbon disulfide, CS_2, used as a solvent in a chemical plant, is evaporated from the product in a dryer into an inert gas (esentially N_2) in order to avoid an explosion hazard. The CS_2-N_2 mixture is to be scrubbed with an absorbent hydrocarbon oil (octadecane, $C_{18}H_{38}$). The gas will flow at the rate of 0.4 m^3/s at 297 K and 1 atm. The partial pressure of CS_2 in the original gas is 50 mm Hg, and the CS_2 concentration in the outlet gas is not to exceed 0.5%. The oil enters the absorber essentially pure at a rate 1.5 times the minimum, and solutions of oil and CS_2 follow Raoult's law. Design a sieve-tray tower for this process. Design for a gas velocity which is 70% of the flooding velocity. Assuming isothermal operation, determine:

(a) Liquid flow rate, kg/s

Answer: 2.24 kg/s

(b) Tower diameter and plate spacing

(c) The details of the tray design

(d) Number of real trays required

(e) Total gas-pressure drop

Data (at 297 K):

Oil average molecular weight = 254   Oil viscosity = 4 cP

Oil density = 810 kg/m^3 Surface tension = 0.030 N/m

Foaming factor = 0.9 CS_2 vapor pressure = 346 mm Hg

Gas viscosity = 1.7×10^{-5} kg/m-s   Gas diffusivity = 0.114 cm^2/s

Liquid diffusivity = 0.765×10^{-5} cm^2/s

5.12^c. Adiabatic absorption of carbon disulfide in a sieve-tray tower.

Determine the number of equilibrium trays and the temperature of the gas and liquid streams leaving the absorber of Problem 5.11, assuming adiabatic operation. The specific heats are:

Substance	J/mole-K
Oil	362.2
CS_2, liquid	76.2
N_2	29.1
CS_2, gas	46.9

the latent heat of vaporization of CS_2 at 297 K is 27.91 kJ/mole. The vapor pressure of CS_2 as a function of temperature is given by

$$\ln P_A = 15.9844 - \frac{2690.85}{T - 31.62}$$

$$P_A = \text{ mm Hg} \qquad T \text{ in K}$$

Answer: 12 ideal stages

5.13[b]. Steam-stripping of benzene in a sieve-plate column.

A straw oil used to absorb benzene from coke-oven gas is to be steam-stripped in a sieve-plate column at atmospheric pressure to recover the dissolved benzene, C_6H_6. Equilibrium conditions at the operating temperature are approximated by Henry's law such that, when the oil phase contains 10 mole % benzene, the equilibrium benzene partial pressure above the oil is 5.07 kPa. The oil may be considered nonvolatile. It enters the stripper containing 8 mole% benzene, 75% of which is to be removed. The steam leaving contains 3 mole% C_6H_6.

(a) How many theorethical stages are required?

Answer: 2.82 ideal stages

(b) How many moles of steam are required per 100 moles of the oil-benzene mixture?

Answer: 194 moles

(c) If 85% of the benzene is to be recovered with the same steam and oil rates, how many theoretical stages are required?

Answer:5 ideal stages

5.14[b]. Relation between N and N_{tOG} for constant absorption factor.

With the help of the Kremser equation (5-3) and equation (5-24), derive the relation between N and N_{tOG} for constant absorption factor. Establish the condition for which $N = N_{tOG}$.

5.15[c,d]. Absorption of carbon disulfide in a random-packed tower.

Design a tower packed with 50-mm ceramic Hiflow rings for the carbon disulfide scrubber of Problem 5.11. Assume isothermal operation and use a liquid rate of 1.5 times the minimum and a gas-pressure drop not exceeding 175 Pa/m of packing. Calculate the tower diameter, packed height, and total gas-pressure drop. Assume that C_h for the packing is 1.0.

Answer: $D = 0.574$ m

5.16[b,d]. Absorption of sulfur dioxide in a random-packed tower.

It is desired to remove 90% of the sulfur dioxide in a flue gas stream at 298 K and 1 atm by countercurrent absorption with pure water at the same temperature, using a packed tower that is 0.7 m in diameter. The tower is packed with 35-mm plastic NORPAC rings. The average gas-pressure drop is 200 Pa/m. Equilibrium is described by Henry's law with $y_i = 8.4x_i$. If the liquid flow is adjusted so that the driving force $(y - y^*)$ is constant, calculate the height of the packed section. Flue gases usually contain less than 1 mole% of SO_2, an air pollutant regulated by law. Assume that the properties of the liquid are similar to those of pure water, and that the properties of the flue gases are similar to those of air.

Answer: 5.77 m

5.17[c]. Benzene vapor recovery system.

Benzene vapor in the gaseous effluent of an industrial process is scrubbed with a wash oil in a countercurrent packed absorber. The resulting benzene-wash-oil solution is then heated to 398 K and stripped in a tray tower, using steam as the stripping medium. The stripped wash oil is then cooled and recycled to the absorber. Some data relative to the operation follow:

Absorption:

> Benzene in entering gas = 1.0 mole%
>
> Operating pressure of absorber = 800 mm Hg
>
> Oil circulation rate = 2 m³/1,000 m³ of gas at STP
>
> Oil specific gravity = 0.88   Molecular weight = 260
>
> Henry's law constant = 0.095 at 293 K;
>
> $$= 0.130 \text{ at } 300 \text{ K}$$
>
> N_{tOG} = 5 transfer units

Stripping:

> Pressure = 1 atm   Steam at 1 atm, 398 K
>
> Henry's law constant = 3.08 at 398 K
>
> Number of equilibrium stages = 5

(a) In the winter, it is possible to cool the recycled oil to 293 K, at which temperature the absorber then operates. Under these conditions 72.0 kg of steam is used in the stripper per 1,000 m³ of gas at STP entering the absorber. Calculate the percent benzene recovery in the winter.

Answer: 92.5%

(b) In the summer it is impossible to cool the recycled wash oil to lower than 300 K with the available cooling water. Assuming that the absorber then operates at 300 K, with the same oil and steam rates, and that N_{tOG} and equilibrium stages remain the same, what summer recovery of benzene can be expected?

Answer: 86.7%

(c) If the oil rate cannot be increased, but the steam rate in the summer is increased by 50% over the winter value, what summer recovery of benzene can be expected?

Answer: 87.9%

5.18^b. Overall transfer units for cocurrent absorption.

  For dilute mixtures and when Henry's law applies, prove that the number of overall transfer units for cocurrent gas absorption in packed towers is given by

$$N_{tOG} = \frac{A}{A+1} \ln\left[\frac{y_1 - mx_1}{y_2 - mx_2}\right] \qquad (5\text{-}55)$$

where subscript 1 indicates the top (where gas and liquid enter) and subscript 2 indicates the bottom of the absorber.

5.19[c]. Absorption of germanium tetrachloride used for optical fibers.

Germanium tetrachloride ($GeCl_4$) and silicon tetrachloride ($SiCl_4$) are used in the production of optical fibers. Both chlorides are oxidized at high temperature and converted to glasslike particles. However, the $GeCl_4$ oxidation is quite incomplete and it is necessary to scrub the unreacted $GeCl_4$ from its air carrier in a packed column operating at 298 K and 1 atm with a dilute caustic solution. At these conditions, the dissolved $GeCl_4$ has no vapor pressure and mass transfer is controlled by the gas phase. Thus, the equilibrium curve is a straight line of zero slope. The entering gas flows at the rate of 23,850 kg/day of air containing 288 kg/day of $GeCl_4$. The air also contains 540 kg/day of Cl_2, which, when dissolved, also will have no vapor pressure.

It is desired to absorb at least 99% of both $GeCl_4$ and Cl_2 in an existing 0.75-m-diameter column that is packed to a height of 3.0 m with 13-mm ceramic Raschig rings. The liquid rate should be set so that the column operates at 75% of flooding. Because the solutions are very dilute, it can be assumed that both gases are absorbed independenttly.

Gas-phase mass-transfer coefficients for GeCl4 and Cl_2 can be estimated from the following empirical equations developed from experimental studies with 13-mm Raschig rings for liquid mass velocities between 0.68–2.0 kg/m^2-s (Shulman, 1971):

$$\frac{k_y}{G_{My}} = 1.195\left[\frac{d_s G_y}{\mu(1-\varepsilon_o)}\right]^{-0.36} Sc^{-2/3} \qquad (5\text{-}56)$$

$$\varepsilon_o = \varepsilon - h_L \qquad (5\text{-}57)$$

$$h_L = 0.0591 G_x^{0.331} \qquad (5\text{-}58)$$

$$a_h = \frac{28.01\left(808 G_y / \rho^{1/2}\right)^n}{G_x^{1.04}} \qquad (5\text{-}59)$$

$$n = 0.2323\,G_x - 0.3 \qquad\qquad (5\text{-}60)$$

where:

$d_s =$ equivalent packing diameter (0.01774 m for 13-mm ceramic Raschig rings)

$\rho =$ gas density, kg/m^3

G_x, G_y mass velocities, $kg/m^2\text{-}s$

For the two diffusing species, take $D_{GeCl4} = 0.06\ cm^2/s$; $D_{Cl2} = 013\ cm^2/s$. For the packing, $F_p = 580\ ft^{-1}$, $\varepsilon = 0.63$. Determine:

(a) Liquid flow rate, in kg/s.

Answer: 0.825 kg/s

(b) The percent absorption of $GeCl_4$ and Cl_2 based on the available 3.0 m of packing.

5.20[c,d]. Absorption of carbon disulfide in a structured-packed tower.

Redesign the absorber of Problem 5.15 using metal Montz B1-300 structured packing. The characteristics of this packing are (Seader and Henley, 1998):

$F_p = 33\ ft^{-1}$ $\qquad a = 300\ m^{-1}$ $\qquad \varepsilon = 0.93$ $\qquad C_h = 0.482$

$C_p = 0.295$ $\qquad C_L = 1.165$ $\qquad C_V = 0.422$

answer: $Z = 3.97$ m

5.21[c,d]. Absorption of ammonia in a random-packed tower.

It is desired to reduce the ammonia content of 0.05 m^3/s of an ammonia-air mixture (300 K and 1 atm) from 5.0 to 0.04% by volume by water scrubbing. There is available a 0.3-m-diameter tower packed with 25-mm ceramic Raschig rings to a depth of 3.5 m. Is the tower satisfactory, and if so, what water rate should be used? At 300 K, ammonia-water solutions follow Henry's law up to 5 mole% ammonia in the liquid, with $m = 1.414$.

Data

Liquid:

Density = 998 kg/m^3 $\qquad\qquad$ Viscosity = 0.8 cP

Diffusivity = $1.64 \times 10^{-5}\ cm^2/s$

Gas:

Viscosity $= 1.84 \times 10^{-5}$ Pa-s   Diffusivity $= 0.28$ cm^2/s

Answer: 4.07 kg/min

5.22^{c,d}. Absorption of sulfur dioxide in a structured-packed tower.

A tower packed with metal Montz B1-300 structured packing is to be designed to absorb SO$_2$ from air by scrubbing with water. The entering gas, at an SO$_2$-free flow rate of 37.44 moles/m^2-s of bed cross-sectional area, contains 20 mole% of SO$_2$. Pure water enters at a flow rate of 1976 moles/m^2-s of bed cross-sectional area. The exiting gas is to contain only 0.5 mole% SO$_2$. Assume that neither air nor water will transfer between the phases and that the tower operates isothermally at 2 atm and 303 K. Equilibrium data for solubility of SO$_2$ in water at 303 K and 1 atm have been fitted by least-squares to the equation (Seader and Henley, 1998):

$$y_i = 12.697x_i + 3148x_i^2 - 4.724 \times 10^5 x_i^3$$
$$+ 3.001 \times 10^7 x_i^4 - 6.524 \times 10^8 x_i^5 \qquad (5\text{-}61)$$

(a) Derive the following operating line equation for the absorber:

$$x = \frac{0.01895\left(\dfrac{y}{1-y}\right) - 9.5213 \times 10^{-5}}{1 + 0.01895\left(\dfrac{y}{1-y}\right)} \qquad (5\text{-}62)$$

(b) If the absorber is to process 1.0 m^3/s (at 2 atm and 303 K) of the entering gas, calculate the water flow rate, the tower diameter, and the gas-pressure drop per unit of packing height at the bottom of the absorber.

Answer: $\Delta P = 167$ Pa/m

(c) Calculate N_{tG}, H_{tG}, and the height of packing required.

Answer: $Z = 1.04$ m

Data:

Liquid:

Density $= 998$ kg/m^3 Viscosity $= 0.7$ cP

Diffusivity = 1.61×10^{-5} cm²/s

Gas:

Viscosity (in) = 1.75×10^{-5} Pa-s   Diffusivity = 0.067 cm²/s

Viscosity (out) = 1.87×10^{-5} Pa-s

Packing characteristics are given in Problem 5.20.

5.23[c,d]. Absorption of sulfur dioxide in a random-packed tower.

Repeat Problem 5.22 using 50-mm ceramic Pall rings as packing material. Compare your results to those of Problem 5.22.

5.24[c,d]. Isothermal absorption of methanol in a random-packed tower.

A system for recovering methanol from a solid product wet with methanol involves evaporation of the alcohol into a stream of inert gas, essentially nitrogen. In order to recover the methanol from the nitrogen, an absorption scheme involving washing the gas countercurrently with water in a packed tower is being considered. The resulting water-methanol solution is then to be distilled to recover the methanol.

The absorption tower will be filled with 50-mm ceramic Pall rings. Design for a gas-pressure drop not to exceed 400 Pa/m of packed depth. Assume cooling coils will allow isothermal operation at 300 K. The gas will enter the column at the rate of 1.0 m³/s at 300 K and 1 atm. The partial pressure of methanol in the inlet gas is 200 mm Hg (Sc_G = 0.783). The partial pressure of methanol in the outlet gas should not exceed 15 mm Hg. Pure water enters the tower at the rate of 0.50 kg/s at 300 K. Neglecting evaporation of water, calculate the diameter and packed depth of the absorber.

Solutions of methanol and water follow Wilson equation. The corresponding parameters can be found in Smith et al. (1996).

5.25[c,d]. Adiabatic absorption of methanol in a random-packed tower.

Consider the methanol absorber of Problem 5.24. Redesign the tower assuming that operation is adiabatic. Do not neglect evaporation of water.

Additional data:

Molar heat capacities of gases:

Gas	N_2	MeOH	H_2O
C_p, J/mole-K	29.14	52.34	33.58

Heat of Solution at 293 K ("International Critical Tables," vol. IV, p. 159):

x_A, mole fraction	0.05	0.10	0.15	0.20	0.25	0.30
$-\Delta H_S$, J/mole solution	341.7	597.8	767.6	872.3	914.1	916.4

The latent heat of vaporization of methanol at 293 K is 1,163 kJ/kg. Heat capacities of liquid solutions of methanol and water are available at "Chemical Engineers' Handbook," 5th ed., p. 3-136 (1973).

5.26[b,d]. Isothermal absorption of acetone in a random-packed tower.

Air at 300 K is used to dry a plastic sheet. The solvent wetting the plastic is acetone. At the end of the dryer, the air leaves containing 2.0 mole% acetone. The acetone is to be recovered by absorption with water in a packed tower. The gas composition is to be reduced to 0.05 mole% acetone. The absorption will be isothermal because of cooling coils inside the tower. For the conditions of the absorber, the equilibrium relationship is $y_i = 1.8x_i$. The rich gas enters the tower at a rate of 0.252 kg/s, and pure water enters the top at the rate of 0.34 kg/s. The tower is packed with 25-mm metal Hiflow rings. Determine the dimensions of the tower.

REFERENCES

Hachmuth, K. H., *Chem. Eng. Prog.,* **47**, 523, 621 (1951).

Holman, J. P., *Heat Transfer*, 7th ed., McGraw-Hill, New York (1990).

"International Critical Tables," vol. V, p. 213, McGraw-Hill, New York, (1929).

King, C. J., *Separation Processes,* McGraw-Hill, New York (1971).

Perry, R. H., and C. H. Chilton (eds.), *Chemical Engineers' Handbook,* 5th ed., McGraw-Hill, New York (1973).

Raal, J. D., and M. K. Khurani, *Can. J. Chem. Eng.,* **51**, 162 (1973).

Reid, R. C., J. M. Prausnitz, and B. E. Poling, *The Properties of Gases and Liquids*, 4th ed., McGraw-Hill, Boston (1987).

Seader, J. D., and E. J. Henley, *Separation Process Principles*, Wiley, New York (1998).

Shulman, H. L., et al., *AIChE J.*, **17**, 631 (1971).

Smith, J. M., H. C. Van Ness, and M. M. Abbott, *Introduction to Chemical Engineering Thermodynamics*, 5th ed., McGraw-Hill, New York (1996).

Taylor, R., and R. Krishna, *Multicomponent Mass Transfer*, Wiley, New York (1993).

Treybal, R. E., *Mass-Transfer Operations*, 3rd ed., McGraw-Hill, New York (1980).

6

Distillation

6.1 INTRODUCTION

Distillation is a method of separating the components of a solution which depends upon the distribution of the substances between a liquid and a gas phase, applied to cases where all components are present in both phases (Treybal, 1980). Instead of introducing a new substance into the mixture in order to provide the second phase, as is done in gas absorption or desorption, the new phase is created from the original solution by vaporization or condensation.

The advantages of distillation as a separation method are clear. In distillation the new phases differ from the original by their heat content, but heat is readily added or removed, although, of course, the cost of doing this must inevitably be considered. Absorption or desorption operations, on the other hand, which depend upon the introduction of a foreign substance, result in a new solution which in turn may have to be separated by one of the diffusional operations unless it happens that the new solution is directly useful.

During the first quarter of the twentieth century, the application of distillation expanded from a tool for enhancing the alcohol content of beverages into the prime separation technique in the chemical industry. During most of the century, distillation was by far the most widely used method for separating liquid mixtures of chemical components (Seader and Henley, 1998). Despite the emergence in recent years of many new separation techniques (e.g., membranes), distillation retains its position of supremacy among chemical engineering unit operations (Taylor and Krishna, 1993). Unfortunately, distillation is a very energy-intensive technique, especially when the relative volatility of the components being separated is low. This is a major drawback of distillation in our times, characterized by spiraling costs of energy.

Distillation is most frequently carried out in multitray columns, although packed columns have long been the preferred alternative when pressure drop is an important consideration. In recent years, the development of highly efficient structured packings has led to increased use of packed columns in distillation.

6.2 SINGLE-STAGE OPERATION—FLASH VAPORIZATION

Your objectives in studying this section are to be able to:

1. Define the concepts of flash vaporization and partial condensation.
2. Relate—through material and energy balances, and equilibrium consid-erations—operating temperature and pressure during flash vaporization to the fraction of the feed vaporized, and to the concentrations of the liquid and vapor products.
3. Calculate the heat requirements of flash vaporization.

Flash vaporization is a single-stage operation wherein a liquid mixture is par-tially vaporized, the vapor is allowed to come to equilibrium with the residual liquid, and the resulting vapor and liquid phases are separated and removed from the appa-ratus. It may be batch or continuous.

A typical flowsheet is shown schematically in Figure 6.1 for continuous oper-ation. The liquid feed is heated in a conventional tubular heat exchanger. The pres-sure is reduced, vapor forms at the expense of the liquid adiabatically, and the mix-ture is introduced into a vapor–liquid separating vessel. The liquid portion of the mixture leaves at the bottom of the separating vessel while the vapor rises and leaves at the top. The vapor may then pass to a condenser, not shown in the figure.

The product D—richer in the more volatile—is, in this case, entirely a vapor. The material and energy balances for the process are

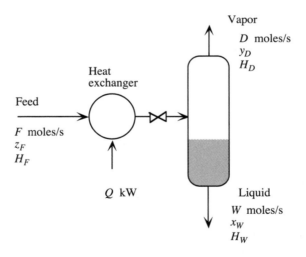

Figure 6.1
Schematic of the
continuous flash
vaporization process.

$$F = D + W \tag{6-1}$$

$$Fz_F = Dy_D + Wx_W \tag{6-2}$$

$$FH_F + Q = DH_D + WH_W \tag{6-3}$$

In equation (6-2), z_F is used to represent the feed composition because the feed could be a liquid or a vapor. Solved simultaneously, equations (6-1) to 6-3) yield

$$-\frac{W}{D} = \frac{y_D - z_F}{x_W - z_F} = \frac{H_D - (H_F + Q/F)}{H_W - (H_F + Q/F)} \tag{6-4}$$

The two left-hand members of equation (6-4) represent the usual single-stage operating line on distribution coordinates, of negative slope as for all single-stage cocurrent operations (see Chapter 3). It passes through compositions representing the influent and effluent streams, points F and M on Figure 6.2. If the effluent streams were in equilibrium, the device would be an ideal stage and the products would be on the equilibrium curve at N on Figure 6.2. In that case, for a given pressure in the separator and a specified degree of vaporization, y_D, x_W, and the separator temperature can be calculated solving simultaneously equation (6-4) and the *vapor/liquid equilibrium (VLE) relationships* for the system. Once the concentrations of the products are known, H_D and H_W can be evaluated. Since the enthalpy of the feed, H_F, is known, the two right-hand members of equation (6-4) can be used to determine the thermal load of the heat exchanger for the degree of vaporization of the feed specified.

The equations developed above apply equally well to the case where the feed is a vapor and Q, the heat removed in the heat exchanger to produce *partial condensation*, is taken as negative.

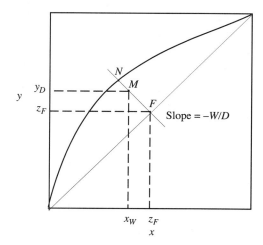

Figure 6.2 Flash vaporization process on the *xy* diagram.

Example 6.1 Flash Vaporization of a Heptane-Octane Mixture.

A liquid mixture containing 50 mole% n-heptane (A), 50 mole% n-octane (B), at
303 K, is to be continuously flash-vaporized at a pressure of 1 atm to vaporize 60
mole% of the feed. What will be the composition of the vapor and liquid and the
temperature in the separator if it behaves as an ideal stage? Calculate the amount of
heat to be added per mole of feed. n-Heptane and n-octane form ideal solutions.

Solution
The VLE relationship that applies under these conditions is Raoult's law. The vapor
pressure of the two components (P_j) is related to temperature through Wagner's
equation (Reid, et al., 1987)

$$\ln\left(\frac{P_j}{P_{c,j}}\right) = \frac{A_j \chi_j + B_j \chi_j^{1.5} + C_j \chi_j^3 + D_j \chi_j^6}{1 - \chi_j}$$

$$\chi_j = 1 - \frac{T}{T_{c,j}}$$

(6-5)

The parameters in equation (6-5) for n-heptane and n-octane are:

Component	T_c, K	P_c, bar	A	B	C	D
n-heptane	540.3	27.4	− 7.675	1.371	− 3.536	− 3.202
n-octane	568.8	24.9	− 7.912	1.380	− 3.804	− 4.501

Figure 6.3 shows a Mathcad computer program to perform flash vaporization calculations. Data supplied to the program include the parameters of the Wagner equation
for both components, the separator pressure, feed concentration, and the desired
fractional vaporization. Initial estimates are given to the program of the separator
temperature (320 K) and the vapor and liquid concentrations ($y_D = 0.6$, $x_W = 0.4$).
The "solve block" includes three relations between the variables: equation (6-4), and
Raoult's law for components A and B. Convergence is achieved at a separator temperature of 385.9 K, and equilibrium compositions of $y_D = 0.576$, $x_W = 0.386$.

The next step is to determine the heat required to vaporize 60% of the feed.
The feed is initially at 303 K and 1 atm. To determine its state of aggregation under
these conditions, estimate the bubble point of the feed. The Mathcad program of
Figure 6.3 can be used to determine both the bubble point (D approaching zero; T_{BP}
= 382.8 K) and the dew point of the mixture ($D = 1.0$; $T_{DP} = 387.9$ K). Therefore, at

Parameters of the Wagner equation:

$T_{cA} := 540.3$ K $P_{cA} := 27.4$ bar $T_{cB} := 568.8$ K $P_{cB} := 24.9$ bar

$A_A := -7.675$ $B_A := 1.371$ $C_A := -3.536$ $D_A := -3.202$

$A_B := -7.912$ $B_B := 1.380$ $C_B := -3.804$ $D_B := -4.501$

$$\chi_A(T) := 1 - \frac{T}{T_{cA}} \qquad \chi_B(T) := 1 - \frac{T}{T_{cB}} \qquad \text{bar} \equiv 10^5 \cdot \text{Pa}$$

$$P_A(T) := P_{cA} \cdot \exp\left[\frac{A_A \cdot \chi_A(T) + B_A \cdot \chi_A(T)^{1.5} + C_A \cdot \chi_A(T)^3 + D_A \cdot \chi_A(T)^6}{1 - \chi_A(T)}\right]$$

$$P_B(T) := P_{cB} \cdot \exp\left[\frac{A_B \cdot \chi_B(T) + B_B \cdot \chi_B(T)^{1.5} + C_B \cdot \chi_B(T)^3 + D_B \cdot \chi_B(T)^6}{1 - \chi_B(T)}\right]$$

$$m_A(T,P) := \frac{P_A(T)}{P} \qquad m_B(T,P) := \frac{P_B(T)}{P}$$

$z_F := 0.5$ $D := 0.6$ $W := 1 - D$ $P := 1 \cdot \text{atm}$

Initial estimates: $T := 320 \cdot K$ $y_D := 0.6$ $x_W := 0.4$

Given

$$\frac{-W}{D} = \frac{y_D - z_F}{x_W - z_F}$$

$$y_D = m_A(T,P) \cdot x_W \qquad 1 - y_D = m_B(T,P) \cdot (1 - x_W)$$

$$\begin{bmatrix} T \\ y_D \\ x_W \end{bmatrix} := \text{Find}\left(T, y_D, x_W\right) \qquad T = 385.891 \cdot K$$

$$y_D = 0.576 \qquad x_W = 0.386$$

Figure 6.3 Flash vaporization calculations with Mathcad.

303 K and 1 atm the feed is a subcooled liquid. Since n-heptane and n-octane form ideal solutions, molar liquid enthalpies with respect to a reference temperature T_0 can be calculated from

$$H_L = \left[x_A C_{L,A} + (1 - x_A) C_{L,B} \right] (T_L - T_0) \tag{6-6}$$

Molar enthalpies for vapors are from

$$H_G = \left[y_A C_{p,A} + (1 - y_A) C_{p,B} \right] (T_G - T_0) + y_A \lambda_A + (1 - y_A) \lambda_B \tag{6-7}$$

where λ_A and λ_B are the latent heats of vaporization at T_0.

The following data were obtained from "Chemical Engineers' Handbook" (1973):

Latent heats of vaporization at $T_0 = 298$ K:

$\lambda_A = 36.5$ kJ/mole $\lambda_B = 41.4$ kJ/mole

Heat capacities of liquids:

$C_{L,A} = 218$ J/mole-K (298 – 303 K) $C_{L,A} = 241$ J/mole-K (298 – 386 K)
$C_{L,B} = 253$ J/mole-K (298 - 303 K) $C_{L,B} = 268$ J/mole-K (298 - 386 K)

Heat capacities of gases, average in the range 298 – 386 K (Smith et al., 1996):

$C_{p,A} = 187$ J/mole-K $C_{p,B} = 247$ J/mole-K

Substituting in equation (6-6), $H_F = 1.18$ kJ/mole of feed, $H_W = 22.64$ kJ/mole of liquid residue. From equation (6-7), $H_D = 57.12$ kJ/mole of vapor. From equation (6-4), $Q = 42.15$ kJ of heat added/mole of feed.

Example 6.2 Flash Vaporization of a Ternary Mixture

A liquid containing 50 mole% benzene (A), 25 mole% toluene (B), and 25 mole% o-xylene (C) is flash-vaporized at 1 atm and 373 K. Compute the amounts of liquid and vapor products and their composition. These components form ideal mixtures. The vapor pressures of the three components at 373 K are: $P_A = 178.8$ kPa, $P_B = 73.6$ kPa, and $P_C = 26.3$ kPa. Therefore, for a total pressure of 1 atm and a temperature of 373 K, $m_A = 1.765, m_B = 0.727, m_C = 0.259$.

Solution
Equation (6-4) and Raoult's law apply for each of the three components in the mixture. Also, the mole fractions must add up to 1.0 in each of the two phases leaving the separator. The Mathcad program of Figure 6.3 is easily modified to include an

additional component and solve for D, W, and the vapor and liquid compositions (see Problem 6.3). The results are: $D = 0.297$, $y_{AD} = 0.719$, $y_{BD} = 0.198$, $y_{CD} = 0.083$, $x_{AW} = 0.408$, $x_{BW} = 0.272$, $x_{CW} = 0.320$.

6.3 BATCH DISTILLATION

Your objectives in studying this section are to be able to:

1. Understand the concept of batch distillation, and how it differs from flash vaporization.
2. Derive and apply *Rayleigh equation* to relate the compositions of the residue solution and composited distillate to the fraction of the feed batch-distilled.

Continuous distillation is a thermodynamically efficient method of producing large amounts of material of constant composition, as will be shown later. When small amounts of material of varying product composition are required, *batch distillation* has several advantages. In batch distillation a charge of feed is heated in a reboiler, and after a short startup period, product can be withdrawn from the top of the equipment. When the distillation is finished, the heat is shut off and the material left in the reboiler is removed. Then a new batch can be started. Usually, the distillate is the desired product.

Batch distillation is versatile; a run may last from a few hours to several days. Batch distillation is the choice when the plant does not run continuously and the batch must be completed in one or two shifts (8 to 16 hours). It is often used when the same equipment distills several different products at different times. If distillation is required only ocassionally, batch distillation would again be the choice. Equipment can be arranged in a variety of configurations. In simple batch distillation (Figure 6.4), the vapor is withdrawn continuously from the reboiler. The system differs from flash vaporization in that there is no continuous feed input and the liquid is drained only at the end of the batch.

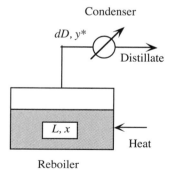

Figure 6.4 Schematic of the simple batch distillation process.

In the simple batch distillation process, the vapor product is in equilibrium with the liquid in the reboiler at any given time but changes continuously in composition. The mathematical approach must therefore be differential. Assume that at any time during the course of the distillation there are L moles of liquid in the still of composition x mole fraction of A and that an amount dD mole of distillate is vaporized, of mole fraction $y*$ in equilibrium with the liquid. Then we have the following differential material balances:

$$dL = -dD \qquad (6\text{-}8)$$

$$0 - y* \, dD = d(L\,x) = L\,dx + x\,dL \qquad (6\text{-}9)$$

Combining these equations

$$y* \, dL = L\,dx + x\,dL \qquad (6\text{-}10)$$

Rearranging and integrating

$$\int_W^F \frac{dL}{L} = \ln\frac{F}{W} = \int_{x_W}^{x_F} \frac{dx}{y*-x} \qquad (6\text{-}11)$$

where F is the moles of charge of composition x_F and W the moles of residual liquid of composition x_W. This is known as the *Rayleigh equation,* after Lord Rayleigh, who first derived it (Treybal, 1980). It can be used to determine F, W, x_F, or x_W when three of these are known. The *composited* distillate composition, $y_{D,av}$ can be determined by a simple material balance,

$$y_{D,av} = \frac{F x_F - W x_W}{D} \qquad (6\text{-}12)$$

Example 6.3 Batch Distillation of a Heptane-Octane Mixture

Suppose the liquid of Example 6.1 [50 mole% n-heptane (A), 50 mole% n-octane (B)] were subjected to a batch distillation at atmospheric pressure, with 60 mole% of the liquid distilled. Compute the composition of the composited distillate and the residue.

Solution
Take as basis for calculation $F = 100$ mole. Then, $D = 60$ moles, $W = 40$ moles, $x_F = 0.50$. Substituting in equation (6-11)

$$\ln\frac{100}{40} = 0.916 = \int_{x_W}^{0.50}\frac{dx}{y*-x} \tag{6-13}$$

The equilibrium-distribution data for this system can be generated using the Mathcad program of Figure 6.3, calculating the liquid composition ($x = x_W$) at the dew point ($D = 1.0$) for different feed compositions ($y^* = z_F$). For example, for $y^* = z_F = 0.5, x = 0.317$. The following table is generated in that manner.

y^*	0.5	0.55	0.60	0.65	0.686	0.70	0.75
x	0.317	0.361	0.409	0.460	0.500	0.516	0.577
$(y^* - x)^{-1}$	5.464	5.291	5.236	5.263	5.376	5.435	5.780

A cubic spline interpolation formula is generated with these data, which is then integrated according to equation (6-13) to determine the residue concentration, x_W, when 60% of the feed has been distilled. Figure 6.5 shows the details of the calculations.

$$x := \begin{bmatrix} .317 \\ .361 \\ .409 \\ .46 \\ .5 \\ .516 \\ .577 \end{bmatrix} \qquad f := \begin{bmatrix} 5.464 \\ 5.291 \\ 5.236 \\ 5.263 \\ 5.376 \\ 5.435 \\ 5.78 \end{bmatrix}$$

$$vs := cspline(x, f)$$

$$x_F := 0.5 \qquad F := 100 \qquad D := 60 \qquad W := F - D$$

$$I(x_W) := \int_{x_W}^{x_F} interp(vs, x, f, \theta)\, d\theta$$

Initial estimate $\qquad x_W := 0.4$

Given

$$\ln\left(\frac{F}{W}\right) = I(x_W)$$

$$x_W := Find(x_W) \qquad x_W = 0.327$$

$$y_{Dav} := \frac{F \cdot x_F - W \cdot x_W}{D} \qquad y_{Dav} = 0.616$$

Figure 6.5 Example 6.3: batch distillation calculations.

6.4 CONTINUOUS RECTIFICATION—BINARY SYSTEMS

Continuous distillation, or fractionation, is a multistage, countercurrent distillation operation. For a binary solution, with certain exceptions, it is ordinarily possible by this method to separate the solution into its components, recovering each in any state of purity desired.

Consider the general countercurrent, multistage, binary distillation operation shown in Figure 6.6. The operation consists of a column containing the equivalent of N theoretical stages arranged in a two-section cascade; a total condenser in which the overhead vapor leaving the top stage is totally condensed to give a liquid distillate product and liquid reflux that is returned to the top stage; a partial reboiler in which liquid from the bottom stage is partially vaporized to give a liquid bottoms product and vapor boilup that is returned to the bottom stage; and an intermediate feed stage.

The feed, which contains a more volatile component—the *light key*, LK—and a less volatile component—the *heavy key*, HK—enters the column at a feed stage, f. At the feed-stage pressure, the feed may be liquid, vapor, or a mixture of liquid and vapor, with its overall mole-fraction composition with respect to the light component denoted z_F. Vapor rising in the section above the feed (called the *enriching* or *rectifying* section) is washed with liquid to remove or absorb the heavy key. In the section below the feed (*stripping* or *exhausting* section), the liquid is stripped of the light key by the rising vapor. The mole fraction of the light key in the distillate is x_D, while the mole fraction of the light component in the bottoms product is x_B. Inside the tower, the liquids and vapors are always at their bubble points and dew points, respectively, so that the highest temperatures are at the bottom, the lowest at the top. The entire device is called a *fractionator*.

6.5 McCABE-THIELE METHOD FOR TRAYED TOWERS

Your objectives in studying this section are to be able to:

1. Understand the fundamentals of the McCabe-Thiele graphical method to analyze binary distillation in trayed towers.
2. Apply the McCabe-Thiele graphical method to determine the number of equilibrium stages required for a given separation, and the optimal location along the cascade for introduction of the feed.

McCabe and Thiele (1925) developed an approximate graphical method for combining the equilibrium-distribution curve for a binary system with operating-line curves for the rectifying and stripping sections of a fractionator to estimate, for a given feed mixture and column operating pressure, the number of equilibrium stages and the amount of reflux required for a desired degree of separation of the feed. The graphical nature of the McCabe-Thiele method greatly facilitates the visualization of many of the important aspects of multistage distillation, and therefore the effort required to learn the method is well justified (Seader and Henley, 1998).

The McCabe-Thiele method hinges upon the fact that, as an approximation, the operating lines on the xy diagram can be considered straight for each section of a fractionator between points of addition or withdrawal of streams. This is true only if the total molar flow rates of liquid and vapor do not vary from stage to stage in each section of the column. This is the case if:

1. The two components have equal and constant molar latent heats of vaporization.
2. Sensible enthalpy changes and heat of mixing are negligible compared to latent heats of vaporization.

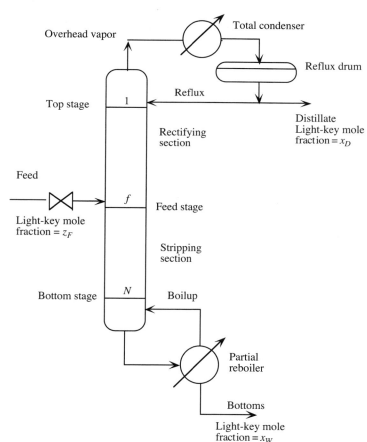

Figure 6.6 Fractionator with total condenser and partial reboiler.

3. Heat losses are negligible.
4. The pressure is uniform throughout the column (negligible pressure drop).

These assumptions are referred to as the *McCabe-Thiele assumptions* leading to the condition of *constant molar overflow*. For constant molar overflow, the analysis of a distillation column is greatly simplified because it is not necessary to consider energy balances in either the rectifying or stripping sections; only material balances and a VLE curve are required.

6.5.1 Rectifying Section

As shown in Figure 6.6, the rectifying section extends from the top stage, 1, to just above the feed stage, f. Consider a top portion of the rectifying stages, including the total condenser. A material balance for the light key over the envelope shown in Figure 6.7a for the total condenser and stages 1 to n is as follows:

$$y_{n+1} = \frac{L_n}{V_{n+1}} x_n + \frac{D}{V_{n+1}} x_D \qquad (6\text{-}14)$$

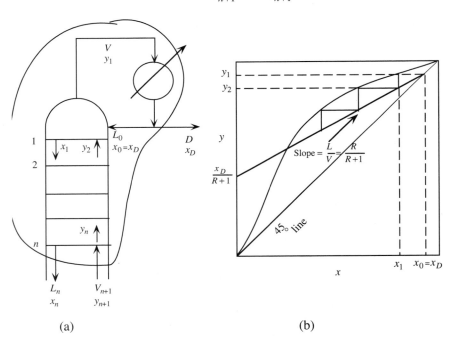

(a)   (b)

Figure 6.7 McCabe-Thiele operating line for the rectifying section.

For constant molar overflow, $V_{n+1} = V =$ constant; $L_n = L =$ constant and equation (6-14) can be written as

$$y_{n+1} = \frac{L}{V} x_n + \frac{D}{V} x_D \tag{6-15}$$

Equation (6-15) is the operating line for the rectifying section of the fractionator.

The liquid entering the top stage is the external reflux rate, L_0, and its ratio to the distillate rate, L_0/D, is the *reflux ratio, R.* for the case of a total condenser, with reflux returned to the column at its bubble point, $L_0 = L$ and $R = L/D$, a constant in the rectifying section. Since $V = L + D$

$$\frac{L}{V} = \frac{L}{L+D} = \frac{L/D}{L/D+1} = \frac{R}{R+1} \tag{6-16}$$

$$\frac{D}{V} = \frac{D}{L+D} = \frac{1}{L/D+1} = \frac{1}{R+1} \tag{6-17}$$

Combining equations (6-15), (6-16), and (6-17) produces the most useful form of the operating line for the rectifying section:

$$y = \left(\frac{R}{R+1}\right) x + \left(\frac{1}{R+1}\right) x_D \tag{6-18}$$

If values of R and x_D are specified, equation (6-18) plots as a straight line on the xy diagram with intersection at $y_1 = x_D$ on the 45° line, slope $= R/(R+1)$, and intersection at $y = x_D/(R+1)$ for $x = 0$, as shown in Figure 6.7b. The equilibrium stages are stepped off in the manner described in Chapter 3.

6.5.2 Stripping Section

As shown in Figure 6.6, the stripping section extends from the feed to the bottom stage. Consider a bottom portion of the stripping stages, including the partial reboiler. A material balance for the light key over the envelope shown in Figure 6.8a for the partial reboiler and stages N to $m + 1$ is as follows:

$$y_{m+1} = \frac{L_{st}}{V_{st}} x_m - \frac{W}{V_{st}} x_W \tag{6-19}$$

or

$$y = \frac{L_{st}}{V_{st}} x - \frac{W}{V_{st}} x_W \tag{6-20}$$

where L_{st} and V_{st} are the total molar flows in the stripping section which, by the constant-molar-overflow assumption, remain constant from stage to stage. The vapor leaving the partial reboiler is assumed to be in equilibrium with the liquid bottoms product. Thus, the partial reboiler acts as an additional equilibrium stage. The vapor rate leaving it is called the boilup, V_{st}, and its ratio to the bottoms product rate, $V_B = V_{st}/W$, is the *boilup ratio*. Since $L_{st} = V_{st} + W$,

$$\frac{L_{st}}{V_{st}} = \frac{V_{st} + W}{V_{st}} = \frac{V_B + 1}{V_B} \tag{6-21}$$

Similarly,

$$\frac{W}{V_{st}} = \frac{1}{V_B} \tag{6-22}$$

Combining equations (6-20), (6-21), and (6-22), the operating line for the stripping section becomes

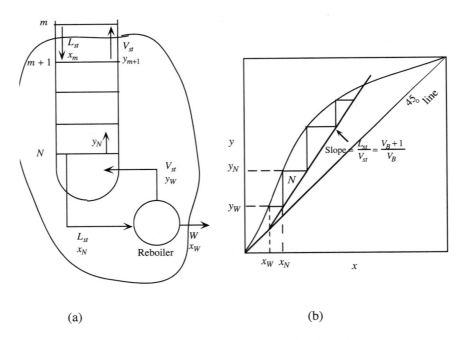

(a)   (b)

Figure 6.8 McCabe-Thiele operating line for the stripping section.

$$y = \left(\frac{V_B + 1}{V_B}\right) x - \frac{1}{V_B} x_W \qquad (6\text{-}23)$$

If values of V_B and x_W are known, equation (6-23) can be plotted on the xy diagram as a straight line with intersection at $y = x_W$ on the 45° line and slope $= (V_B + 1)/V_B$, as shown in Figure 6.8. The equilibrium stages are stepped off in a manner similar to that described for the rectifying section. The very last stage at the bottom corresponds to the partial reboiler.

6.5.3 Feed Stage

Thus far, the McCabe-Thiele construction has not considered the feed to the column. In determining the operating lines for the rectifying and stripping sections, it is very important to note that although x_D and x_W can be selected independently, R and V_B are related by the feed phase condition.

Consider the section of the column at the tray where the feed is introduced. The quantities of the liquid and vapor streams change abruptly at this tray since the feed may consist of liquid, vapor, or a mixture of both (fraction vaporized $= V_F/F$). If, for example, the feed is a saturated liquid, L_{st} will exceed L by the amount of the added feed liquid. To establish a general relationship, an overall material balance around the feed plate is

$$F + L + V_{st} = V + L_{st} \qquad (6\text{-}24)$$

and an energy balance,

$$F H_F + L H_{L,f-1} + V_{st} H_{G,f+1} = V H_{G,f} + L_{st} H_{L,f} \qquad (6\text{-}25)$$

The vapors and liquids inside the tower are all saturated, and the molal enthalpies of all saturated vapors at this section are esentially identical since the temperature and composition changes over one tray are small. Therefore, $H_{G,f+1} = H_{G,f}$. The same is true of the molal enthalpies of the saturated liquids, therefore, $H_{L,f-1} = H_{L,f}$. Equation (6-25) then becomes

$$\left(L_{st} - L\right) H_{L,f} = \left(V_{st} - V\right) H_{G,f} + F H_F \qquad (6\text{-}26)$$

Combining this with equation (6-24):

$$\frac{L_{st} - L}{F} = \frac{H_{G,f} - H_F}{H_{G,f} - H_{L,f}} = q \qquad (6\text{-}27)$$

Since there are only small composition changes across the feed plate, $H_{G,f}$ is basically the molal enthalpy that the feed would have if it were a saturated vapor, while $H_{L,f}$ is basically the molal enthalpy that the feed would have if it were a saturated liquid. Therefore, the quantity q in equation (6-27) is the energy required to convert 1 mole of feed from its condition H_F to a saturated vapor, divided by the molal latent heat of evaporation $(H_{G,f} - H_{L,f})$.

The feed may be introduced under five different thermal conditions ranging from a liquid well below its bubble point to a superheated vapor. For each condition, the value of q will be different, as the next table shows:

Feed condition	q-Value
Subcooled liquid	> 1
Bubble-point liquid	1
Partially vaporized	$L_F/F = 1 - V_F/F$
Dew-point vapor	0
Superheated vapor	< 0

Combining equations (6-24) and (6-27) yields

$$V_{st} = V + F(q - 1) \qquad (6\text{-}28)$$

which provides a convenient method for determining V_{st}.

The point of intersection of the two operating lines will help locate the exhausting-section operating line. This can be established as follows. Subtracting equation (6-20) from (6-15) gives

$$y(V - V_{st}) = (L - L_{st})x + Dx_D + Wx_W \qquad (6\text{-}29)$$

Further, by an overall material balance

$$Fz_F = Dx_D + Wx_W \qquad (6\text{-}30)$$

Combining equations (6-27) to (6-30) gives

$$y = \frac{q}{q-1}x - \frac{z_F}{q-1} \qquad\qquad (6\text{-}31)$$

Equation (6-31), representing the locus of intersection of operating lines (the q-line), is a straight line on the xy diagram of slope $= q/(q-1)$ and it passes through the point $x = y = z_F$ on the 45° line. Figure 6.9 shows the graphical interpretation of the q-line for typical cases. Here the operating-lines intersection is shown for a particular case of feed as a mixture of vapor and liquid. Following the placement of the rectifying section operating line and the q-line, the stripping-section operating line is located by drawing a straight line from the point $x = y = x_W$ on the 45° line to and through the point of intersection of the q-line and the rectifying-section operating line as shown in Figure 6.9. The point of intersection must lie somewhere between the equilibrium curve and the 45° line. It is clear from this analysis that, for a given feed condition, fixing the reflux ratio automatically establishes the liquid/vapor ratio in the stripping section and the reboiler heat load as well.

As q changes from a value greater than 1 (subcooled liquid) to a value less than 0 (superheated vapor) the slope of the q-line changes from positive to negative, and back to positive as shown in Figure 6.9. For a saturated liquid feed, the q-line is vertical; for a saturated vapor it is horizontal.

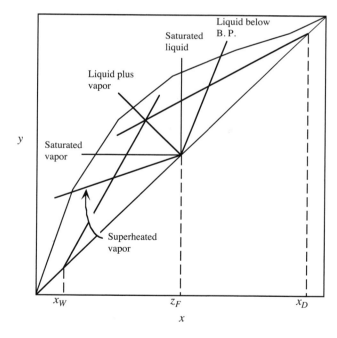

Figure 6.9 Location of the q-line for typical feed conditions.

6.5.4 Number of Equilibrium Stages and Feed-Stage Location

The q-line is useful in simplifying the graphical location of the stripping-section line, but the point of intersection of the two operating lines does not necessarily establish the demarcation between the stripping and rectifying sections of the fractionator. Rather, it is the introduction of the feed which governs the change from one operating line to the other. At least in the design of a new column some latitude in the introduction of the feed is available to the designer.

Consider the separation shown in Figure 6.10, for example. For a given feed, z_F and the q-line are fixed. For particular overhead and residue products, x_D and x_W are fixed. If the reflux ratio is specified, the location of the rectifying line is fixed and the stripping line must pass through the q-line at E. If the feed is introduced upon the fourth stage from the top (Figure 6.10a) the rectifying line is used for stages 1 through 3, and, beginning with stage 4, the stripping line must be used. The total number of ideal stages required is approximately 5.3, including the partial reboiler. If, on the other hand, the feed is introduced upon the sixth stage from the top (Figure 6.10b) the rectifying line is used for stages 1 through 5. The total number of ideal stages required this time is approximately 6.5. If the feed is introduced upon the third stage from the top (Figure 6.10c) the rectifying line is used for stages 1 and 2, and, beginning with stage 3, the stripping line must be used, for a total of approximately 6 stages. The least total number of trays will result if the steps on the diagram are kept as large as possible, a condition accomplished if the transition from one line to the other is made at the first opportunity after passing the operating-line intersection at E, as shown in Figure 6.10a. In the design of a new column, this is the practice to be followed.

In the adaptation of an existing column to a new separation, the point of introducing the feed is limited to the location of existing nozzles in the column wall. The slope of the operating lines and the product compositions to be realized must then be determined by trial and error, in order to obtain numbers of theoretical stages in the two sections of the fractionator consistent with the number of real trays in each section and the expected tray efficiency.

6.5.5 Limiting Conditions

For a given specification, a reflux ratio can be selected anywhere from the minimum, R_{min}, to an infinite value (total reflux) where all the overhead vapor is condensed and returned to the top stage (thus, no distillate is withdrawn!). The minimum reflux corresponds to an infinite number of stages, while an infinite reflux ratio corresponds to the minimum number of stages.

As the reflux ratio increases, the operating lines of both sections of the tower move toward the 45° line until, eventually, at total reflux they will coincide. Because the operating lines are located as far away as possible from the equilibrium curve, a minimum number of stages is required, as Figure 6.11 shows.

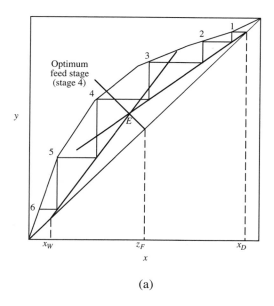

(a)

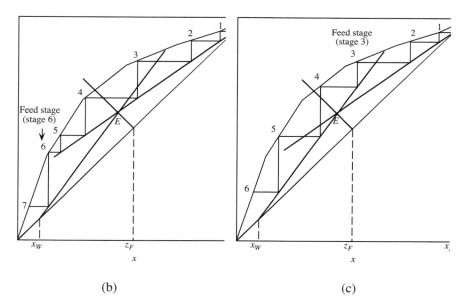

(b)   (c)

Figure 6.10 Location of feed stage: (a) optimum location; (b) location below optimum stage; (c) location above optimum stage.

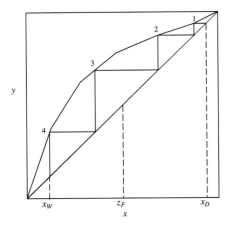

Figure 6.11 Total reflux and minimum stages.

As the reflux ratio decreases from the limiting case of total reflux, the intersection of the two operating lines and the q-line moves from 45° line toward the equilibrium curve. The number of ideal stages required increases because the operating lines move closer and closer to the equilibrium curve, thus requiring more and more steps to move from the top of the column to the bottom. Finally, a limiting condition is reached when the point of intersection is on the equilibrium curve, as shown in Figure 6.12. For binary mixtures that are not highly nonideal, the typical case is shown in Figure 6.12a where the intersection, P, is at the feed stage. To reach that stage from either the rectifying section or the stripping section, an infinite number of stages is required. The point P is called a *pinch point* because the two operating lines each pinch the equilibrium curve.

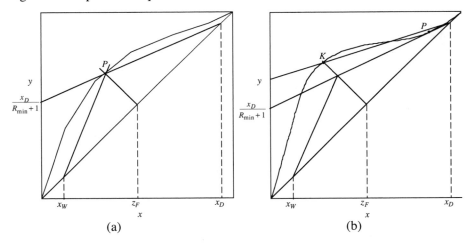

(a)   (b)

Figure 6.12 Minimum reflux ratio construction: (a) ideal or near ideal system; (b) nonideal system, pinch point above the feed stage.

For a highly nonideal binary mixture, the pinch may occur at a stage above or below the feed stage. A pinch point above the feed stage is illustrated in Figure 6.12b, where the operating line for the rectifying section is tangent to the equilibrium curve at point P, well before the feed stage is reached. The slope of this tangent operating line cannot be reduced any further because it would then cross over the equilibrium curve, as shown on Figure 6.12b for the operating line through point K. The line through K clearly represents too small a reflux ratio. Because of the interdependence of the liquid/vapor ratios in the two sections of the column, a tangent operating line in the exhausting section may also set the minimum reflux ratio (see Problem 6.9).

6.5.6 Optimum Reflux Ratio

Any reflux ratio between the minimum and infinity will provide the desired separation, with the corresponding number of theoretical trays required varying from infinity to the minimum number. The reflux ratio to be used for a new design should be the optimum, the one for which the total cost of the operation will be the least. At the minimum reflux ratio the column requires an infinite number of stages and, consequently, the fixed cost is infinite, but the operating costs (heat for the reboiler, condenser cooling water, power for the reflux pump) are least. As R increases, the number of trays rapidly decreases, but the column diameter increases owing to the larger quantities of recycled liquid and vapor per unit quantity of feed. The condenser, reboiler, and reflux pump must also be larger. The fixed costs therefore fall through a minimum value and rise to infinity again at total reflux. On the other hand, the operating costs increase almost directly with reflux ratio. The total cost, which is the sum of the fixed and operating costs, must therefore pass through a minimum at the optimum reflux ratio, as Figure 6.13 shows. This frequently occurs at a value of $R = R_{opt}$ in the range of $1.2R_{min}$ to $1.5R_{min}$ (Treybal, 1980).

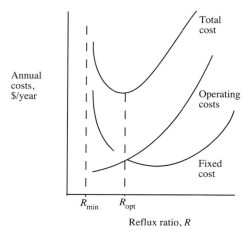

Figure 6.13 Optimum reflux ratio.

Example 6.4 Rectification of a Benzene-Toluene Mixture

A trayed tower operating at 1 atm is to be designed to continuously distill 200 kmoles/hour (55.6 moles/s) of a binary mixture of 60 mole% benzene, 40 mole% toluene. A liquid distillate and a liquid bottoms product of 95 mole% and 5 mole% benzene, respectively, are to be produced. Before entering the column, the feed—originally at 298 K—is flash-vaporized at 1 atm to produce an equimolal vapor/liquid mixture ($V_F/F = L_F/F = 0.5$). A reflux ratio 30% above the minimum is specified. Calculate: (a) quantity of the products; (b) minimum number of theoretical stages, N_{min}; (c) minimum reflux ratio; (d) number of equilibrium stages and the optimal location of the feed stage for the reflux ratio specified; and (e) thermal load of the condenser, reboiler, and feed preheater.

Solution

(a) Calculate D and W. An overall material balance on benzene gives

$$0.60(200) = 0.95D + 0.05W$$

A total balance gives

$$200 = D + W$$

Combining these two balances gives $D = 122.2$ kmoles/hr, $W = 77.8$ kmoles/hr.

(b) Benzene and toluene form ideal solutions, therefore the VLE data for this system at 1 atm is generated in the manner illustrated in Example 6.1. The parameters in equation (6-5) for benzenene and toluenene are (Reid, et al., 1987):

Component	T_c, K	P_c, bar	A	B	C	D
Benzene	562.2	48.9	− 6.983	1.332	− 2.629	− 3.333
Toluene	591.8	41.0	− 7.286	1.381	−2.834	− 2.792

Applying Raoult's law at 1 atm, the following VLE data are generated:

x	0.10	0.20	0.30	0.40	0.50	0.60	0.70	0.80	0.90
y*	0.21	0.37	0.51	0.64	0.72	0.79	0.86	0.91	0.96
T	379.4	375.5	371.7	368.4	365.1	362.6	359.8	357.7	355.3

In Figure 6.14, where y and x refer to benzene—the light key—with $x_D = 0.95$ and $x_W = 0.05$, the minimum number of equilibrium stages is stepped off between the equilibrium curve and the 45° line, starting from the top, giving $N_{min} = 6.7$ stages.

(c) Calculate the slope of the q-line: For this example, $q = L_F/F = 0.5$. From equation (6-31), the slope of the q-line $= q/(q − 1) = − 1.0$. In Figure 6.14, a q-line is drawn

that has a slope of − 1.0 and passes through the point $x = y = z_F = 0.6$ on the 45°
line. For the minimum reflux ratio, an operating line for the rectifying section passes
through the point $x = y = x_D = 0.95$ on the 45° line and through the point of intersection of the q-line and the equilibrium curve, to a y-intercept of 0.457. Then,

$$\frac{x_D}{R_{min} + 1} = \frac{0.95}{R_{min} + 1} = 0.457$$

Therefore, $R_{min} = 1.079$ moles reflux/mole distillate.

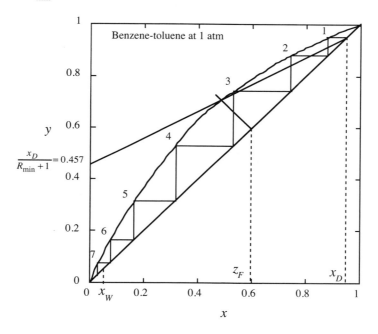

Figure 6.14 Determination of minimum stages and minimum reflux ratio for
Example 6.4.

(d) For $R = 1.3R_{min} = 1.403$, the y-intercept of the rectifying-section operating line is
$(0.95)/(1.403 + 1) = 0.395$. Figure 6.15 shows the location of both operating lines.
The operating line for the stripping section is drawn to pass through the point $x = y$
$= x_W = 0.05$ on the 45° line and the point of intersection of the q-line and the rectifying-section operating line. The number of equilibrium stages is stepped off between
first, the rectifying-section operating line and the equilibrium curve, and then, the
equilibrium curve and the stripping-section operating line. For the optimal feed-
stage location, the transition from one operating line to the other occurs at the first
opportunity after passing the operating-line intersection. Therefore, Figure 6.15
shows that the feed is to be introduced on the sixth ideal stage from the top. A total

of 13 equilibrium stages, including the reboiler, is required, and the tower must then contain 12 ideal stages.

(e) Calculate the molal flows of liquid and vapor throughout the column:

$$L = L_0 = RD = (1.403)(122.2) = 171.4 \text{ kmoles/hr}$$
$$V = L + D = 171.4 + 122.2 = 293.6 \text{ kmoles/hr}$$

From equation (6-27),

$$L_{st} = L + qF = 171.4 + (0.5)(200) = 271.4 \text{ kmoles/hr}$$

From equation (6-28),

$$V_{st} = V + (q - 1)F = 293.6 - (0.5)(200) = 193.6 \text{ kmoles/hr}$$

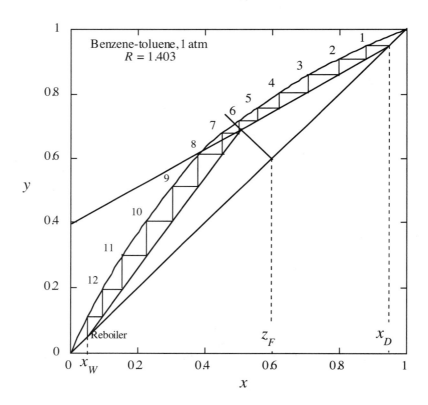

Figure 6.15 Determination of number of equilibrium stages and feed-stage location for Example 6.4.

Consider the feed preheater, a flash-vaporization unit, shown schematically in Figure 6.16. For 50% vaporization of the feed ($z_F = 0.60$), from calculations similar to those illustrated in Example 6.1, the separator temperature is $T_F = 365.6$ K and the equilibrium compositions are $y_F = 0.707$, $x_F = 0.493$. Since benzene (A) and toluene (B) form ideal solutions, molar liquid and vapor enthalpies with respect to a reference temperature T_0 can be calculated from equations (6-6) and (6-7). The following data were obtained from Perry and Chilton (1973):

Latent heats of vaporization at $T_0 = 298$ K:

$$\lambda_A = 33.9 \text{ kJ/mole} \qquad \lambda_B = 38.0 \text{ kJ/mole}$$

Heat capacities of liquids (298 – 366 K):

$$C_{L,A} = 0.147 \text{ kJ/mole-K} \qquad C_{L,B} = 0.174 \text{ kJ/mole-K}$$

Heat capacities of gases, average in the range 298 – 366 K (Smith et al., 1996):

$$C_{p,A} = 0.094 \text{ kJ/mole-K} \qquad C_{p,B} = 0.118 \text{ kJ/mole-K}$$

Substituting in equation (6-6), $H_F = 0$,
$$H_{LF} = [x_F C_{L,A} + (1 - x_F)C_{L,B}](T_F - T_0) = 10.86 \text{ kJ/mole of liquid feed}$$

From equation (6-7),

$$H_{VF} = [y_F C_{p,A} + (1 - y_F)C_{p,B}](T_F - T_0) + y_F \lambda_A + (1 - y_F)\lambda_B$$
$$= 41.93 \text{ kJ/mole of vapor feed}$$

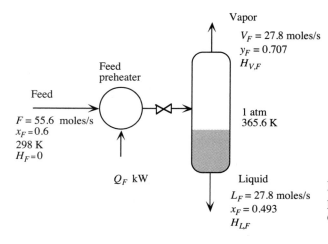

Figure 6.16 Feed preheater for Example 6.4.

From equation (6-3),

$$Q_F = V_F H_{VF} + L_F H_{LF} - F H_F = 1,450 \text{ kW}$$

Consider now the total condenser. Saturated vapor containing 95 mole% benzene at a dew-point temperature of 355.8 K will enter the condenser at the rate of 293.6 kmoles/hr (81.6 moles/s). It will leave the condenser as saturated liquid of the same composition and a bubble-point temperature of 354.3 K. An energy balance around the condenser is as follows:

$$Q_c = V(H_{V,1} - H_{L,0})$$

From equation (6-7),

$$H_{V,1} = [y_1 C_{p,A} + (1 - y_1) C_{p,B}](T_1 - T_0) + y_1 \lambda_A + (1 - y_1) \lambda_B$$
$$= 40.0 \text{ kJ/mole of vapor}$$

From equation (6-6),

$$H_{L,0} = [x_0 C_{L,A} + (1 - x_0) C_{L,B}](T_{L,0} - T_0)$$
$$= 8.3 \text{ kJ/mole of liquid}$$
$$Q_c = V(H_{V,1} - H_{L,0}) = 2,590 \text{ kW}$$

Consider now the partial reboiler shown schematically in Figure 6.17. Saturated liquid leaving the last equilibrium stage in the tower enters the reboiler at a rate of 271.4 kmoles/hr (75.4 moles/s). Saturated vapor leaves the reboiler and returns to the column at the rate of 193.6 kmoles/hr (53.8 moles/s), while the liquid residue is withdrawn as the bottoms product at the rate of 77.8 kmoles/hr (21.6 moles/s). The bottoms product is a saturated liquid with a composition of 5 mole % benzene. A flash-vaporization calculation is done in which the fraction vaporized is known ($53.8/75.4 = 0.714$) and the concentration of the liquid residue is fixed at $x_W = 0.05$. The calculations yield the following results: $T_R = 381.6$ K, $x_{12} = 0.093$, $y_{13} = 0.111$. The liquid entering the reboiler is at its bubble point, which is $T_{12} = 379.7$ K. An energy balance around the reboiler is

$$Q_R = V_{st} H_{V,13} + W H_{L,W} - L_{st} H_{L,12}$$

From equation (6-7),

$$H_{V,13} = [y_{13} C_{p,A} + (1 - y_{13}) C_{p,B}](T_R - T_0) + y_{13} \lambda_A + (1 - y_{13}) \lambda_B$$
$$= 47.19 \text{ kJ/mole of vapor feed}$$

From equation (6-6),

$$H_{L,12} = [x_{12} C_{L,A} + (1 - x_{12}) C_{L,B}](T_{12} - T_0)$$
$$= 14.01 \text{ kJ/mole of liquid}$$
$$H_{L,W} = [x_W C_{L,A} + (1 - x_W) C_{L,B}](T_R - T_0)$$

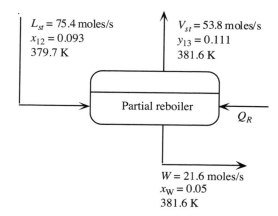

$L_{st} = 75.4$ moles/s
$x_{12} = 0.093$
379.7 K

$V_{st} = 53.8$ moles/s
$y_{13} = 0.111$
381.6 K

Partial reboiler

Q_R

$W = 21.6$ moles/s
$x_W = 0.05$
381.6 K

Figure 6.17 Partial reboiler for Example 6.4.

$$= 14.43 \text{ kJ/mole of liquid}$$
$$Q_R = (53.8)(47.19) + (21.6)(14.43) - (75.4)(14.01) = 1,790 \text{ kW}$$

6.5.7 Large Number of Stages

The McCabe-Thiele graphical construction is difficult to apply when conditions of *relative volatility* and/or product purities are such that a large number of stages must be stepped off. In that event, one of the following techniques can be used to determine the stage requirements.

1. Separate plots of expanded scales and/or larger dimensions are used for stepping off stages at the ends of the *xy* diagram. For example, the additional plots may cover just the regions (1) 0.95 to 1.0, and (2) 0 to 0.05.
2. The stages are determined by combining the McCabe-Thiele graphical construction, for a suitable region in the middle, with the Kremser equations for the low and/or high ends, where absorption and stripping factors are almost constant.
3. If the equilibrium data are given in analytical form, a McCabe-Thiele computer program can be used. Appendix F is an example of Mathcad programs to implement the McCabe-Thiele method (Hwalek, 2001). They generate the required VLE data from the Antoine equation for vapor pressure, and the NRTL equation for liquid-phase activity coefficients. Appendix F-1 is for column feed as saturated liquid; Appendix F-2 is for column feed as saturated vapor.

Example 6.5 Rectification of a Methanol-Water Solution

A methanol (A)–water (B) solution containing 36 mole% methanol is to be continuously rectified at 1 atm pressure at a rate of 216.8 kmoles/hr (60.22 moles/s) to provide a distillate containing 99.9 mole% methanol and a residue containing 0.1 mole% methanol. The feed to the column is to be preheated to its bubble point. The distillate is to be totally condensed and the reflux returned at the bubble point. A reflux ratio of 1.5 times the minimum will be used. Determine the number of theoretical stages required if the feed is introduced at the optimal location:

(a) using the Mathcad program of Appendix F-1,

(b) combining the McCabe-Thiele graphical method with the Kremser equations.

Solution
(a) The Antoine constants for methanol (1) and water (2) are (Smith, et al., 1996): A_1 = 16.5938, B_1 = 3,644.3 K, C_1 = 239.76 K; A_2 = 16.2620, B_2 = 3799.89 K, C_2 = 226.35 K. The NRTL constants are b_{12} = 253.88 cal/mole, b_{21} = 845.21 cal/mole, α = 0.2994 (Smith, et al., 1996). Table 6.1 presents the VLE data for the system generated by the computer program in Appendix F-1.

Table 6.1 VLE for the System Methanol-Water at 1 atm

CH$_3$OH mole fraction in liquid, x	CH$_3$OH mole fraction in gas, y^*	Temp (K)
0.000	0.000	373.1
0.100	0.416	361.2
0.200	0.576	355.3
0.300	0.667	351.5
0.400	0.731	348.8
0.500	0.784	346.5
0.600	0.832	344.5
0.700	0.876	342.7
0.800	0.919	340.9
0.900	0.960	339.3
1.000	1.000	337.7

Source: Generated by the computer program in Appendix F-1 (Hwalek, 2001).

A summary of the results obtained in Appendix F-1 for $R = 1.5\, R_{min}$ is as follows:
Distillate flow rate = 21.66 moles/s   Bottoms flow rate = 38.56 moles/s
Minimum reflux ratio = 0.84   Reflux ratio = 1.26
N = 22 ideal stages   Feed introduced in stage 16 from top
L = 27.29 moles/s V = 48.95 moles/s
L_{st} = 87.51 moles/s $V_{st} = V$ = 48.95 moles/s

(b) Figure 6.18 shows the McCabe-Thiele construction for the region of x from 0.017 to 0.961, where the stages have been stepped off in two directions starting from the feed stage. Notice that, because of the high purity of the products, the operating lines originate basically from the corners of the xy diagram where the 45° line ends. In this middle region, eight stages are stepped off above the feed stage and four below it, for a total of 13 stages including the feed stage. The Kremser equations can now be applied to determine the remaining stages needed to achieve the desired high purities for the distillate and bottoms.

Let us consider now the number of additional ideal stages required in the stripping section (N_S) to achieve the desired bottoms concentration $x_W = 0.001$. This portion of the column behaves like a methanol stripper with very dilute solutions of methanol in water. From the construction in Figure 6.18, the liquid entering the "stripper" contains 1.7 mole% of methanol, while the liquid leaving the section contains only 0.1 mole% methanol. Since the operating line for the stripping section of the distillation column ends on the 45° line at $y = x = x_W$, the "stripper" operates as if the vapor entering it had a concentration $y = x_W = 0.001$. To calculate the absorption factor for the "stripper," we must estimate the slope of the equilibrium curve in

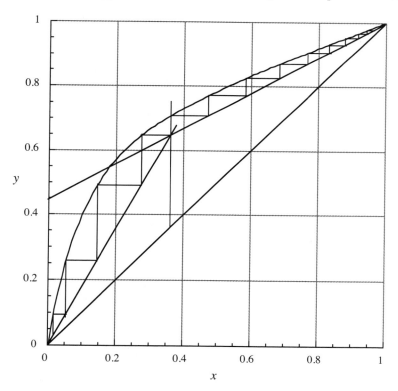

Figure 6.18 McCabe-Thiele construction for Example 6.5, from $x = 0.017$ to $x = 0.961$.

the limit as x_A tends to zero (m_{st}). In that portion of the column, the temperature is very close to the normal boiling point of pure water, then $T_{st} = 373.1$ K. From the modified Raoult's law,

$$m_{st} = \frac{P_A(T_{st})\gamma_A^\infty(T_{st})}{P} \qquad (6\text{-}32)$$

The infinite-dilution activity coefficients for liquid solutions described by the NRTL equation are given by the equations (Smith, et al., 1996)

$$\ln\gamma_A^\infty(T) = \tau_{21}(T) + \tau_{12}(T)\exp\left[-\alpha\,\tau_{12}(T)\right] \qquad (6\text{-}33)$$

$$\ln\gamma_B^\infty(T) = \tau_{12}(T) + \tau_{21}(T)\exp\left[-\alpha\,\tau_{21}(T)\right] \qquad (6\text{-}34)$$

where
$$\tau_{12}(T) = \frac{b_{12}}{RT} \qquad \tau_{21}(T) = \frac{b_{21}}{RT} \qquad (6\text{-}35)$$

Substituting in equations (6-32) to (6-35), $m_{st} = 7.452$. Then, the absorption factor for the "stripper" is $A_{st} = L_{st}/(m_{st}V_{st}) = 87.51/(7.542 \times 48.95) = 0.24$. Substituting in equation (5-1), $N_S = 1.9$ stages.

Let us consider now the number of additional ideal stages required in the rectifying section (N_R) to achieve the desired distillate concentration $x_D = 0.001$. This portion of the column behaves like a water absorber with very dilute solutions of water in methanol. From the construction in Figure 6.18, the vapor entering the "absorber" contains 2.3 mole% water (97.7 mole% methanol), while the vapor leaving it contains only 0.1 mole% water. The liquid entering it has a concentration $x = x_D = 0.001$. To calculate the absorption factor for the "absorber," we must estimate the slope of the equilibrium curve in the limit as x_B tends to zero (m_{ab}). In that portion of the column, the temperature is very close to the normal boiling point of pure methanol, then $T_{ab} = 337.7$ K. From the modified Raoult's law,

$$m_{ab} = \frac{P_B(T_{ab})\gamma_B^\infty(T_{ab})}{P} \qquad (6\text{-}36)$$

Substituting in equations (6-34) to (6-36), $m_{ab} = 0.393$. Then, the absorption factor for the "absorber" is $A_{ab} = L/(m_{ab}V) = 27.29/(0.393 \times 48.95) = 1.418$. Substituting in equation (5-3), $N_R = 7.0$ stages.

Combining these results with those obtained by the McCabe-Thiele graphical construction for the intermediate portion of the column, we have $8 + 7 = 15$ stages above the feed stage, and $4 + 1.9 = 5.9$ stages below the feed stage, for a total of 21.9 ideal stages, and feed introduced in stage 16 from the top. These results agree with those obtained in part a using the computer program in Appendix F-1.

6.5.8 Use of Open Steam

Ordinarily, heat is applied at the base of the fractionator by means of a reboiler. However, when an aqueous solution is distilled to give the nonaqueous solute as the distillate and the water is removed as the bottoms product, the heat required may be provided by the use of open steam at the bottom of the tower. The reboiler is then dispensed with. For a given reflux ratio and overhead composition, however, more trays will be required in the tower.

Figure 6.19 shows the effects of using open steam instead of a reboiler. Overall material balances are

$$F + V_{st} = D + W \tag{6-37}$$

$$F z_F = D x_D + W x_W \tag{6-30}$$

where V_{st} is the molal flow of steam used, assumed saturated at the fractionator pressure. The rectifying operating line is located as usual, and the slope of the stripping line, L_{st}/V_{st}, is related to L/V and the feed conditions in the same manner as before. However, the stripping line now ends at the point $y = 0, x = x_W$, as shown in Figure

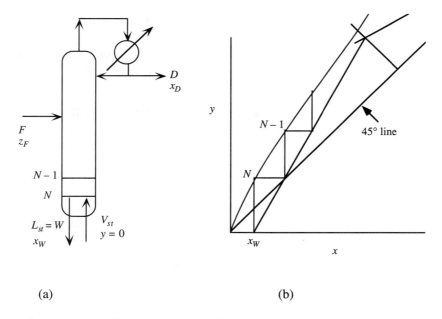

(a)   (b)

Figure 6.19 Use of open steam instead of a reboiler.

6.19b. The graphical tray construction must therefore be continued to the x axis of the diagram.

If the steam entering the tower, V_{N+1}, is superheated, it will vaporize liquid on tray N to the extent necessary to bring it to saturation. Then,

$$V_{st} = V_{N+1}\left[1 + \frac{H_{G,N+1} - H_{G,sat}}{\lambda_B}\right]$$

$$L_{st} = V_{st} - V_{N+1} + W \qquad\qquad (6\text{-}38)$$

where $H_{G,sat}$ is the molar enthalpy of steam saturated at the column pressure, and λ_B is the molar heat of vaporization of water at the saturation temperature corresponding to the column pressure.

6.5.9 Tray Efficiencies

Methods for estimating tray efficiencies were discussed in Chapter 4. Murphree vapor efficiencies are most simply used graphically on the xy diagram. Overall efficiencies $\mathbf{E}_O$ strictly have meaning only when the Murphree efficiency of all trays is the same, and the equilibrium and operating line are both straight over the concentration range considered. Nevertheless, empirical correlations to estimate $\mathbf{E}_O$ are easy to use and are useful for rough estimates, if not for final designs.

The most widely used empirical approach to estimate $\mathbf{E}_O$ is the O'Connell correlation, although it has been pointed out that it usually underpredicts the overall efficiency in the distillation of water solutions (Wankat, 1988; Kister, 1992):

$$\mathbf{E}_O = 0.52782 - 0.27511\log_{10}(\alpha\mu_L) + 0.044923\left[\log_{10}(\alpha\mu_L)\right]^2$$

$$0.1\text{ cP } \leq \alpha\mu_L \leq 10.0\text{ cP} \qquad\qquad (6\text{-}39)$$

where:

μ_L = viscosity of the feed as liquid at the average temperature of the tower, expressed in cP

α = average relative volatility. The relative volatility is defined, for a binary system, as the ratio of the equilibrium concentration ratio of A and B in one phase to that in the other; then

$$\alpha = \frac{y^*(1-x)}{x(1-y^*)} \qquad\qquad (6\text{-}40)$$

Example 6.6 Rectification of an Ethanol-Water Solution

In the development of a new process, it will be necessary to fractionate 34.43 kmoles/hr of an ethanol-water solution containing 30 mole% ethanol, available at the bubble point. It is desired to recover 99.9% of the ethanol in a distillate containing 80 mole% ethanol. The tower will be designed for operation at 1 atm, with a reflux ratio twice the minimum. Open steam at 1 atm and 523 K is available for heating. Calculate:

(a) Product rates, kmoles/hr, and rate of steam use, kg/hr.

(b) Number of theoretical trays required.

(c) For a sieve-tray tower of "conventional design," determine the tower diameter for the gas velocity not to exceed 70% of the flooding velocity.

d) Estimate the number of real trays combining the Murphree efficiency and the McCabe-Thiele method, and calculate the total tower height.

Solution

(a) Table 6.2 shows the VLE data for the system ethanol-water at 1 atm. A minimum-boiling azeotropic mixture is formed at 89.43 mole% ethanol. The distillate concentration of 80 mole% is safely below the azeotropic composition. To recover

Table 6.2 VLE for the System Ethanol-Water at 1 atm

Ethanol mole fraction in liquid, x	Ethanol mole fraction in gas, $y*$	Temp (K)
0.000	0.000	373.10
0.0721	0.3891	362.10
0.1238	0.4704	358.40
0.2377	0.5445	355.80
0.2608	0.5580	355.40
0.3965	0.6122	353.80
0.5198	0.6599	352.80
0.5732	0.6841	352.40
0.6763	0.7385	351.84
0.7472	0.7815	351.51
0.8943	0.8943	351.25
1.000	1.000	351.40

Source: Wankat, 1988.

99.9% of the alcohol in the overhead product,

$$D = \frac{0.999 F z_F}{x_D} = \frac{0.999 \times 34.43 \times 0.3}{0.8} = 12.90 \text{ kmoles / hr}$$

Figure 6.20 is the xy diagram for this example. Since the feed is a saturated liquid, the q-line is vertical ($q = 1.0$). The pinch point that determines the minimum reflux ratio is above the feed stage. From the y-intercept of the pinch-point tangent rectifying operating line, $x_D/(R_{min} + 1) = 0.387$, $R_{min} = 1.067$. For a reflux ratio that is twice the minimum, $R = 2.134$. Calculate the molal flows of liquid and vapor throughout the column:

$$L = L_0 = RD = (2.134)(12.90) = 27.53 \text{ kmoles/hr}$$
$$V = L + D = 27.53 + 12.90 = 40.43 \text{ kmoles/hr}$$
$$L_{st} = L + qF = 27.53 + (1.0)(34.43) = 61.96 \text{ kmoles/hr}$$
$$V_{st} = V + (q - 1)F = V = 40.43 \text{ kmoles/hr}$$

From the Steam Tables (Smith, et al., 1996), the enthalpy of saturated steam at 101.3 kPa is $H_{G,sat} = 2,676$ kJ/kg; the enthalpy of superheated steam at 101.3 kPa and 523 K is $H_{G,N+1} = 2825.8$ kJ/kg; the latent heat of vaporization of water at 373.1 K is $\lambda_B = 2,256.9$ kJ/kg. Substituting in equation (6-38), the flow rate of open steam required is $V_{N+1} = 37.93$ kmoles/hr (or 682.74 kg/hr), and the flow rate of the bottoms product is $W = 59.46$ kmoles/hr. Since 0.1% of the ethanol in the feed will be lost in the bottoms product

$$x_W = \frac{0.001 F z_F}{W} = \frac{0.001 \times 34.43 \times 0.3}{59.46} = 1.74 \times 10^{-4}$$

(b) For $R = 2.134$, the rectifying operating line y-intercept in Figure 6.20 is $x_D/(R + 1) = 0.255$. The operating line for the stripping section, on this scale of plot, for all practical purposes passes through the origin. Ideal stages are stepped off as usual starting at the top of the column. The feed should be introduced at ideal stage number 9 from the top. Below ideal stage 10, Kremser equations should be used to determine the additional number of trays required to reach the bottoms concentration, N_S. From Figure 6.20, the concentration of the liquid leaving ideal stage 10 is $x = 0.025$, and from this concentration down. to $x_W = 1.74 \times 10^{-4}$ the equilibrium curve is esentially straight ($m = 8.95$). The absorption factor $A = L_{st}/mV_{st} = 61.96/[(8.95)(40.43)]$ $= 0.171$. When open steam is used, the concentration of the entering vapor is $y = 0$. Substituting in equation (5-1), $N_S = 2.71$ stages. Therefore, the total number of ideal stages required is 12.71.

(c) For a single-feed, two-product column, there is generally a need to carry out the column sizing calculations for the top tray, bottom tray, tray just above the feed, and tray just below the feed. The column is then designed for the more severe conditions (Kister, 1992). Consider first the bottom tray. In this part of the tower both the liquid and the vapor are virtually pure water ($M_G = M_L = 18$). Therefore, the liquid mass flow rate is $L_{st}' = (61.96)(18)/3{,}600 = 0.3098$ kg/s; the vapor mass flow rate is $V_{st}' = 0.2022$ kg/s. The gas density, ρ_G, is from the ideal gas law,

$$\rho_G = \frac{PM_G}{RT} = \frac{101.3 \times 18}{8.314 \times 373.1} = 0.588 \text{ kg / m}^3$$

The density of the liquid is from the Steam Tables, $\rho_L = 958$ kg/m³. The viscosity of the vapor is $\mu_G = 1.28 \times 10^{-5}$ kg/m-s (Holman, 1990). The foaming factor for aqueous solutions of alcohols is close to 1.0 (Kister, 1992). The surface tension of liquid water at 293 K is 72.8 dyn/cm. It can be corrected to a different temperature from

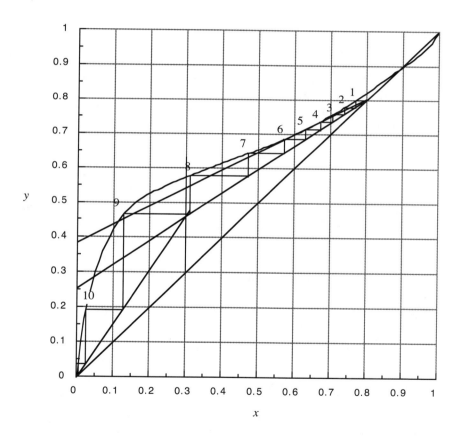

Figure 6.20 Ideal stages for the fractionator of Example 6.6.

(Reid, et al., 1987)

$$\sigma_2 = \sigma_1 \left[\frac{1 - T_{r2}}{1 - T_{r1}} \right]^{1.22} \tag{6-41}$$

For water, $T_c = 647.3$ K, therefore, the surface tension at 373.1 K is

$$\sigma = 72.8 \left[\frac{1 - \left(\dfrac{293}{647.3} \right)}{1 - \left(\dfrac{373.1}{647.3} \right)} \right]^{1.22} = 53.3 \text{ dyn / cm}$$

The slope of the equilibrium curve is $m = 8.95$. The gas-phase diffusivity is from the Wilke-Lee equation, $D_G = 0.177$ cm²/s. The liquid-phase diffusivity is from the Hayduk-Minhas correlation for aqueous solutions, $D_L = 5.54 \times 10^{-5}$ cm²/s. Take $d_o = 4.5$ mm on an equilateral-triangular pitch 12.0 mm between hole centers, punched in stainless steel sheet metal 2 mm thick. Use a weir height of 50 mm. Design for a 70% approach to the flood velocity. From the Mathcad program in Appendix E we get the following results for conditions at the bottom tray:

$$D = 0.423 \text{ m} \qquad t = 0.5 \text{ m} \qquad E_{MGE} = 0.567$$

Consider now conditions at the top tray. Both the liquid entering the tray and the vapor leaving it contain 80 mole% ethanol. The temperature there is around 351.4 K (see Table 6.2). The average molecular weight of both streams is 40.5 kg/kmole. Therefore, the liquid mass flow rate is $L' = (27.53)(40.5)/3{,}600 = 0.310$ kg/s; the vapor mass flow rate is $V' = (40.43)(40.5)/3{,}600 = 0.455$ kg/s. For the conditions at the top tray, the following data are available from Wankat (1988):

$$\rho_G = 1.393 \text{ kg/m}^3 \qquad \rho_L = 772 \text{ kg/m}^3 \qquad \sigma = 18.2 \text{ dyn/cm}$$
$$m = 0.63 \qquad D_L = 3.64 \times 10^{-5} \text{ cm}^2/\text{s}$$

From the Lucas method: $\mu_G = 1.04 \times 10^{-5}$ kg/m-s. From the Wilke-Lee equation, $D_G = 0.157$ cm²/s. From the Mathcad program in Appendix E we get the following results for conditions at the top tray:

$$D = 0.60 \text{ m} \qquad t = 0.5 \text{ m} \qquad E_{MGE} = 0.80$$
$$\Delta P = 336 \text{ Pa/tray} \qquad Fr_o = 1.03 \qquad E = 0.0193$$

Similar calculations for the trays just above, and just below the feed tray show that the conditions at the top tray govern the design of the tower. Therefore, a tower diameter of 0.6 m is specified, with tray spacing of 0.5 m.

(d) The Murphree vapor efficiency corrected for entrainment remains relatively constant at $E_{MGE} = 0.80$ for the rectifying section. For the stripping section, the efficiency is recalculated for the specified tower diameter $D = 0.60$ m. For this diameter, flow conditions at the bottom tray correspond to only a 34% approach to flooding, and $E_{MGE} = 0.78$. Therefore, it is appropriate to assume that the Murphree vapor efficiency for this problem remains basically constant throughout the column at an average value of $E_{MGE} = 0.79$.

Figure 6.21 shows the xy diagram for this problem in which a pseudoequilibrium curve (dotted-line) has been added such that the vertical distance from the operating lines to the pseudoequilibrium curve is 79% of the distance from the operating lines to the original equilibrium curve. Real stages are stepped off between the operating lines and the pseudoequilibrium curve. Feed is introduced in tray 11 from the top. The liquid leaving tray 13 contains 0.025 mole fraction of ethanol. It was shown in part b above that 2.71 additional ideal stages are required to reduce the liquid concentration from 0.025 mole fraction to $x_W = 1.74 \times 10^{-4}$. In that part of the column, $E_{MGE} = 0.78$, $A = 0.171$. The overall efficiency for that part of the column can be estimated from equation (5-5):

$$E_O = \frac{\ln\left[1 + E_{MGE}\left(A^{-1} - 1\right)\right]}{\ln\left(A^{-1}\right)} = 0.886$$

Therefore, the number of additional real stages is 2.71/0.886 = 3.06 stages, which must be rounded up to the next integer, in this case, 4 additional stages. A total of 17 real stages is required, with the feed introduced in stage 11 from the top. In this case, all 17 stages must be in the column since open steam is used instead of a partial reboiler. For a tray spacing of 0.5 m, the total height between the bottom and top tray is 8.5 m. Add 1 m above the top tray for entrainment separation, andadd 3 m beneath the bottom tray for bottoms surge capacity (Seader and Henley, 1998). The total column height is 12.5 m.

Example 6.7 Overall Efficiency of a Benzene-Toluene Fractionator

A binary distillation operation separates 78.1 moles/s of a mixture of 46 mole% benzene and 54 mole% toluene. The purpose of the sieve-tray column (equivalent to 20 theoretical stages and a partial reboiler) is to separate the feed into a liquid distillate of 99 mole% benzene and a liquid bottoms product of 98 mole% toluene. The pres-

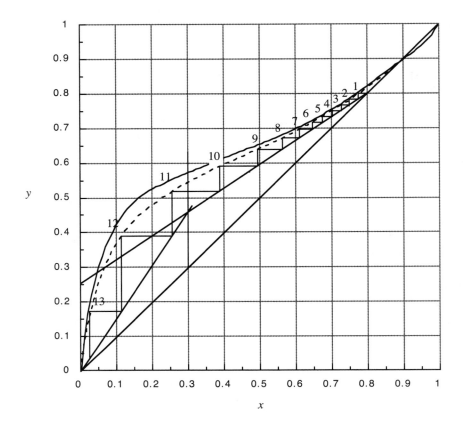

Figure 6.21 Real stages for the fractionator of Example 6.6.

sure in the reboiler is 141 kPa. In this range of pressure, benzene and toluene form ideal solutions with a relative volatility of 2.26 at the bottom tray (395 K) and 2.52 at the top tray (360 K). The feed to the column is a mixture of vapor and liquid; 23.4 mole % of the feed is vaporized. A reflux ratio 30% above the minimum is used. The diameter of the column is 1.53 m, which corresponds to 84% of flooding at the top tray, 81% of flooding at the bottom tray, and an average gas-pressure drop of 700 Pa/tray.

(a) Estimate the overall tray efficiency using the O'Connell correlation.

(b) Calculate the number of real trays and the total tower height.

(c) Estimate the total gas-pressure drop through the column.

Solution

(a) The average temperature in the column is $(365 + 390)/2 = 377.5$ K. The average relative volatility is $\alpha = (2.26 + 2.52)/2 = 2.39$. The viscosity of an ideal binary solution can be estimated in terms of the viscosities of the pure components at the solution temperature and the solution composition from (Reid, et al., 1987)

$$\mu_L = \mu_A^{x_A} \mu_B^{x_B} \tag{6-41}$$

The viscosity of liquid benzene and liquid toluene as functions of temperature are given by (Reid, et al., 1987)

$$\ln \mu = A + \frac{B}{T} + CT + DT^2$$

μ in cP; T in K

$$\tag{6-42}$$

where:

Component	A	B	C	$D \times 10^5$
Benzene	4.612	148.9	−0.0254	2.222
Toluene	−5.878	1,287	0.00458	− 0.450

Source: Reid, et al., 1987.

At the average column temperature of 377.5 K, $\mu_A = 0.242$ cP, $\mu_B = 0.251$ cP, $\mu_L = 0.247$ cP. Then, $\alpha \mu_L = 0.5903$; $\log_{10}(\alpha \mu_L) = -0.229$. From the O'Connell correlation, $E_O = 0.593$.

(b) Number of real trays = number of ideal trays/E_O = 33.7, or call it 34 trays. For a tower diameter of 1.53 m, Table 4.3 recommends a tray spacing of 0.6 m. Adding 1 m over the top tray as entrainment separator and 3 m beneath the bottom tray for bottoms surge capacity, the total column height is $4 + (34)(0.6) = 24.4$ m.

(c) Total gas-pressure drop = (0.700 kPa/tray)(34 trays) = 23.8 kPa.

6.6 BINARY DISTILLATION IN PACKED TOWERS

Your objectives in studying this section are to be able to:

1. Understand the advantages of using packed versus trayed towers for certain distillation applications.
2. Combine material-balances information from the McCabe-Thiele graphical method with interphase mass-transfer considerations to determine the packed height of the rectifying and stripping sections of a continuous-contact binary distillation tower.

With the availability of economical and efficient packings, packed towers are finding increasing use in new distillation processes and for retrofitting existing trayed towers. They are particularly useful in applications where pressure drop must be low, as in low-pressure distillation, and where liquid holdup must be small, such as when distilling heat-sensitive materials whose exposure to high temperatures must be minimized.

As in the case of packed absorbers, the changes in concentration with height produced by these towers are continuous rather than stepwise as for tray towers, and the computation procedure must take this into consideration. Figure 6.22a shows a schematic diagram of a packed-tower fractionator. Like tray towers, it must be provided with a reboiler at the bottom (or open steam may be used if an aqueous residue is produced), a condenser, means of returning reflux and reboiled vapor, as well as means for introducing feed. The last can be accomplished by providing a short unpacked section at the feed entry, with adequate distribution of liquid over the top of the exhausting section.

The operating diagram, Figure 6.22b, is determined exactly as for tray towers using the McCabe-Thiele method. Equations for operating lines already derived for trays are also applicable, except that tray-number subscripts are omitted. The operating lines are then simply the relation between x and y, the bulk liquid and gas compositions, prevailing at each horizontal section of the tower. As before, the change from rectifying- to stripping-section operating lines is made at the point where the feed is actually introduced, and for new designs a shorter column results, for a given reflux ratio, if this is done at the intersection of the operating lines. In what follows, this practice is assumed.

By material balance over an incremental section of packed height in the rectifying section, assuming equimolar counterdiffusion ($N_B = -N_A$)

$$Vdy = k'_y a_h (y_i - y)Sdz = Ldx = k'_x a_h (x - x_i)Sdz \qquad (6\text{-}43)$$

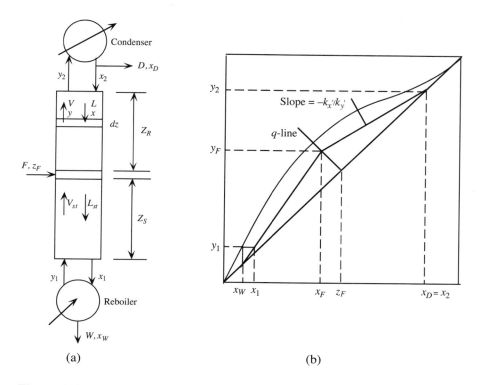

Figure 6.22 Binary distillation in a packed tower.

Integrating over the entire rectifying section

$$Z_R = \int_{y_F}^{y_2} \frac{V dy}{k_y' a_h S(y_i - y)} = \int_{x_F}^{x_D} \frac{L dx}{k_x' a_h S(x - x_i)} \qquad (6\text{-}44)$$

Integrating over the stripping section

$$Z_S = \int_{y_1}^{y_F} \frac{V_{st}\, dy}{k_y' a_h S(y_i - y)} = \int_{x_1}^{x_F} \frac{L_{st}\, dx}{k_x' a_h S(x - x_i)} \qquad (6\text{-}45)$$

For any point (x, y) on the operating lines, the corresponding point (x_i, y_i) on the equilibrium curve is obtained at the intersection with a line of slope $-k_x'/k_y'$ drawn from (x, y), as shown in Figure 6.22b. For $k_x' > k_y'$, so that the principal resistance to mass transfer lies within the vapor, $(y_i - y)$ is more accurately read than $(x -$

x_i). The middle integral of equations (6-44) and (6-45) is then best used. For $k_x' < k_y'$, it is better to use the last integral. In this manner, variations in V, L, the mass-transfer coefficients, and the interfacial area with location on the operating lines are readily dealt with.

Example 6.8 Benzene-Toluene Fractionator Using a Structured Packed Tower

Determine suitable dimensions of packed sections of a tower for the benzene-toluene separation of Example 6.4, using Montz B1-300 metal structured packing. Design for a gas-pressure drop not to exceed 400 Pa/m. The hydraulic and mass-transfer characteristics for this packing are (Seader and Henley, 1998):

$$F_p = 33 \text{ ft}^{-1} \quad a = 300 \text{ m}^{-1} \quad \varepsilon = 0.930 \quad C_h = 0.482$$
$$C_p = 0.295 \quad C_L = 1.165 \quad C_V = 0.422$$

Solution
From Example 6.4, $L = 171.4$ kmole/hr, $V = 293.6$ kmole/hr, $L_{st} = 271.4$ kmole/hr, and $V_{st} = 193.6$ kmole/hr. Vapor and liquid mass-flow rates throughout the tower are:

Rectifying Section

x	T_L, K	y	T_G, K	M_G	V', kg/s	M_L	L', kg/s
0.900	355.4	0.922	357.1	79.2	6.459	79.5	3.785
0.800	357.6	0.861	359.8	80.0	6.524	80.9	3.852
0.700	360.0	0.802	362.1	80.9	6.598	82.3	3.918
0.600	362.6	0.743	364.3	81.7	6.663	83.7	3.985
0.505	365.2	0.691	366.1	82.4	6.720	85.0	4.047

Stripping Section

x	T_L, K	y	T_G, K	M_G	V_{st}', kg/s	M_L	L_{st}', kg/s
0.505	365.2	0.691	366.1	82.4	4.431	85.0	6.408
0.400	368.4	0.5401	370.9	84.5	4.544	86.5	6.521
0.300	371.7	0.401	374.4	86.5	4.652	87.9	6.627
0.200	375.4	0.261	378.2	88.4	4.802	89.3	6.732
0.100	379.4	0.120	381.4	90.4	4.862	90.7	6.838

The x and y values are from the operating lines (Figure 6.15). The temperatures are bubble and dew points for the liquids and vapors, respectively. The operating lines intersect at $x = x_F = 0.505$, $y = y_F = 0.691$, the dividing point between the rectifying and stripping sections.

The tower diameter will be set by the conditions at the bottom of the rectifying section, right above the feed entrance, because of the large vapor flow at this point. The gas and liquid properties at this point are estimated as follows. The gas density is from the ideal gas law

$$\rho_G = \frac{PM_G}{RT_G} = \frac{101.3 \times 82.4}{8.314 \times 366.1} = 2.742 \text{ kg / m}^3$$

Since benzene and toluene form virtually ideal solutions, to estimate the liquid density only the molar fractions and the densities of the pure components as liquids at the solution temperature and pressure are required. Use the Rackett equation to estimate the molar volumes of the pure components as saturated liquids (Smith, et al., 1996):

$$V^{sat} = V_c Z_c^{(1-T_r)^{0.2857}} \tag{6-46}$$

The values predicted by equation (6-46) can be used, with no further correction for pressure, as good estimates of the molar volumes of the pure components at the solution temperature and pressure. Then, $V_A = 98.4$ cm^3/mole; $V_B = 114.5$ cm^3/mole;

$$V_L = (0.505)(98.4) + (0.495)(114.5) = 106.4 \text{ cm}^3/\text{mole}$$
$$\rho_L = M_L/V_L = 799.4 \text{ kg/m}^3$$

The liquid viscosity is estimated from equations (6-41) and (6-42), $\mu_L = 0.274$ cP. The gas viscosity is from the method of Lucas, $\mu_G = 8.92 \times 10^{-6}$ kg/m-s. Gas diffusivity is from the Wilke-Lee equation, $D_G = 0.0456$ cm^2/s. The liquid diffusivity is from equation (1-56), with $\Gamma = 1.0$, using the corresponding Hayduk-Minhas correlation to estimate the infinite dilution diffusion coefficients: $D_L = 5.43 \times 10^{-5}$ cm^2/s.

The Mathcad program in Appendix D is used to determine the tower diameter that satisfies the gas-pressure drop criterion at the bottom of the rectifying section. The result is $D = 1.41$ m, which corresponds to a 67% approach to flooding. The effective specific interfacial area is $a_h = 130.2$ m^2/m^3. The mass-transfer coefficients are $k_x' = 3.705$ moles/m^2-s; $k_y' = 2.125$ moles/m^2-s.

To determine the conditions at the interface for this point in the tower, solve simultaneously the equilibrium relationship (Raoult's law, in this case) and the interface mass-transfer condition given by an equation similar to (3-10):

$$\frac{y - y_i}{x - x_i} = -\frac{k_x'}{k_y'} \tag{6-47}$$

$$y_i = \frac{P_A(T)x_i}{P} \qquad 1 - y_i = \frac{P_B(T)(1 - x_i)}{P} \tag{6-48}$$

These equations are easily solved using the "solve block" capabilities of Mathcad®. The solution is $T = 365.5$ K, $x_i = 0.495$, $y_i = 0.709$.

Similarly, values at other concentrations in the rectifying section are:

y	k_y' mole/m²-s	k_x' mole/m²-s	a_h m⁻¹	y_i	$[k_y' a_h(y_i-y)]^{-1}$ m³-s/mole	z m	dP/dz Pa/m
0.691	2.125	3.705	130.2	0.709	0.201	0	400
0.743	2.119	3.7461	128.7	0.777	0.108	0.393	393
0.802	2.115	3.804	126.8	0.842	0.093	0.691	385
0.861	2.107	3.853	125.1	0.900	0.097	0.981	376
0.922	2.095	3.914	123.4	0.952	0.131	1.335	368
0.950	2.097	3.944	122.5	0.975	0.158	1.546	364

The height, z, for a given gas composition along the rectifying section is calculated from equation (6-44), written here for constant molar overflow as

$$z = \int_{y_F}^{y} \frac{V \, dy}{k_y' a_h S(y_i - y)} = G_{My} \int_{y_F}^{y} \frac{dy}{k_y' a_h(y_i - y)}$$

$$Z_R = G_{My} \int_{y_F}^{y_2} \frac{dy}{k_y' a_h(y_i - y)} \tag{6-46}$$

For the rectifying section, $G_{My} = V/(\pi D^2/4) = 52.25$ moles/m²-s. The integrals in equation (6-46) are estimated numerically using Mathcad, as Example 5.3 illustrated while calculating N_{tG}. The total height of the rectifying section is $Z_R = 1.546$ m. The gas-pressure drop per unit packed height (dP/dz), estimated at different positions along the section, is integrated in a similar manner to yield the total pressure drop for the rectifying section, $\Delta P_R = 591$ Pa.

As Figure 6.17 shows, conditions at the bottom of the tower are: $y_1 = 0.111$, $x_1 = 0.093$. This is one extreme of the stripping-section operating line. The other extreme is at the point right underneath the introduction of the feed: $x = x_F = 0.505$, $y = y_F = 0.691$. For the rectifying section, $G_{My} = V_{st}/(\pi D^2/4) = 34.45$ moles/m²-s. Calculations of mass-transfer coefficients, interfacial area, interfacial gas and liquid concentrations, and gas-pressure drop at different points on the stripping-section operating line yield the following results:

y	k_y' mole/m^2-s	k_x' mole/m^2-s	a_h m^{-1}	y_i	$[k_y'a_h(y_i-y)]^{-1}$ m^3-s/mole	z m	dP/dz Pa/m
0.111	1.601	4.192	170.1	0.160	0.098	0	240
0.120	1.602	4.197	170.1	0.172	0.093	0.030	240
0.261	1.608	4.151	167.4	0.333	0.068	0.395	234
0.401	1.587	4.113	165.2	0.475	0.068	0.723	221
0.540	1.577	4.079	162.8	0.599	0.087	1.069	212
0.691	1.565	4.032	160.2	0.711	0.261	1.869	202

The total height of the stripping section is $Z_S = 1.859$ m; the total pressure drop for the rectifying section is $\Delta P_S = 409$ Pa. Therefore, the total packed height is $Z = Z_R + Z_S = 3.415$ m. The total gas-pressure drop is $\Delta P = \Delta P_R + \Delta P_S = 1.0$ kPa.

6.7 INTRODUCTION TO MULTICOMPONENT DISTILLATION

Your objectives in studying this section are to be able to:

1. Explain why analysis of multicomponent distillation problems is always by trial-and-error.
2. Make appropriate assumptions and solve the overall material balances.

Many of the distillations of industry involve more than two components. While the principles established for binary solutions generally apply to such distillations, new problems of design are introduced which require special consideration.

An important principle to be emphasized is that a single fractionator cannot separate more than one component in reasonably pure form from a multicomponent solution, and that a total of $C - 1$ fractionators will be required for complete separation of a system of C components. Consider, for example, the continuous separation of a ternary solution consisting of components A, B, and C whose relative volatilities are in that order (A most volatile). In order to obtain the three substances in substantially pure form, the following two-column scheme can be used. The first column is used to separate C as a residue from the rest of the solution. This residue is necessarily contaminated with a small amount of B and an even smaller amount of A. The distillate, which is necessarily contaminated with a small amount of C, is then fractionated in the second column to give nearly pure A and B.

Another significant difference between multicomponent and binary distillation problems arises from a degree-of-freedom analysis around the column (Wankat, 1988). Assuming constant pressure and negligible heat losses in the column, the number of degrees of freedom is $C + 6$. For ternary distillation, then, there are 9 variables that can be specified by the designer; the most likely are listed in Table 6.3.

Table 6.3 Specified Design Variables; Ternary Distillation

Number of variables	Variable
1	Feed rate, F
2	Feed composition, z_1, z_2
1	Feed quality, q (or H_F or T_F)
1	Distillate, $x_{1,D}$ (or D or one fractional recovery)
1	Bottoms, $x_{2,W}$ (or W or one fractional recovery)
1	Reflux ratio
1	Saturated liquid reflux or T_0
1	Optimum feed plate
9	Total

Source: Wankat (1988).

In multicomponent distillation neither the distillate nor the bottoms composition is completely specified because there are not enough degrees of freedom to allow complete complete specification. This inability to completely specify the distillate and bottoms compositions has major effects on the calculation procedure. The components that do have their distillate and bottoms fractional recoveries specified (such as component 1 in the distillate and component 2 in the bottoms in Table 6.3) are called *key components*. The most volatile of the keys is called the *light key* (LK), and the least volatile the *heavy key* (HK). The other components are *non-keys* (NK). If a non-key is more volatile than the light key, it is a *light non-key* (LNK); if it is less volatile than the heavy key, it is a *heavy non-key* (HNK).

Consider the overall mass balances around a ternary distillation column. They are:

$$F z_i = W x_{i,W} + D x_{i,D} \qquad i = 1,2,3 \qquad (6\text{-}47)$$

$$\sum_{i=1}^{3} x_{i,D} = 1.0 \qquad \sum_{i=1}^{3} x_{i,W} = 1.0 \qquad (6\text{-}48)$$

The unknowns are six: D, W, $x_{2,D}$, $x_{3,D}$, $x_{1,W}$, and $x_{3,W}$. There are only five independent equations. Additional equations (energy balances and equilibrium expressions) always add additional variables, so we cannot start out by solving the overall mass and energy balances.

Can we do the internal stage-by-stage calculations first and then solve the overall balances? To begin the stage-by-stage calculation procedure in a distillation column, we need to know all the compositions at one end of the column. For ternary systems with the variables specified as in Table 6.3, these compositions are unknown. To begin the analysis we would have to assume that one of them is known. Therefore, internal calculations for multicomponent distillation problems are necessarily trial-and-error. In a ternary system, once an additional composition is assumed, both the overall and internal calculations are easily done. The results can then be compared and the assumed composition modified as needed until convergence is achieved.

Fortunately, in many cases it is easy to make an excellent first guess. If a sharp separation of the keys is required, then almost all of the heavy non-keys will appear only in the bottoms, and almost all of the light non-keys will appear only in the distillate. If there are only light non-keys or only heavy non-keys, then an accurate first guess of compositions can be made.

Example 6.9 Overall Mass Balances Using Fractional Recoveries

We wish to distill 2,000 kmoles/hr of a saturated liquid feed of composition 45.6 mole% propane, 22.1 mole% n-butane, 18.2 mole% n-pentane, and 14.1 mole% n-hexane at a total pressure of 101.3 kPa. A fractional recovery of 99.4% of propane is desired in the distillate and 99.7% of the n-butane in the bottoms. Estimate distillate and bottoms compositions and flow rates.

Solution

This appears to be a straightforward application of overall material balances, except that there are two variables too many. Thus, we will have to assume the recoveries or concentrations of two of the components. The normal boiling point of propane is 231.1 K, that of n-butane is 272.7 K, 309.2 for n-pentane, and 341.9 for n-hexane. Thus, the order of volatilities is propane > n-butane > n-pentane > n-hexane. This makes propane the light key, n-butane the heavy key, and n-pentane and n-hexane the heavy non-keys (HNKs). Since the recoveries of the keys are quite high, it is reasonable to assume that all of the HNKs appear only in the bottoms. Based on this assumption, the overall material balances yield: $D = 907.9$ kmoles/hr, $W = 1,092.1$ kmoles/hr. The estimated compositions are as follows.

Component	Mole fraction in D	Mole fraction in W
Propane	0.9985	0.0060
n-Butane	0.0015	0.4030
n-Pentane	0.0000	0.3330
n-Hexane	0.0000	0.2580

The composition calculated in the bottoms is quite accurate. Thus in this case we can step off stages from the bottoms upward and be confident that the results are reliable. On the other hand, if only light non-keys are present, the distillate composition can be estimated with high accuracy, and stage-by-stage calculations should proceed from the top downward. When both light and heavy non-keys are present, stage-by-stage calculation methods are difficult and other design procedures should be used.

6.8 FENSKE-UNDERWOOD-GILLILAND METHOD

Your objectives in studying this section are to be able to:

1. Derive the Fenske equation and use it to determine the number of stages required at total reflux and the splits of non-key components.
2. Use the Underwood equations to determine the minimum reflux ratio for multicomponent distillation.
3. Use the Gilliland correlation to estimate the actual number of stages in a multicomponent column, and the optimum feed stage location.

Although rigorous computer methods are available for solving multicomponent separation problems, approximate methods continue to be used in practice for various purposes, including preliminary design, parametric studies to establish optimum design conditions, and process synthesis studies to determine optimal separation sequences (Seader and Henley, 1998). A widely used approximate method is commonly referred to as the *Fenske-Underwood-Gilliland (FUG)* method.

6.8.1 Total Reflux: Fenske Equation

Fenske (1932) derived a rigorous solution for binary and multicomponent distillation at total reflux. The derivation assumes that the stages are equilibrium stages. Consider a multicomponent distillation column operating at total reflux. For an equilibrium partial reboiler, for any two components A and B,

$$\left[\frac{y_{A,R}}{y_{B,R}}\right] = \alpha_R \left[\frac{x_{A,W}}{x_{B,W}}\right] \tag{6-49}$$

Equation (6-49) is just the definition of the relative volatility applied to the conditions in the reboiler. Material balances for these components around the reboiler are

$$V_R y_{A,R} = L_N x_{A,N} - W x_{A,W} \qquad (6\text{-}50a)$$

$$V_R y_{B,R} = L_N x_{B,N} - W x_{B,W} \qquad (6\text{-}50b)$$

However, at total reflux, $W = 0$ and $L_N = V_R$. Thus, the mass balances become

$$y_{A,R} = x_{A,N} \qquad y_{B,R} = x_{B,N} \quad \text{(at total reflux)} \qquad (6\text{-}51)$$

For a binary system this means, naturally, that the operating line is the $y = x$ line. Combining equations (6-49) and (6-51),

$$\left[\frac{x_{A,N}}{x_{B,N}}\right] = \alpha_R \left[\frac{x_{A,W}}{x_{B,W}}\right] \qquad (6\text{-}52)$$

If we now move up the column to stage N, combining the equilibrium equation and the mass balances, we obtain

$$\left[\frac{x_{A,N-1}}{x_{B,N-1}}\right] = \alpha_N \left[\frac{x_{A,N}}{x_{B,N}}\right] \qquad (6\text{-}53)$$

Then equations (6-52) and (6-53) can be conbined to give

$$\left[\frac{x_{A,N-1}}{x_{B,N-1}}\right] = \alpha_N \alpha_R \left[\frac{x_{A,W}}{x_{B,W}}\right] \qquad (6\text{-}54)$$

We can repeat this procedure until we reach the top stage. The result is

$$\left[\frac{x_{A,D}}{x_{B,D}}\right] = \alpha_1 \alpha_2 \alpha_3 \cdots \alpha_{N-1} \alpha_N \alpha_R \left[\frac{x_{A,W}}{x_{B,W}}\right] \qquad (6\text{-}55)$$

If we define α_{AB} as the geometric average relative volatility,

$$\alpha_{AB} = \left[\alpha_1 \alpha_2 \alpha_3 \cdots \alpha_{N-1} \alpha_N\right]^{1/N_{min}} \qquad (6\text{-}56)$$

Equation (6-55) becomes

$$\left[\frac{x_{A,D}}{x_{B,D}}\right] = \alpha_{AB}^{N_{min}}\left[\frac{x_{A,W}}{x_{B,W}}\right] \tag{6-57}$$

Solving equation (6-57) for N_{min},

$$N_{min} = \frac{\ln\left[\dfrac{x_{A,D}\,x_{B,W}}{x_{B,D}\,x_{A,W}}\right]}{\ln \alpha_{AB}} \tag{6-58}$$

which is one form of the Fenske equation. In this equation, N_{min} is the number of equilibrium stages required at total reflux, including the partial reboiler.

An alternative form of the Fenske equation that is very convenient for multicomponent calculations is easily derived. Equation (6-58) can also be written as

$$N_{min} = \frac{\ln\left[\dfrac{(Dx_{A,D})(Wx_{B,W})}{(Dx_{B,D})(Wx_{A,W})}\right]}{\ln \alpha_{AB}} \tag{6-59}$$

The amount of substance A recovered in the distillate is $(Dx_{A,D})$ and is also equal to the fractional recovery of A in the distillate, $FR_{A,D}$, times the amount of A in the feed

$$Dx_{A,D} = \left(FR_{A,D}\right)Fz_A \tag{6-60}$$

From the definition of the fractional recovery,

$$Wx_{A,W} = \left[1 - FR_{A,D}\right]Fz_A \tag{6-61}$$

Substituting equations (6-60) and (6-61) and the corresponding equations for component B into equation (6-59) gives

$$N_{min} = \frac{\ln\left[\frac{(FR_{A,D})(FR_{B,W})}{(1-FR_{A,D})(1-FR_{B,W})}\right]}{\ln\alpha_{AB}} \qquad (6\text{-}62)$$

For multicomponent systems calculations with the Fenske equation are straightforward if fractional recoveries of the two keys, A and B, are specified. If the relative volatility is not constant, the average defined in equation (6-56) can be approximated by

$$\alpha_{AB} \approx \left(\alpha_R\alpha_D\right)^{1/2} \qquad (6\text{-}63)$$

where α_D is determined at the distillate composition. Once N_{min} is known, the fractional recovery of the non-keys can be found by writing equation (6-62) for a non-key component, C, and either key component. Then, solve the resulting equation for $FR_{C,W}$ or $FR_{C,D}$, depending on the key component chosen. If the key component chosen is B, the result is

$$FR_{C,D} = \frac{\alpha_{CB}^{N_{min}}}{\frac{FR_{B,W}}{1-FR_{B,W}} + \alpha_{CB}^{N_{min}}} \qquad (6\text{-}64)$$

Example 6.10 Use of Fenske Equation for Ternary Distillation

A distillation column with a partial reboiler and a total condenser is being used to separate a mixture of benzene, toluene, and 1,2,3-trimethylbenzene. The feed, 40 mole% benzene, 30 mole% toluene, and 30 mole% 1,2,3-trimethylbenzene, enters the column as a saturated vapor. We desire 95% recovery of the toluene in the distillate and 95% of the 1,2,3-trimethylbenzene in the bottoms. The reflux is returned as a saturated liquid, and constant molar overflow can be assumed. The column operates at a pressure of 1 atm. Find the number of equilibrium stages required at total reflux, and the recovery fraction of benzene in the distillate. Solutions of benzene, toluene, and 1,2,3-trimethylbenzene are ideal.

Solution
In this case, toluene (A) is the light key, 1,2,3-trimethylbenzene (B) is the heavy key, and benzene (C) is the light non-key. Since the liquid solutions are ideal, and the gases are ideal at the pressure of 1 atm, the VLE can be described by Raoult's law. The relative volatilities will be simply the ratio of the vapor pressures. To estimate the relative volatilities, we need an initial estimate of the conditions at the bottom and top of the column. For a 95% recovery of the LK in the distillate, we may

assume initially that virtually all of the LNK will be recovered in the distillate. Therefore, the bottoms will be almost pure 1,2,3-trimethylbenzene, and the reboiler temperature will be close to the normal boiling point of 1,2,3-trimethylbenzene, 449.3 K. For a 95% recovery of the HK in the bottoms, we may assume initially that the distillate will contain all of the LK, all of the LNK, and virtually none of the HK. Therefore a first estimate of the distillate composition is x_C = 40/70 = 0.571; x_A = 30/70 = 0.429; x_B = 0.0. The bubble point temperature for this solution is 390 K, a good preliminary estimate of the conditions at the top of the column.

The vapor pressures of the pure components as functions of temperature are given by equation (6-5). The corresponding parameters for benzene and toluene are given in Example 6.4. For 1,2,3-trimethylbenzene, they are: T_c = 664.5 K, P_c = 34.5 bar, A = –8.442, B = 2.922, C = –5.667, and D = 2.281 (Reid, et al., 1987). At the estimated reboiler temperature of 449.3 K, the vapor pressures are : P_A = 3.389 atm, P_B = 1.0 atm, P_C = 6.274 atm. The relative volatilities are α_{AB} = 3.389, α_{CB} = 6.274. At the estimated distillate temperature of 390 K, the vapor pressures are : P_A = 0.779 atm, P_B = 0.173 atm, P_C = 1.666 atm. The relative volatilities at this temperature are α_{AB} = 0.779/0.173 = 4.503, α_{CB} = 1.666/0.173 = 9.63. The geometric-average relative volatilities are α_{AB} = 3.91, α_{CB} = 7.77. Equation (6-62) gives

$$N_{min} = \frac{\ln\left[\dfrac{0.95 \times 0.95}{0.05 \times 0.05}\right]}{\ln(3.91)} = 4.32$$

Equation (6-63) gives the desired benzene fractional recovery in the distillate

$$FR_{C,D} = \frac{7.77^{4.32}}{\dfrac{0.95}{1-0.95} + 7.77^{4.32}} = 0.997$$

Thus, the assumption that virtually all of the LNK will be recovered in the distillate is justified.

6.8.2 Minimum Reflux: Underwood Equations

For binary distillation at minimum reflux, most of the stages are crowded into a constant-composition zone that bridges the feed stage. In this zone, all liquid and vapor streams have compositions essentially identical to those of the flashed feed. This zone constitutes a single pinch point as shown in Figure 6.12. This is also true for multicomponent distillation when all the components in the feed distribute to both the distillate and bottoms product. When this occurs, an analytical solution for the limiting flows can be derived (King, 1980). Unfortunately, for multicomponent systems, there will be separate pinch points in both the stripping and rectifying sections if one or more of the components appear in only one of the products. In this case, an alternative analysis procedure developed by Underwood is used to find the minimum reflux ratio (Wankat, 1988).

If there are nondistributing heavy non-keys present, a pinch point of constant composition will occur at minimum reflux in the rectifying section above where the heavy non-keys are fractionated out. With nondistributing light non-keys present, a pinch point will occur in the stripping section. Consider the case where the pinch point is in the rectifying section. The mass balance for component i around the top portion of the rectifying section as illustrated in Figure 6.7a is

$$V_{min}y_{i,n+1} = L_{min}x_{i,n} + Dx_{i,D} \tag{6-65}$$

At the pinch point where compositions are constant

$$x_{i,n-1} = x_{i,n} = x_{i,n+1} , \quad \text{and} \quad y_{i,n-1} = y_{i,n} = y_{i,n+1} \tag{6-66}$$

The equilibrium expression can be written as

$$y_{i,n+1} = m_i x_{i,n+1} \tag{6-67}$$

Combining equations (6-65) to (6-67) we obtain a simplified balance valid in the region of constant composition

$$V_{min}y_{i,n+1} = \frac{L_{min}}{m_i} y_{i,n+1} + Dx_{i,D} \tag{6-68}$$

Defining the relative volatility $\alpha_i = m_i/m_{HK}$ and combining terms in equation (6-68)

$$V_{min}y_{i,n+1}\left[1 - \frac{L_{min}}{V_{min}\alpha_i m_{HK}}\right] = Dx_{i,,} \tag{6-69}$$

Rearranging,

$$V_{min}y_{i,n+1} = \frac{\alpha_i Dx_{i,D}}{\alpha_i - \dfrac{L_{min}}{V_{min}m_{HK}}} \tag{6-70}$$

Equation (6-70) can be summed over all components to give the total vapor flow in the enriching section at minimum reflux:

$$V_{min} = \sum_i V_{min} y_{i,n+1} = \sum_i \frac{\alpha_i D x_{i,D}}{\alpha_i - \dfrac{L_{min}}{V_{min} m_{HK}}} \qquad (6\text{-}71)$$

In the stripping section, a similar analysis can be used to derive

$$-V_{st,min} = \sum_i \frac{\alpha_{i,st} W x_{i,W}}{\alpha_{i,st} - \dfrac{L_{st,min}}{V_{st,min} m_{HK,st}}} \qquad (6\text{-}72)$$

Defining

$$\phi = \frac{L_{min}}{V_{min} m_{HK}} \quad \text{and} \quad \phi_{st} = \frac{L_{st,min}}{V_{st,min} m_{HK,st}} \qquad (6\text{-}73)$$

equations (6-71) and (6-72) become polynomials in ϕ and ϕ_{st} and have C roots. The equations are now

$$V_{min} = \sum_i \frac{\alpha_i D x_{i,D}}{\alpha_i - \phi}, \quad \text{and} \quad -V_{st,min} = \sum_i \frac{\alpha_i W x_{i,W}}{\alpha_i - \phi_{st}} \qquad (6\text{-}74)$$

Assuming constant molar overflow and constant relative volatilities, Underwood showed there are common values of $\phi = \phi_{st}$ that satisfy both equations. Adding both equations in (6-74)

$$\Delta V_{feed} = V_{min} - V_{st,min} = \sum_i \left[\frac{\alpha_i D x_{i,D}}{\alpha_i - \phi} + \frac{\alpha_i W x_{i,W}}{\alpha_i - \phi} \right] \qquad (6\text{-}75)$$

where ΔV_{feed} is the change in vapor flow at the feed stage, and α_i is now an average relative volatility. Equation (6-75) is easily simplified, combining it with the overall column mass balance for component i to give

$$\Delta V_{feed} = \sum_i \left[\frac{\alpha_i F z_i}{\alpha_i - \phi} \right] \qquad (6\text{-}76)$$

If q is known

$$\Delta V_{feed} = F(1 - q) \qquad (6\text{-}77)$$

Combining equations (6-76) and (6-77),

$$1 - q = \sum_i \left[\frac{\alpha_i z_i}{\alpha_i - \phi} \right] \qquad (6\text{-}78)$$

Equation (6-78) is known as the first Underwood equation. It can be used to calculate appropriate values of ϕ. Equation (6-74) is known as the second Underwood equation and is used to calculate V_{min}. Once V_{min} is known, L_{min} is calculated from the mass balance:

$$L_{min} = V_{min} - D \qquad (6\text{-}79)$$

The exact method for using the Underwood equations depends on what can be assumed about the distillation process. Three cases will be considered.

Case A. Assume that none of the non-keys distribute. In this case the amounts of non-keys in the distillate are:

$$Dx_{HNK,D} = 0 \quad \text{and} \quad Dx_{LNK,D} = Fz_{LNK}$$

while the amounts for the keys are:

$$Dx_{LK,D} = FR_{LK,D} Fz_{LK}$$

$$Dx_{HK,D} = [1 - FR_{HK,W}] Fz_{HK}$$

Equation (6-78) can now be solved for the one value of ϕ between the relative volatilities of the two keys, $\alpha_{HK} < \phi < \alpha_{LK}$. This value of ϕ can now be substituted into equation (6-74) to immediately calculate V_{min}. Then

$$D = \sum_{i=1}^{C}\left(Dx_{i,D}\right) \tag{6-80}$$

and L_{min} is found from mass balance equation (6-79).

Case B. Assume that the distribution of the non-keys determined from the Fenske equation at total reflux are also valid at minimum reflux. In this case, the $Dx_{HNK,D}$ values are obtained from the Fenske equation as described before. The rest of the procedure is similar to the one described for *Case A.*

Case C. Exact solution without further assumptions. Equation (6-78) is a polynomial with C roots. Solve this equation for all values of ϕ lying between the relative volatilities of all components. This gives $C - 1$ valid roots. Now, write equation (6-74) $C - 1$ times, once for each value of ϕ. There are now $C - 1$ equations in $C - 1$ unknowns (V_{min}, and $Dx_{i,D}$ for all non-keys). Solve these simultaneous equations; calculate D from equation (6-80); calculate L_{min} from equation (6-79).

Example 6.11 Underwood Equations for Ternary Distillation

For the distillation problem of Example 6-10 find the minimum reflux ratio. Use a basis of 100 kmoles/hr of feed.

Solution

This problem fits into *Case B.* Since the feed is a saturated vapor, $q = 0$ and equation (6-78) becomes

$$1 = \frac{(3.91)(0.3)}{3.91 - \phi} + \frac{(1)(0.3)}{1 - \phi} + \frac{(7.77)(0.4)}{7.77 - \phi}$$

Solving for the value of ϕ between 1 and 3.91, we obtain $\phi = 2.2085$. From the problem statement (A = toluene, B = 1,2,3-trimethylbenzene, C = benzene),

$$Dx_{A,D} = (100)(0.3)(0.95) = 28.5 \text{ kmoles/hr}$$

$$Dx_{B,D} = (100)(0.3)(0.05) = 1.5 \text{ kmoles/hr}$$

From the results of Example 6-10,

$$Dx_{C,D} = (100)(0.4)(0.997) = 39.9 \text{ kmoles/hr}$$

Summing the three distillate flows, $D = 69.9$ kmoles/hr. Equation (6-74) becomes

$$V_{min} = \frac{(3.91)(28.5)}{3.91 - 2.2085} + \frac{(1)(1.5)}{1 - 2.2085} + \frac{(7.77)(39.88)}{7.77 - 2.2085}$$
$$= 120 \text{ kmoles / hr}$$

From the mass balance, $L_{min} = 120.0 - 69.9 = 50.1$ kmoles/hr. the minimum reflux ratio is $R_{min} = L_{min}/D = 0.717$.

Example 6.12 Underwood Equations for a Depropanizer

The feed to a depropanizer is 66% vaporized at the column inlet. The feed composition and average relative volatilities are given in the table below. It is required that 98% of the propane in the feed is recovered in the distillate, and 99% of the pentane is to be recovered in the bottoms product. Calculate the minimum reflux ratio for this case using Underwood's method.

Component	Mole Fraction in Feed	Relative Volatility
Methane (C)	0.26	39.47
Ethane (D)	0.09	10.00
Propane (A)	0.25	4.08
Butane (E)	0.17	2.11
Pentane (B)	0.11	1.00
Hexane (F)	0.12	0.50

Solution

In this case, propane and pentane are the light key and heavy key, respectively. Methane and ethane are LNK, hexane is a HNK, while butane is a "sandwich component," meaning that it has a volatility intermediate between the keys. This problem fits into *Case C*, where the distribution of the non-keys at minimum reflux must be determined, simultaneously with the minimum reflux ratio. Shiras et al. (1950) developed the following equation to determine whether or not a component is distributed at minimum reflux

$$D_{i,R} = \frac{\alpha_i - 1}{\alpha_{LK} - 1} FR_{LK,D} + \frac{\alpha_{LK} - \alpha_i}{\alpha_{LK} - 1} FR_{HK,D} \qquad (6\text{-}81)$$

where the relative volatilities are based on a reference value of 1.0 for the heavy key component. The Shiras et al. criterion applies at minimum reflux as follows:

$D_{i,R} > 1.0$ Component is nondistributed; contained entirely in distillate.

$0 < D_{i,R} < 1.0$ Component is distributed; appears both in distillate and bottoms.

$D_{i,R} < 0$ Component is nondistributed; contained entirely in bottoms.

We will determine which of the non-key components distribute according to the Shiras criterion. From the statement of the problem, $FR_{LK,D} = 0.98$, $FR_{HK,D} = 0.01$.

For methane,

$$D_{C,R} = \frac{39.47-1}{4.08-1} \times 0.98 + \frac{4.08-39.47}{4.08-1} \times 0.01 = 12.13$$

For ethane,

$$D_{D,R} = \frac{10.00-1}{4.08-1} \times 0.98 + \frac{4.08-10.00}{4.08-1} \times 0.01 = 2.84$$

For butane,

$$D_{E,R} = \frac{2.11-1}{4.08-1} \times 0.98 + \frac{4.08-2.11}{4.08-1} \times 0.01 = 0.36$$

For hexane,

$$D_{F,R} = \frac{0.50-1}{4.08-1} \times 0.98 + \frac{4.08-0.50}{4.08-1} \times 0.01 = -0.15$$

Then, according to the Shiras criterion, methane and ethane appear only in the distillate, hexane appears only in the bottoms, and butane is the only distributed non-key.

Equation (6-78) is now solved for two values of ϕ such that $1.0 < \phi_1 < 2.11$, and $2.11 < \phi_2 < 4.08$. Since the feed is 66% vaporized, $1 - q = 0.66$. The two values of ϕ in the given intervals are $\phi_1 = 1.263$, $\phi_2 = 2.846$. For the purpose of calculating the minimum reflux ratio, choose a basis:

Basis: 100 mole of feed

$$Dx_{A,D} = (100)(0.25)(0.98) = 24.5 \text{ moles (propane)}$$

$$Dx_{B,D} = (100)(0.11)(0.01) = 0.11 \text{ moles (pentane)}$$

$$Dx_{C,D} = (100)(0.26) = 26 \text{ moles (methane)}$$

$$Dx_{D,D} = (100)(0.09) = 9 \text{ moles (ethane)}$$

$$Dx_{E,D} = \text{unknown (butane)}$$

$$Dx_{F,D} = (100)(0.12)(0) = 0 \text{ moles (hexane)}$$

Applying equation (6-74) for each value of ϕ,

$$V_{min} = \frac{39.47 \times 26}{39.47 - 1.263} + \frac{10.00 \times 9}{10 - 1.263} + \frac{4.08 \times 24.5}{4.08 - 1.263}$$
$$+ \frac{2.11 \times \left(Dx_{E,D}\right)}{2.11 - 1.263} + \frac{0.11}{1.00 - 1.263}$$

$$V_{min} = \frac{39.47 \times 26}{39.47 - 2.846} + \frac{10.00 \times 9}{10 - 2.846} + \frac{4.08 \times 24.5}{4.08 - 2.846}$$
$$+ \frac{2.11 \times \left(Dx_{E,D}\right)}{2.11 - 2.846} + \frac{0.11}{1.00 - 2.846}$$

or

$$V_{min} = 72.243 + 2.494\left(Dx_{E,D}\right)$$
$$V_{min} = 121.614 - 2.863\left(Dx_{E,D}\right)$$

Solving simultaneously, $V_{min} = 95.23$ moles, $Dx_{E,D} = 9.22$ moles. From equation (6-80), $D = 68.83$ moles. From the mass balance, $L_{min} = 95.23 - 68.83 = 26.40$ moles, and the minimum reflux ratio is $R_{min} = L_{min}/D = 26.40/68.83 = 0.384$.

6.8.3 Gilliland Correlation for Number of Stages at Finite Reflux

A general shortcut method for determining the number of stages required for a multicomponent distillation at finite reflux ratios would be extremely useful. Unfortunately, such a method has not been developed. However, Gilliland (1940) noted that he could empirically relate the number of stages N at a finite reflux ratio L/D to the minimum number of stages and to the minimum reflux ratio. Gilliland did a series of accurate stage-by-stage calculations and found that he could correlate the variable

$$Y = \frac{N - N_{\min}}{N + 1} \tag{6-82}$$

with the variable

$$X = \frac{R - R_{\min}}{R + 1} \tag{6-83}$$

The original Gilliland correlation was graphical. Molkanov et al. (1972) fit the Gilliland correlation to the equation

$$Y = 1 - \exp\left[\left(\frac{1 + 54.4X}{11 + 117.2X}\right)\left(\frac{X - 1}{X^{0.5}}\right)\right] \tag{6-84}$$

Implicit in the application of the Gilliland correlation is the specification that the theoretical stages be distributed optimally between the rectifying and stripping sections. A reasonably good approximation of optimum feed-stage location, according to Seader and Henley (1998), can be made by employing the empirical equation of Kirkbride (1944)

$$\frac{N_R}{N_S} = \left[\left(\frac{z_{HK}}{z_{LK}}\right)\left(\frac{x_{LK,W}}{x_{HK,D}}\right)^2\left(\frac{W}{D}\right)\right]^{0.206} \tag{6-85}$$

where N_R and N_S are the number of stages in the rectifying and stripping sections, respectively. Application of equation (6-85) requires knowledge of the distillate and bottoms composition at the specified reflux ratio. Seader and Henley (1998) suggest that the distribution of the non-key components at a finite reflux ratio is close to that estimated by the Fenske equation at total reflux conditions.

Example 6.13 Application of the Gilliland Correlation

Estimate the total number of equilibrium stages and the optimum feed-stage location for the distillation problem presented in Examples 6.10 and 6.11 if the actual reflux ratio is set at $R = 1.0$.

Solution

From Example 6.10, $N_{\min} = 4.32$; from Example 6.11, $R_{\min} = 0.717$. For $R = 1.0$, $X = (1.0 - 0.717)/(1 + 1) = 0.142$. Substituting in equation (6-84), $Y = 0.513$. From equation (6.82), $N = 9.93$ equilibrium stages.

To use the Kirkbride equation to determine the feed-stage location, we must first estimate the composition of the distillate and bottoms from the results of Example 6.10. We found there that, at total reflux, 99.7% of the LNK (benzene) is recovered in the distillate, 95% of the light key is in the distillate, and 95% of the heavy key is in the bottoms. For a basis of 100 moles of feed, the material balances for the three components are:

Component	Distillate		Bottoms	
	Moles	Molar Fraction	Moles	Molar Fraction
Benzene (LNK)	39.88	0.571	0.12	0.004
Toluene (LK)	28.50	0.408	1.50	0.050
Trimethylbenzene (HK)	1.50	0.021	28.50	0.946
Total	$69.88 = D$		$30.12 = W$	

From the problem statement, $z_{LK} = z_{HK} = 0.30$. Substituting in equation (6-85), $N_R/N_S = 1.202$. On the other hand, $N = N_R + N_S = 9.93$. Solving simultaneously, $N_R = 5.42$, $N_S = 4.51$. Rounding the estimated equilibrium stage requirement leads to 1 stage as a partial reboiler, 4 stages below the feed, and 5 stages above the feed.

6.9 RIGOROUS CALCULATION PROCEDURES FOR MULTICOMPONENT DISTILLATION

Your objectives in studying this section are to be able to:

1. Understand the importance of the introduction of computers as tools for rigorous analysis and design of multicomponent distillation equipment.
2. Understand the difference between the equilibrium-efficiency approach and the rate-based approach to multicomponent distillation problems.
3. Mention some of the computer programs available for implementation of the equilibrium- and rate-based models.

Before the 1950s, distillation column calculations were performed by hand. Although rigorous calculation procedures were available, they were difficult to apply for all but very small columns. Shortcut methods were therefore the primary design tool. Rigorous procedures were seldom used. Inaccuracies and uncertainties in the shortcut procedures were usually accommodated by overdesign.

The introduction of computers has entirely reversed the design procedure. Rigorous calculations can now be performed quickly and efficiently using a computer. In modern distillation practice, rigorous methods are the primary design tool. The role of shortcut calculations is now restricted to eliminating the least-desirable design options, providing the designer with an initial estimate for the rigorous step and for troubleshooting the final design (Kister, 1992).

Two different approaches have evolved for the simulation and design of multicomponent distillation columns. The conventional approach is through the use of an *equilibrium stage model* together with methods for estimating the *tray efficiency*. An alternative approach, the *nonequilibrium, rate-based model,* applies rigorous multicomponent mass- and heat-transfer theory to distillation calculations. This non-equilibrium stage model is also applicable, with only minor modifications, to gas absorption, liquid extraction, and to operations in trayed or packed columns (Taylor and Krishna, 1993).

6.9.1 Equilibrium Stage Model

A schematic diagram of a single equilibrium stage is shown in Figure 6.23. The key assumption of this model is that the vapor and liquid streams leaving a stage are in equilibrium with each other. A complete multicomponent distillation or absorption column may be modeled as a sequence of these stages. The equations that model equilibrium stages have been termed the MESH equations after Wang

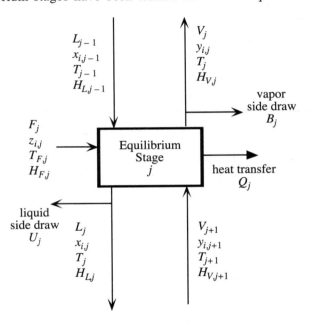

Figure 6.23 Equilibrium stage model.

and Henke (1966). MESH is an acronym referring to the different types of equations that form the mathematical model:

1. *M* equations—material balance for each component (*C* equations for each stage):

$$M_{i,j} = L_{j-1}x_{i,j-1} + V_{j+1}y_{i,j+1} + F_j z_{i,j}$$
$$- (L_j + U_j)x_{i,j} - (V_j + B_j)y_{i,j} = 0 \qquad (6\text{-}86)$$

2. E equations—phase equilibrium relation for each component, here modified to include the Murphree efficiency defined by equation 4-55 (C equations for each stage):

$$E_{i,j} = \mathbf{E}_{MG,i,j}m_{i,j}x_{i,j} - y_{i,j} - \left(1 - \mathbf{E}_{MG,i,j}\right)y_{i,j+1} = 0 \qquad (6\text{-}87)$$

3. S equations—mole fractions summation (2 for each stage):

$$\left(S_y\right)_j = \sum_{i=1}^{C} y_{i,j} - 1 = 0 \qquad\qquad \left(S_x\right)_j = \sum_{i=1}^{C} x_{i,j} - 1 = 0 \qquad (6\text{-}88)$$

4. H equation—enthalpy balance (one for each stage):

$$H_j = L_{j-1}H_{L,j-1} + V_{j+1}H_{V,j+1} + F_j H_{F,j} - (L_j + U_j)H_{L,j}$$
$$- (V_j + B_j)H_{V,j} - Q_j = 0 \qquad (6\text{-}89)$$

The MESH equations can be applied to all of the equilibrium stages in the column, including the reboiler and condenser. The result is a set of nonlinear equations that must be solved by iterative techniques.

   A wide variety of iterative solution procedures for solving the MESH equations has appeared in the literature. Current practice is based mainly on the bubble-point (BP) method, the simultaneous correction (SC) method, and the inside-out method. The BP method is usually restricted to distillation problems involving narrow-boiling feed mixtures. The SC and inside-out methods are designed to solve any type of column configuration for any type of feed mixture. Because of its computational efficiency, the inside-out method is often the method of choice; however, it may fail to converge when highly nonideal liquid mixtures are involved, in which case the slower SC method should be tried (Seader and Henley, 1998). Computer implementations of these methods are found in the following widely available programs and simulators: (1) ASPEN PLUS of Aspen Technology, (2) ChemCAD of Chemstations, (3) HYSIM of Hyprotech, and (4) PRO/II of Simulation Sciences.

6.9.2 Nonequilibrium, Rate-based Model

   Although the equilibrium-based model, modified to incorporate stage efficiency, is adequate for binary mixtures and for the major components in nearly ideal multicomponent mixtures, that model has serious deficiencies for more general multicomponent vapor-liquid mixtures. Murphree himself stated clearly the limitations of his development for multicomponent mixtures. He even stated that the theoretical plate should not be the basis of calculation for ternary systems.

When the equilibrium-based model is applied to multicomponent mixtures, a number of problems arise. Values of E_{MG} differ from component to component and vary from stage to stage. But, at each stage, the number of independent values of E_{MG} must be determined so as to force the mole fractions in the vapor phase to sum 1. This introduces the possibility that negative values of E_{MG} can result. This is in contrast to binary mixtures for which the values of E_{MG} are always positive and are identical for the two components.

Krishna et al. (1977) showed that when the vapor mole-fraction driving force of a component (call it A) is small compared to the other components in the mixture, the transport rate of A is controlled by the other components, with the result that E_{MG} for A is anywhere in the range from minus infinity to plus infinity. They confirmed this theoretical prediction by conducting experiments with the ethanol/*tert*-butanol/water system and obtained values of E_{MG} for *tert*-butanol ranging from –2,978% to +527%. In addition, the observed values of E_{MG} for ethanol and water sometimes differed significantly.

Krishna and Standardt (1979) were the first to show the possibility of applying rigorous multicomponent mass- and heat-transfer theory to calculations of simultaneous transport. The availability of this theory led to the development by Krishnamurthy and Taylor (1985) of the first general rate-based, computer-aided model for application to trayed and packed columns for distillation and other continuous, countercurrent, vapor-liquid separation operations. This model applies the two-resistance theory of mass-transfer discussed in Chapter 3, with equilibrium assumed at the interface of the two phases, and provides options for vapor and liquid flow configurations in trayed columns, including plug flow and perfectly mixed flow, on each tray. Although the model does not require tray efficiencies, correlations of mass- and heat-transfer coefficients are needed for the particular type of tray or packing employed. The theory was further developed by Taylor and Krishna (1993), and the model was extended by Taylor et al. (1994). The 1994 version, unlike the 1985 model, includes a design mode that estimates column diameter for a specified approach to flooding or pressure drop. Baur et al. (2001) further modified the model to simulate the dynamic, nonequilibrium behavior of reactive distillation tray columns.

In the rate-based models, the mass and energy balances around each equilibrium stage are each replaced by separate balances for each phase around a stage, which can be a tray, a collection of trays, or a segment of a packed section. Rate-based models use the same *m*-value and enthalpy correlations as the equilibrium-based models. However, the *m*-values apply only at the equilibrium interphase between the vapor and liquid phases. The accuracy of enthalpies and, particularly, *m*-values is crucial to equilibrium-based models. For rate-based models, accurate predictions of heat-transfer rates and, particularly, mass-transfer rates are also required. These rates depend upon transport coefficients, interfacial area, and driving forces. It is important that mass-transfer rates account for component-coupling effects through binary pair coefficients.

The general forms for component mass-transfer rates across the vapor and liquid films, ($N_{i,j}{}^V$, $N_{i,j}{}^L$) respectively, on a tray or in a packed segment are as follows, where both diffusive and bulk-flow contributions are included:

$$N_{i,j}^V = a_j^{\mathbf{I}} J_{i,j}^V + y_{i,j} N_{T,j}^V \tag{6-90}$$

and

$$N_{i,j}^L = a_j^{\mathbf{I}} J_{i,j}^L + x_{i,j} N_{T,j}^L \tag{6-91}$$

where $a_j^{\mathbf{I}}$ is the total interfacial area for stage j, and $N_{T_j}{}^V$, $N_{T_j}{}^L$ are total molar rates.

As discussed in detail by Taylor and Krishna (1993), the general multicomponent case for mass transfer is considerably more complex than the binary case because of component-coupling effects. For example, for a ternary system the fluxes for the first two components are

$$J_1^V = k_{11}^V \left(y_1^V - y_1^{\mathbf{I}}\right)_{avg} + k_{12}^V \left(y_2^V - y_2^{\mathbf{I}}\right)_{avg} \tag{6-92}$$

$$J_2^V = k_{21}^V \left(y_1^V - y_1^{\mathbf{I}}\right)_{avg} + k_{22}^V \left(y_2^V - y_2^{\mathbf{I}}\right)_{avg} \tag{6-93}$$

The flux for the third component is not independent of the other two, but is obtained from equation (1-18):

$$J_3^V = -J_1^V - J_2^V \tag{6-94}$$

In these equations, the multicomponent mass-transfer coefficients, $k_{ij}{}^V$, are complex functions related to inverse rate functions described below.

For the general multicomponent system (1, 2, 3, ..., C), the independent fluxes for the first $C - 1$ components are given in matrix equation form as

$$\mathbf{J}^V = \left[\mathbf{k}^V\right]\left(\mathbf{y}^V - \mathbf{y}^{\mathbf{I}}\right)_{avg} \tag{6-95}$$

$$\mathbf{J}^L = \left[\mathbf{k}^L\right]\left(x^{\mathbf{I}} - x^L\right)_{\text{avg}} \qquad (6\text{-}96)$$

where $\mathbf{J}^V$, $\mathbf{J}^L$, $(\mathbf{y}^V - \mathbf{y}^{\mathbf{I}})_{\text{avg}}$, and $(x^{\mathbf{I}} - x^L)_{\text{avg}}$ are column vectors of length $C - 1$ and $[\mathbf{k}^V]$ and $[\mathbf{k}^L]$ are $(C - 1) \times (C - 1)$ square matrices.

From the Maxwell-Stefan theory for multicomponent diffusion, Taylor and Krishna (1993) developed the following scheme to estimate the matrices of *zero-flux* multicomponent mass-transfer coefficients from binary-pair mass-transfer coefficients, F_{ij}. These are obtained from correlations of experimental data, with the Chan and Fair (1984) correlations being the most widely used. For an ideal gas solution:

$$\left[\mathbf{k}^V\right] = \left[\mathbf{R}^V\right]^{-1} \qquad (6\text{-}97)$$

For a nonideal liquid solution:

$$\left[\mathbf{k}^L\right] = \left[\mathbf{R}^L\right]^{-1}[\Gamma] \qquad (6\text{-}98)$$

where the thermodynamic factor matrix Γ is as defined in equation (1-32), and the elements of $\mathbf{R}^V$ are:

$$R_{ii}^V = \frac{y_i}{F_{iC}^V} + \sum_{\substack{k=1 \\ k \neq i}}^{C} \frac{y_k}{F_{ik}^V} \qquad (6\text{-}99)$$

$$R_{ij}^V = -y_i\left(\frac{1}{F_{ij}^V} - \frac{1}{F_{iC}^V}\right) \qquad (6\text{-}100)$$

where here, j refers to the jth component and not the jth stage. Similar expressions are used to calculate $\mathbf{R}^L$.

When mass-transfer rates are moderate to high, an additional correction term is needed in equations (6-97) and (6-98) to correct for distortion of the composition profiles. This correction, which can have a serious effect on the results, is discussed in detail by Taylor and Krishna (1993). An alternative approach would be to numerically solve the Maxwell-Stefan equations, as illustrated in Examples 1.15 and 1.16. The calculation of the low mass-transfer fluxes according to equations (6-90) to (6-100) is illustrated in the following example.

Example 6.14 Rate-Based Ternary Distillation Calculations.

The following results were obtained for tray n from a rate-based calculation of a ternary distillation at 1 atm involving acetone (1), methanol (2), and water (3). Vapor and liquid phases are assumed to be completely mixed.

Component	y_n	y_{n+1}	y_n^I	m_n^I	x_n
1	0.2971	0.1700	0.3521	2.759	0.1459
2	0.4631	0.4290	0.4677	1.225	0.3865
3	0.2398	0.4010	0.1802	0.3673	0.4676

The gas-phase binary mass-transfer coefficients were estimated as follows:

$$F_{12}^V = 4.927 \text{ moles/m}^2\text{-s}; \ F_{13}^V = 6.066 \text{ moles/m}^2\text{-s}; \ F_{23}^V = 7.048 \text{ moles/m}^2\text{-s}$$

The total interfacial area of the tray was estimated as $a^I = 50.0$ m^2. The total vapor flow rate entering the plate was estimated as $V_{n+1} = 188.0$ moles/s; the vapor rate leaving the plate was estimated from an energy balance as $V_n = 194.8$ moles/s.

(a) Compute the molar diffusion rates.

(b) Compute the mass-transfer rates.

(c) Calculate the Murphree vapor tray efficiencies.

Solution
(a) Compute the reciprocal rate functions, **R**, from equations (6-99) and (6-100). For low total flux conditions, we may assume linear mole fraction gradients, such that the mole fractions in equations (6-99) and (6-100) can be assumed constant at their average value $(y_i + y_i^I)/2$. Thus:

$$y_1 = (0.2971 + 0.3521)/2 = 0.3246$$
$$y_2 = (0.4631 + 0.4677)/2 = 0.4654$$
$$y_3 = (0.2398 + 0.1802)/2 = 0.2100$$

$$R_{11}^V = \frac{y_1}{F_{13}^V} + \frac{y_2}{F_{12}^V} + \frac{y_3}{F_{13}^V} = \frac{0.3246}{6.066} + \frac{0.4654}{4.927} + \frac{0.2100}{6.066} = 0.1830$$

$$R_{22}^V = \frac{y_2}{F_{23}^V} + \frac{y_1}{F_{12}^V} + \frac{y_3}{F_{23}^V} = \frac{0.4654}{7.048} + \frac{0.3246}{4.927} + \frac{0.2100}{7.048} = 0.1617$$

$$R_{12}^V = -y_1\left(\frac{1}{F_{12}^V} - \frac{1}{F_{13}^V}\right) = -0.3246\left(\frac{1}{4.927} - \frac{1}{6.066}\right) = -0.0124$$

$$R_{21}^V = -y_2\left(\frac{1}{F_{12}^V} - \frac{1}{F_{23}^V}\right) = -0.4654\left(\frac{1}{4.927} - \frac{1}{7.048}\right) = -0.0284$$

Thus, in matrix form:

$$[\mathbf{R}^V] = \begin{bmatrix} 0.1830 & -0.0124 \\ -0.0284 & 0.1617 \end{bmatrix}$$

From equation (6-97), by matrix inversion using Mathcad

$$[\mathbf{k}^V] = [\mathbf{R}^V]^{-1} = \begin{bmatrix} 5.5269 & 0.4024 \\ 0.9217 & 6.2588 \end{bmatrix}$$

From equation (6-95),

$$\begin{bmatrix} J_1^V \\ J_2^V \end{bmatrix} = \begin{bmatrix} k_{11}^V & k_{12}^V \\ k_{21}^V & k_{22}^V \end{bmatrix}\begin{bmatrix} (y_1 - y_1^I) \\ (y_2 - y_2^I) \end{bmatrix} = \begin{bmatrix} 5.5269 & 0.4024 \\ 0.9217 & 6.2588 \end{bmatrix}\begin{bmatrix} (0.2971 - 0.3521) \\ (0.4631 - 0.4677) \end{bmatrix}$$

Then,

$$J_1^V = -0.3061 \text{ moles/m}^2\text{-s} \qquad J_2^V = -0.0822 \text{ moles/m}^2\text{-s}$$

From equation (6-94),

$$J_3^V = -J_1^V - J_2^V = 0.3883 \text{ moles/m}^2\text{-s}$$

(b) To determine the component mass-transfer rates, it is necessary to know the total mass-transfer rate for the tray, N_T^V. The problem of determining this quantity when the diffusion rates, J, are known is referred to as the *bootstrap problem* (Taylor and Krishna, 1993). In distillation, N_T^V is determined by an energy balance which gives the change in vapor molar rate across a tray. For the assumption of constant molar overflow, $N_T^V = 0$. In this example, that assumption is not valid. Since mass transfer from the vapor to the liquid is taken as positive in the derivation of the nonequilibrium model,

$$N_T^V = V_{n+1} - V_n = -6.8 \text{ moles/s}$$

From equation (6-90),

$$N_1^V = a^I J_1^V + y_1 N_T^V = (50) \times (-0.3061) + (0.3246) \times (-6.8)$$
$$= -17.5 \text{ moles/s}$$

$$N_2^V = a^I J_2^V + y_2 N_T^V = (50) \times (-0.0822) + (0.4654) \times (-6.8)$$
$$= -7.3 \text{ moles/s}$$

$$N_3^V = a^I J_3^V + y_3 N_T^V = (50) \times (0.3883) + (0.2100) \times (-6.8)$$
$$= 18.0 \text{ moles/s}$$

(c) Approximate values of the Murphree vapor tray efficiency are obtained from

$$E_{MG,i} = \frac{y_{i,n} - y_{i,n+1}}{m_n^I x_{i,n} - y_{i,n+1}} \qquad (6\text{-}101)$$

Then

$$E_{MG,1} = \frac{0.2971 - 0.1700}{2.759 \times 0.1459 - 0.1700} = 0.547$$

$$E_{MG,2} = \frac{0.4631 - 0.4290}{1.225 \times 0.3865 - 0.4290} = 0.767$$

$$E_{MG,3} = \frac{0.2398 - 0.4010}{0.3673 \times 0.4676 - 0.4010} = 0.703$$

Rate-based models are implemented in several computer programs, including RATEFRAC of Aspen Technology and ChemSep v4.20. Both programs offer considerable flexibility in user specifications. Computing time for these rate-based models is not generally more than an order of magnitude greater than for an equilibrium-based model.

ChemSep Program

ChemSep is a suite of programs for performing multicomponent separation process calculations on PCs. It was created specifically for use by students in university courses on stagewise separation processes, but it is also used by professionals in industry. It includes ChemProp, a program that allows estimation of physical properties such as densities, viscosities, heat capacities and conductivities, and surface tensions. Thermodynamic quantities such as m-values, vapor pressures, activity coefficients, or fugacity coefficients can also be calculated. Included are thermodynamic databases for commonly used chemicals as well as group contribution methods, such as UNIFAC.

The ChemSep program applies the transport equations to trays or short heights (called segments) of packing. The condenser and reboiler are treated as equilibrium stages. Calculations of transport coefficients and pressure drop require column diameter and dimensions of column internals. These may be specified (simulation mode) or computed (design mode). In the design mode, default dimensions are selected for the internals, with column diameter computed from a specified fractional approach to flooding for a trayed or packed column, or a specified pressure drop per unit height for a packed column. ChemSep can also perform rate-based calculations for liquid–liquid extraction.

Example 6.15 Equilibrium Distillation Calculations: ChemSep

The following feed, at 80 °C and 1,035 kPa, is to be fractionated at the rate of 1.0 kmoles/s at the given pressure so that the vapor contains 98% of the propane, but only 1% of the n-butane:

Component	CH_4	C_2H_6	C_3H_8	$n\text{-}C_4H_{10}$	$n\text{-}C_5H_{12}$	$n\text{-}C_6H_{14}$
z_F, mole fraction	0.03	0.07	0.15	0.33	0.30	0.12

A number of simulations were carried out using the equilibrium model of ChemSep for different column configurations (number of equilibrium stages and feed stage location) and for differing operation specifications (reflux ratio and bottom product rate). Only the specifications and results of our final simulation are reported here.

Solution

The problem specifications presented in Table 6.4 were entered via the ChemSep menu and the program was executed. A converged solution was achieved in 5 iterations on a PC running the Windows 2000 operating system. Initialization of all the variables was done by the program. The predicted separation is as follows:

Component	Feed (moles/s)	Distillate (moles/s)	Bottoms (moles/s)
CH_4	30	30.0	0.0
C_2H_6	70	70.0	0.0
C_3H_8	150	147.2	2.8
$n\text{-}C_4H_{10}$	330	129.7	200.3
$n\text{-}C_5H_{12}$	300	3.1	296.9
$n\text{-}C_6H_{14}$	120	0.0	120.0
Total	1,000	380.0	620.0

Predicted column profiles for temperature, total vapor and liquid flow rates, and component vapor and liquid mole fractions are shown in Figure 6.24, where stages are numbered from the top down (stage 1 is the partial condenser, stage 14 is the partial reboiler). Computed heat exchanger duties are as follows:

Condenser: 8.94 MW
Reboiler: 11.77 MW

Table 6.4 Specifications for Example 6.15.

Operation:
        Simple distillation
        Partial (vapor product) condenser
        Partial (liquid product) reboiler
        14 equilibrium stages
        Feed to stage 6
Properties:
        EOS *m*-model
        Soave-RK Cubic EOS
        Excess enthalpy from EOS
Specifications:
        Column pressure 10.21 atm (constant)
        Feed
                Pressure 10.21 atm
                Temperature 80 °C
        Condenser
                Reflux ratio = 1.2
        Reboiler
                Bottom product flow rate = 620 moles/s

Example 6.16 Nonequilibrium Distillation Calculations Using ChemSep; Sieve-Tray Column Design

A sieve-tray distillation column with a partial reboiler and a total condenser is to be designed to separate a mixture of benzene, toluene, and cumene (a common name for isopropyl benzene). The feed flows at the rate of 40 moles/s benzene, 30 moles/s toluene, and 30 moles/s cumene. It enters the column at a pressure of 1.05 atm and 50 mole% vaporized. We desire at least 98.5% recovery of the toluene in the distillate, and at least 98.5% recovery of the cumene in the bottoms. Design the column for a reflux ratio of 2.0.

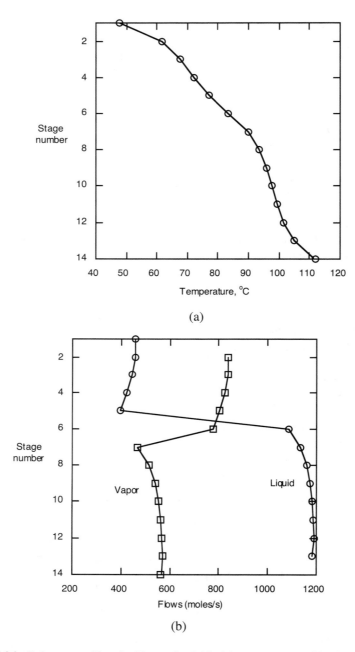

(a)

(b)

Figure 6.24 Column profiles for Example 6.15: (a) temperature; (b) vapor and liquid flow rates; (c) vapor mole fractions; (d) liquid mole fractions.

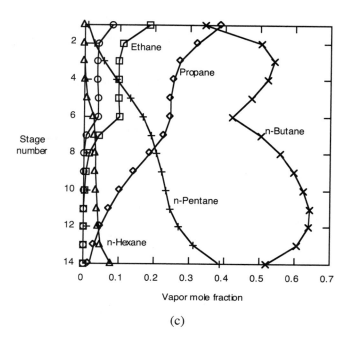

(c)

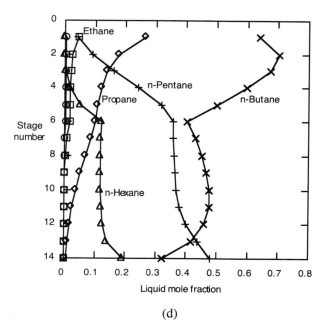

(d)

Figure 6.24 (continued)

Solution

A number of simulations were carried out using the nonequilibrium model of ChemSep for different column configurations (number of equilibrium stages and feed stage location). Only the specifications and results of our final simulation are reported here.The problem specifications presented in Table 6.5 were entered via the ChemSep menu and the program was executed, converging in 15 iterations.

Table 6.5 Specifications for Example 6.16

Operation:

       Simple distillation

       Total (liquid product) condenser

       Partial (liquid product) reboiler

       10 sieve trays

       Feed to stage 6 (stage 1 is the condenser; stage 12 is the reboiler)

Properties:

       DECHEMA m-model

       UNIFAC activity coefficients

       Equation of state—ideal gas law

       Vapor pressure—Antoine

       Excess enthalpy from UNIFAC model

Specifications:

       Condenser pressure—1 atm

       Column top pressure—1 atm

       Feed

              Pressure 1.05 atm, 50 mole% vapor

              Flowrates (moles/s)

Benzene	40
Toluene	30
Cumene	30

       Condenser

              Reflux ratio = 2.0

       Reboiler

              Bottom product flow rate = 30 moles/s

Design:

	Section 1	Section2
Stages	2–5	6–11
Mass-transfer coefficient	Chan Fair	Chan Fair
Vapor flow model	Plug	Plug
Liquid flow model	Mixed	Mixed
Fraction of flooding	0.61	0.70
System factor	0.90	0.90

The results of the simulation are summarized in Table 6.6. Predicted column profiles for pressure and back-calculated Murphree tray efficiencies are shown in Figure 6.25.

Table 6.6 Simulation Results for Example 6.16

Streams	Feed	Top	Bottom
Stage	6	1	12
Pressure, atm	1.05	1.0	1.049
Vapor fraction	0.5	0.0	0.0
Temperature, °C	111.4	90.9	153.2
Mole flows (moles/s)			
Benzene	40.00	40.00	0.00
Toluene	30.00	29.55	0.45
Cumene	30.00	0.45	29.55
Total	100.00	70.00	30.00
Internals design			
Section	1		2
Column diameter (m)	2.94		2.94
Tray spacing (m)	0.61		0.61
Downcomer area (% total)	5.30		7.65
Hole diameter (mm)	4.76		4.76
Hole pitch (mm)	13.0		13.0
Weir height (mm)	50.0		50.0
Weir length (m)	6.59		6.94
Condenser duty (MW)	7.11		
Reboiler duty (MW)	5.58		

Example 6.17 Non Equilibrium Distillation Calculations Using ChemSep; Packed Tower Design

Repeat Example 6.16 for a tower packed with 50-mm metal Hiflow rings. Design for a 70% approach to flooding.

Solution

A number of simulations were carried out using the nonequilibrium model of ChemSep, trying different combinations of packing heights above and below the feed until the desired separation was achieved. Table 6.7 presents the final results.

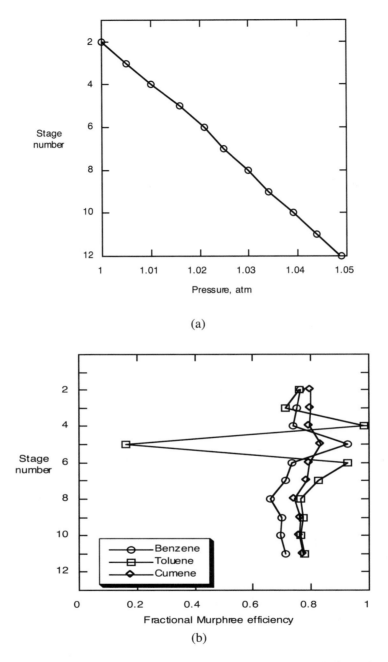

(a)

(b)

Figure 6.25 Column profiles for Example 6.16: (a) pressure; (b) Murphree vapor tray efficiencies.

Table 6.7 Simulation Results for Example 6.17

Mole flows (moles/s)	Feed	Top	Bottom
Benzene	40.00	40.00	0.00
Toluene	30.00	29.55	0.45
Cumene	30.00	0.45	29.55
Total	100.00	70.00	30.00
Section		Above feed	Below feed
Packing		50-mm Hiflow rings	50-mm Hiflow rings
Column diameter (m)		2.20	2.10
Packed height (m)		2.01	2.01
Total gas pressure drop	1.20 kPa		
Convergence in 8 iterations			
Segment height 3.7 cm			

RATEFRAC Program

The RATEFRAC program differs somewhat from the ChemSep program in that RATEFRAC is designed to model columns used for reactive distillation. The reactions can be equilibrium-based or kinetics-based, including reactions among electrolytes. For kinetically controlled reactions, built-in power-law expressions are selected, or the user supplies FORTRAN subroutines for the rate laws. For equilibrium-based reactions, the user supplies a temperature-dependent equilibrium constant, or RATEFRAC computes reaction-equilibrium constants from free-energy values stored in the data bank. The user specifies the phase in which the reaction takes place. Flow rates of side streams and the column pressure profile must be provided. Mass-transfer correlations are built into RATEFRAC for bubble-cap trays, valve trays, sieve trays, and random packings. Users may provide their own FORTRAN subroutines for transport coefficients and interfacial area.

PROBLEMS

The problems at the end of each chapter have been grouped into four classes (designated by a superscript after the problem number).

Class a: Illustrates direct numerical application of the formulas in the text.
Class b: Requires elementary analysis of physical situations, based on the subject material in the chapter.
Class c: Requires somewhat more mature analysis.
Class d: Requires computer solution.

6.1ᵃ. Flash vaporization of a heptane-octane mixture.

A liquid mixture containing 50 mole% n-heptane (A), 50 mole% n-octane (B), at 303 K, is to be continuously flash-vaporized at a pressure of 1 atm to vaporize 30 mole% of the feed. What will be the composition of the vapor and liquid and the temperature in the separator if it behaves as an ideal stage?

Answer: $x_W = 0.443$

6.2ᵃ. Flash vaporization of a heptane-octane mixture.

A liquid mixture containing 50 mole% n-heptane (A), 50 mole% n-octane (B), at 303 K, is to be continuously flash-vaporized at a temperature of 350 K to vaporize 30 mole% of the feed. What will be the composition of the vapor and liquid and the pressure in the separator if it behaves as an ideal stage?

Answer: $y_D = 0.654$

6.3ᵈ. Flash vaporization of a ternary mixture.

Modify the Mathcad program of Figure 6.3 for ternary mixtures. Test your program with the data presented in Example 6.2 for a mixture of benzene (A), toluene (B), and o-xylene (C). Critical temperatures and pressures, and the parameters of the Wagner equation for estimating vapor pressure (equation 6-5), are included in the following table (Reid, et al., 1987).

Component	T_c, K	P_c, bar	A	B	C	D
benzene	562.2	48.9	– 6.983	1.332	– 2.629	– 3.334
toluene	591.8	41.0	– 7.286	1.381	– 2.834	– 2.792
o-xylene	630.3	37.3	– 7.534	1.410	– 3.110	– 2.860

6.4[b]. Flash vaporization of a ternary mixture.

Consider the ternary mixture of Example 6.2 and Problem 6.3. Estimate the temperature, and composition of the liquid and vapor phases when 60% of the mixture has been vaporized at a constant pressure of 1 atm.

Answer: $y_{AD} = 0.630$

6.5[b]. Flash vaporization of a ternary mixture.

Consider the ternary mixture of Example 6.2 and Problem 6.3. It is desired to recover in the vapor 75% of the benzene in the feed, and to recover in the liquid 70% of the o-xylene in the feed. Calculate the temperature, pressure, fraction of the feed vaporized, and the concentration of the liquid and gas phases.

Answer: $P = 0.756$ atm

6.6[b]. Batch distillation of a heptane-octane mixture.

Repeat the calculations of Example 6.3, but for 80 mole% of the liquid distilled.

Answer: $y_{D,av} = 0.573$

6.7[b]. Binary batch distillation with constant relative volatility.

If equation (6-40) describes the equilibrium relation at constant pressure by use of some average relative volatility α over the concentration range involved, show that equation (6-11) becomes

$$\ln \frac{F}{W} = \frac{1}{\alpha - 1} \ln \frac{x_F(1 - x_W)}{x_W(1 - x_F)} + \ln \frac{(1 - x_W)}{(1 - x_F)} \qquad (6\text{-}102)$$

6.8ᵃ. Binary batch distillation with constant relative volatility.

Consider the binary batch distillation of Example 6.3. For this system, n-heptane with n-octane at 1 atm, the average relative volatility is $\alpha = 2.16$ (Treybal, 1980). Using equation (6-102) derived in Problem 6.7, compute the composition of the residue after 60 mole% of the feed is batch distilled.

6.9ᵇ. Binary batch distillation with constant relative volatility.

Consider the binary batch distillation of Example 6.3. For this system, n-heptane with n-octane at 1 atm, the average relative volatility is $\alpha = 2.16$ (Treybal, 1980). The mixture will be batch distilled until the average concentration of the distillate is 65 mole% heptane. Using equation (6-102) derived in Problem 6.7, compute the composition of the residue, and the fraction of the feed that is distilled.

Answer: $x_W = 0.419$

6.10ᵇ. Mixtures of light hydrocarbons: *m*-value correlations.

Because of the complex concentration functionality of the *m*-values, VLE calculations in general require iterative procedures suited only to computer solutions. However, in the case of mixtures of light hydrocarbons, we may assume as a reasonable approximation that both the liquid and the vapor phases are ideal. This allows *m*-values for light hydrocarbons to be calculated and correlated as functions of T and P. Approximate values can be determined from the monographs prepared by DePriester (1953). The DePriester charts have been fit to the following equation (McWilliams, 1973):

$$\ln m = \frac{a_{T1}}{T^2} + a_{T2} + a_{P1} \ln P + \frac{a_{P2}}{P^2} + \frac{a_{P3}}{P} \tag{6-103}$$

where T is in K and P is in kPa. The constants $a_{T1}, a_{T2}, a_{P1}, a_{P2}$, and a_{P3} are given in Table 6.8. This equation is valid for temperatures from 200 K to 473 K, and pressures from 101.3 kPa to 6,000 kPa.

What is the bubble point of a mixture that is 15 mole% isopentane (A), 30 mole% n-pentane (B), and 55 mole% n-hexane (C)? Calculate the composition of the first bubble of vapor. The pressure is 1.0 atm.

Answer: $y_A = 0.282$

Table 6.8 Constants in Equation (6-103); T in K, P in kPa

Compound	$-a_{T1}$	a_{T2}	$-a_{P1}$	a_{P2}	a_{P3}
Methane	90,389	9.9730	0.89510	2,846.4	0
Ethylene	185,209	9.5411	0.84677	2,042.6	0
Ethane	212,114	9.6178	0.88600	2,331.8	0
Propylene	285,026	9.4141	0.87871	2,267.6	0
Propane	299,595	8.6372	0.76984	0	47.6017
Isobutane	360,138	9.5073	0.92213	0	0
n-Butane	395,234	9.8124	0.96455	0	0
Isopentane	457,279	9.3796	0.93159	0	0
n-Pentane	470,645	9.0527	0.89143	0	0
n-Hexane	549,044	8.6021	0.84634	0	0
n-Heptane	621,544	8.0651	0.79543	0	0

Source: McWilliams, M. L., *Chem. Eng.*, **80**, 138 (1973).

6.11[b]. Mixtures of light hydrocarbons: *m*-value correlations.

A solution has the following composition, expressed as mole percent: ethane, 0.25%; propane, 25%; isobutane, 18.5%; n-butane, 56%; isopentane, 0.25%. In the following, the pressure is 10 bars. Use equation (6-103) and Table 6.8 to calculate equilibrium distribution coefficients.

(a) Calculate the bubble point.

Answer: 331.2 K

(b) Calculate the dew point.

Answer: 341.9 K

(c) The solution is flash-vaporized to vaporize 40 mol% of the feed. Calculate the temperature and composition of the products.

Answer: 63.2 mole% n-butane in liquid

6.12[b]. Binary batch distillation with constant relative volatility.

A 30 mole% feed of benzene in toluene is to be distilled in a batch operation. A product having an average composition of 45 mole% benzene is to be produced. Calculate the amount of residue left, assuming that $\alpha = 2.5$ and $F = 100$ moles.

Answer: $W = 57.4$ moles

6.13[b]. Batch distillation of a mixture of isopropanol in water.

A mixture of 40 mole% isopropanol in water is to be batch-distilled at 1 atm until 70 mole% of the charge has been vaporized. Calculate the composition of the liquid residue in the still pot, and the average composition of the collected distillate. VLE data for this system, in mole fraction of isopropanol, at 1 atm are (Seader and Henley, 1998):

T, K	366	357	355.1	354.3	353.6	353.2	353.3	354.5
y	0.220	0.462	0.524	0.569	0.593	0.682	0.742	0.916
x	0.012	0.084	0.198	0.350	0.453	0.679	0.769	0.944

Composition of the azeotrope is $x = y = 0.685$; boiling point of the azeotrope = 353.2 K.

Answer: $y_{D,av} = 0.543$

6.14[c]. Batch steam distillation.

An open kettle contains 50 kmoles of a dilute aqueous solution of methanol (2 mole% of methanol), at the bubble point, into which steam is continuously sparged. The entering steam agitates the kettle contents so that they are always of uniform composition, and the vapor produced, always in equilibrium with the liquid, is led away. Operation is adiabatic. For the concentrations encountered it may be assumed that the enthalpy of the steam and evolved vapor are the same, the enthalpy of of the liquid in the kettle is essentially constant, and the relative volatility is constant at 7.6.

a) Show that, under these conditions:

$$V = \frac{F}{\alpha}\left[\ln\left(\frac{x_F}{x_W}\right) + (\alpha-1)(x_F - x_W)\right] \qquad (6\text{-}104)$$

where V = moles of steam required.

(b) Compute the quantity of steam to be introduced in order to reduce the concentration of methanol to 0.1 mole%.

Answer: 20.55 kmoles

6.15^a. Continuous distillation of a binary mixture of constant relative volatility.

For continuous distillation of a binary mixture of constant relative volatility, Fenske equation (6-58) can be used to estimate the minimum number of equilibrium stages required for the given separation, N_{min}.

Use the Fenske equation to estimate N_{min} for distillation of the benzene-toluene mixture of Example 6.4. Assume that, for this system at 1 atm, the relative volatility is constant at $\alpha = 2.5$.

Answer: 6.43 stages

6.16^a. Continuous distillation of a binary mixture of constant relative volatility.

For continuous distillation of a binary mixture of constant relative volatility, the minimum reflux ratio can be determined analytically from the following equation (Treybal, 1980):

$$\frac{R_{min}z_F + qx_D}{R_{min}(1-z_F)+q(1-x_D)} = \frac{\alpha\left[x_D(q-1)+z_F(R_{min}+1)\right]}{(R_{min}+1)(1-z_F)+(q-1)(1-x_D)} \quad (6\text{-}105)$$

Use equation (6-105) to estimate R_{min} for distillation of the benzene-toluene mixture of Example 6.4. Assume that, for this system at 1 atm, the relative volatility is constant at $\alpha = 2.5$.

Answer: $R_{min} = 1.123$

6.17^b. Continuous rectification of a water-isopropanol mixture.

A water-isopropanol mixture at its bubble point containing 10 mole% isopropanol is to be continuously rectified at atmospheric pressure to produce a distillate containing 67.5 mole% isopropanol. Ninety-eight percent of the isopropanol in the feed must be recovered. VLE data are given in Problem 6.13. If a reflux ratio of 1.5 times the minimum is used, how many theoretical stages will be required:

(a) If a partial reboiler is used?

(b) If no reboiler is used and saturated steam at 101.3 kPa is introduced below the bottom plate?

6.18[b]. Rectification of a binary mixture; overall efficiency.

A solution of carbon tetrachloride and carbon disulfide containing 50 wt% each is to be continuously fractionated at standard atmospheric pressure at the rate of 5,000 kg/h. The distillate product is to contain 95 wt% carbon disulfide, the residue 0.5 wt%. The feed will be 30 mole% vaporized before it enters the tower. A total condenser will be used, and reflux will be returned at the bubble point. VLE data at 1 atm (Treybal, 1980), x, y = mole fraction of CS_2, are as follows:

T, K	x	y^*	T, K	x	y^*
349.7	0.0	0.0	332.3	0.391	0.634
347.9	0.030	0.082	328.3	0.532	0.747
346.1	0.062	0.156	325.3	0.663	0.829
343.3	0.111	0.266	323.4	0.757	0.878
341.6	0.144	0.333	321.5	0.860	0.932
336.8	0.259	0.495	319.3	1.000	1.000

(a) Determine the product rates, in kg/h.

(b) Determine the minimum number of theoretical stages required.

(c) Determine the minimum reflux ratio.

(d) Determine the number of theoretical trays required, and the location of the feed tray, at a reflux ratio equal to 1.5 the minimum.

(e) Estimate the overall tray efficiency of a sieve-tray tower of "conventional design" and the number of real trays. The viscosity of liquid CCl_4 and liquid CS_2 as functions of temperature are given by equation (6-42) with:

Component	A	B	C	$D \times 10^5$
CCl_4	− 13.03	2,290	0.0234	− 2.011
CS_2	− 3.442	713.8	0.00	0.0

Source: Reid, et al., 1987.

6.19[b]. Rectification of an ethanol-water mixture; Murphree efficiency.

A distillation column is separating 1,000 moles/h of a 32 mole% ethanol, 68 mole% water mixture at atmospheric pressure. The feed enters as a subcooled liquid that will condense 1 mole of vapor in the feed plate for every four moles of feed. The column has a total condenser and uses open steam heating. We desire a distillate with an ethanol content of 75 mole%, and a bottoms product with 10 mole% ethanol. The steam used is saturated at 1 atm. Table 6.2 gives VLE data for this system.

(a) Find the minimum reflux ratio.

(b) Find the number of real stages and the optimum feed location for $R = 1.5$ R_{min} if the Murphree vapor efficiency is 0.7 for all stages.

(c) Find the steam flow rate used.

6.20[d]. Rectification of an acetone-methanol mixture; large number of stages.

A distillation column is separating 100 moles/s of a 30 mole% acetone, 70 mole% methanol mixture at atmospheric pressure. The feed enters as a saturated liquid. The column has a total condenser and a partial reboiler We desire a distillate with an acetone content of 72 mole%, and a bottoms product with 99.9 mole% methanol. A reflux ratio of 1.25 the minimum will be used. Calculate the number of ideal stages required and the optimum feed location. VLE for this system is described by the modified Raoult's law, with the NRTL equation for calculation of liquid-phase activity coefficients, and the Antoine equation for estimation of the vapor pressures.

The Antoine constants for acetone (1) and methanol (2) are (P_i in kPa, T in °C): $A_1 = 14.3916$, $B_1 = 2,795.92$, $C_1 = 230.0$; $A_2 = 16.5938$, $B_2 = 3,644.30$, $C_2 = 239.76$. The NRTL constants are $b_{12} = 184.70$ cal/mole, $b_{21} = 222.64$ cal/mole, $\alpha = 0.3084$ (Smith, et al., 1996).

Answer: 28 ideal stages, including the reboiler

6.21[c]. Binary distillation with two feeds.

We wish to separate ethanol from water from two different feed solutions in a distillation column with a total condenser and a partial reboiler, at a column pressure of 1 atm. One of the feed solutions, a saturated vapor, flows at the rate of 200 kmoles/h and contains 30 mole% ethanol. The other feed, a subcooled liquid, flows at the rate of 300 kmoles/h and contains 40 mole% ethanol. One mole of vapor must condense inside the column to heat up 4 moles of the second feed to its boiling point. We desire a bottoms product that is 2 mole% ethanol and a distillate product that is 72 mole% ethanol. The reflux ratio is 1.0; VLE data are available in Table 6.2. The feeds are to be input at their optimum feed locations. Find the optimum feed locations and the total number of equilibrium stages required.

Answer: 6.5 ideal stages, including the reboiler

6.22[c]. Binary distillation with a liquid side stream.

A mixture of ethanol and methanol enters a distillation column at the rate of 100 kmoles/h. The overall concentration of the mixture is 75 mole% methanol, 25 mole% ethanol, and it is 25 mole% vaporized. A distillate containing 96 mole% methanol, and a bottoms product containing 95 mole% ethanol are produced. A liquid side stream is removed from the distillation column at the rate of 15 kmoles/h, containing 80 mole% ethanol. A reflux ratio 1.2 the minimum is specified. The column operates at a pressure of 1 atm. Methanol and ethanol form ideal solutions. Determine the number of theoretical stages required and the optimal location of the feed and side stream.

The Antoine constants for methanol (1) and ethanol (2) are (P_i in kPa, T in °C): A_1 = 14.3916, B_1 = 2,795.92, C_1 = 230.0; A_2 = 16.5938, B_2 = 3,644.30, C_2 = 239.76. The NRTL constants are b_{12} = 184.70 cal/mole, b_{21} = 222.64 cal/mole, α = 0.3084 (Smith, et al., 1996).

6.23[c]. Binary distillation in a random packed column.

Redesign the fractionator of Example 6.8 using a random packing. The column is to be packed with 50-mm metal Pall rings. Determine the diameter of the tower, the height of packing in the stripping and rectifying sections, and the total gas-pressure drop. Design for a gas-pressure drop not to exceed 400 Pa/m.

6.24[c]. Binary distillation in a structured packed column.

Redesign the fractionator of Example 6.8 for a reflux ratio that is twice the minimum. Determine the diameter of the tower, the height of packing in the stripping and rectifying sections, and the total gas-pressure drop. Design for a gas-pressure drop not to exceed 400 Pa/m.

6.25^c. A distillation-membrane hybrid for ethanol dehydration.

Many industrially important liquid systems are difficult or impossible to separate by simple continuous distillation because the phase behavior contains an azeotrope, a tangent pinch, or an overall low relative volatility. One solution is to combine distillation with one or more complementary separation technologies to form a hybrid. An example of such a combination is the dehydration of ethanol using a distillation-membrane hybrid, as shown in Figure 6.26.

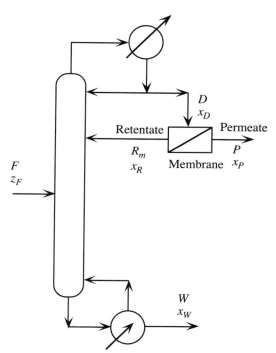

Figure 6.26 Distillation-membrane hybrid.

The membrane performance is specified by defining a *membrane separation factor*, α_m, and *membrane cut*, θ:

$$\alpha_m = \frac{x_P(1-x_R)}{x_R(1-x_P)} \tag{6-106}$$

$$\theta = \frac{P}{D} \tag{6-107}$$

In a given application, 100 moles/s of a saturated liquid containing 37 mole% ethanol and 63 mole% water must be separated to yield a product which is 99 mole% ethanol, and a residue containing 99 mole% water. The solution will be fed to a distillation column operating at atmospheric pressure, with a partial reboiler and a total condenser. The reflux ratio will be 1.5 the minimum. The distillate will enter a membrane with parameters $\alpha_m = 70$ and $\theta = 0.6$. The membrane boosts the concentration so that the permeate stream is the ethanol-rich product ($x_P = 0.99$). The retentate stream is returned as a saturated liquid to the column to the tray at the nearest liquid concentration.

(a) Calculate the molar flow rate of the product and of the residue.

Answer: $P = 36.7$ moles/s

(b) Calculate the molar flow rate and composition of the distillate coming out of the column.

Answer: $x_D = 0.828$

(c)Calculate the molar flow rate and composition of the retentate returned to the column.

Answer: $x_R = 0.586$

(d) Calculate the number of ideal stages required, and the optimal location of the two feeds (the original feed and the retentate recycle) to the distillation column.

6.26[a]. Gilliland correlation applied to a binary system.

Use the Gilliland correlation to estimate the number of ideal stages required for the separation of Example 6.4. Assume that, for the system benzene-toluene at atmospheric pressure, the relative volatility is constant at $\alpha = 2.5$. Use the results of Problems 6.15 and 6.16. Use the Kirkbride equation to estimate the optimum feed-stage location.

6.27[b]. Fenske-Underwood-Gilliland method.

A distillation column has a feed of 100 kmoles/h. The feed is 10 mole% LNK, 55 mole% LK, and 35 mole% HK and is a saturated liquid. The reflux ratio is 1.2 the minimum. We desire 99.5% recovery of the light key in the distillate. The mole fraction of the light key in the distillate should be 0.75. Use the FUG approach to estimate the number of ideal stages required and the optimal location of the feed-stage. Equilibrium data:

$$\alpha_{LNK} = 4.0 \qquad \alpha_{LK} = 1.0 \qquad \alpha_{HK} = 0.75$$

Answer: $N = 49$ ideal stages

6.28[b]. Fenske-Underwood-Gilliland method.

One hundred kmoles/h of a ternary bubble-point mixture to be separated by distillation has the following composition:

Component	Mole fraction	Relative volatility
A	0.4	5
B	0.2	3
C	0.4	1

(a) For a distillate rate of 60 kmoles/h, five theoretical stages, and total reflux, calculate the distillate and bottoms composition by the Fenske equation.

Answer: $x_{AD} = 0.662, x_{CW} = 0.954$

(b) Using the separation in part a) for components A and C, determine the minimum reflux ratio by the Underwood equation.

Answer: $R_{min} = 0.367$

(c) For a reflux ratio of 1.2 times the minimum, determine the number of theoretical stages required, and the optimum feed location.

Answer: 14.3 ideal stages

6.29[c]. Fenske-Underwood-Gilliland method.

The following feed at 355 K and 1,035 kPa is to be fractionated at the given pressure so that the distillate contains 98% of the C_3H_8, but only 1% of the C_5H_{12}:

Component	CH_4	C_2H_6	$n\text{-}C_3H_8$	$n\text{-}C_4H_{10}$	$n\text{-}C_5H_{12}$	$n\text{-}C_6H_{14}$
z_F, mole %	3.0	7.0	15.0	33.0	30.0	12.0

(a) Compute the number of theoretical stages at total reflux.

(b) Estimate the minimum reflux ratio.

(c) Estimate the number of theoretical stages and the optimal location of the feed tray for a reflux ratio $R = 0.8$.

See Problem 6.10 for VLE data.

6.30[d]. Extractive distillation (Use of ChemSep is suggested).

Extractive distillation is used for the separation of azeotropes and close-boiling mixtures. In this process, a solvent is added to the distillation column. This solvent is selected so that one of the components, B, is selectively attracted to it. Since the solvent is usually chosen to have a significantly higher boiling point than the component being separated, the attracted component, B, has its volatility reduced. Thus, the other component, A, becomes more volatile and is easy to remove in the distillate. An additional column is required to separate the solvent and component B (Wankat, 1988).

The following bubble-point mixture at 1.4 atm is distilled by extractive distillation with the following phenol-rich solvent at 1.4 atm and at the same temperature as the main feed (Seader and Henley, 1998):

Component	Feed, kmoles/h	Solvent, kmoles/h
Methanol	50	0
n-Hexane	20	0
n-Heptane	180	0
Toluene	150	10
Phenol	0	800

The column has 30 sieve trays, with a total condenser and a partial reboiler. The solvent enters the 5th tray and the feed enters tray 15, from the top. The pressure in the condenser is 1.1 atm; the pressure at the top tray is 1.2 atm, and the pressure at the bottom is 1.4 atm. The reflux ratio is 5 and the bottoms rate is 960 kmoles/h. Use the nonequilibrium model of the ChemSep program to estimate the separation achieved. Assume that the vapor and the liquid are both well mixed, and that the trays operate at 75% of flooding. In addition, determine from the tray-by-tray results the average Murphree tray efficiency for each component.

6.31[d]. Equilibrium- versus rate-based model for distillation of a ternary mixture (The use of ChemSep is suggested).

A ternary mixture of methanol, ethanol, and water is distilled in a sieve-tray column to obtain a distillate with not more than 0.01 mole% water. The feed to the column is as follows:

Flow rate, kmoles/h	142.46
Pressure, atm	1.3
Temperature, K	316
Mole fractions:	

Methanol	0.6536
Ethanol	0.0351
Water	0.3113

For a distillate rate of 93.1 kmoles/h, a reflux ratio of 1.2, a condenser outlet pressure of 1.0 atm, and a top tray pressure of 1.1 atm, determine:

(a) The number of equilibrium stages required and the corresponding split, if the feed enters at the optimal stage.

(b) The number of actual stages required if the column operates at about 85% of flooding and the feed is introduced to the optimal tray. Compare the split to that of part a.

(c) Compute the component Murphree tray efficiencies.

6.32[d]. Equilibrium- versus rate-based model for distillation of a ternary mixture (Use of ChemSep is suggested).

Repeat Problem 6.30 for a column packed with 50-mm stainless steel Pall rings.

6.33[d]. Depropanizer column design (Use of ChemSep is suggested).

A depropanizer is a distillation operation encountered in almost all oil refineries. Our task here is to design a sieve-tray column to separate 1,000 moles/s of a mixture containing 100 moles/s of ethane, 300 moles/s of propane, 500 moles/s of n-butane, and 100 moles/s of n-pentane at 298 K and 15 atm. The distillate should contain no more than 3.5 moles/s of n-butane, and the bottoms should contain no more than 3.5 moles/s of propane. The trays should operate at about 70% of flooding (Taylor and Krishna, 1993).

6.34[d] Nonequilibrium distillation in packed tower (Use of ChemSep is suggested).

Repeat Example 6.17 for a tower packed with 50-mm ceramic Berl saddles.

6.35^d Equilibrium distillation calculations with multiple feeds (Use of ChemSep is suggested).

Calculate the product compositions, stage temperatures, interstage vapor and liquid flow rates and compositions, reboiler duty, and condenser duty for the following multiple-feed distillation column, which has 30 equilibrium stages exclusive of a partial condenser and a partial reboiler and operates at 17 atm.

Feeds (both bubble-point liquids at 17 atm)

Flow rates (moles/s)	Feed 1 to stage 15 from the bottom	Feed 2 to stage 6 from the bottom
Ethane	0.189	0.063
Propane	3.024	1.260
n-Butane	2.079	2.772
n-Pentane	0.945	1.827
n-Hexane	0.063	0.378

Distillate rate = 4.536 moles/s
Reflux rate = 18.9 moles/s

6.36^c Flash Calculations: the Rachford-Rice Method for Ideal Mixtures.

(a) Show that the problem of flash vaporization of a multicomponent ideal mixture can be reformulated as suggested by Rachford and Rice (Doherty and Malone, 2001):

$$f(\phi) = \sum_{i=1}^{C} \frac{z_i(m_i - 1)}{1 + \phi(m_i - 1)} = 0 \qquad (6\text{-}108)$$

where

$$m_i = P_i/P$$

$$x_i = \frac{z_i}{1 + \phi(m_i - 1)} \qquad i = 1, 2, \cdots, C \qquad (6\text{-}109)$$

$$y_i = m_i x_i \qquad i = 1, 2, \cdots, C \qquad (6\text{-}110)$$

$$D = \phi F \qquad W = (1 - \phi)F \qquad (6\text{-}111)$$

Equation (6-108) is solved iteratively for ϕ; all other variables are then calculated explicitly from equations (6-109) to (6-111).

(b) Solve Example 6.2 using the Rachford-Rice method.

REFERENCES

Baur, R., et al., *Chem. Eng. Sci.*, **56**, 1721, 2085 (2001).

Chan, H., and J. R. Fair, *Ind. Eng. Chem. Process Des. Dev.*, **23**, 814 (1984).

DePriester, C. L., *Chem. Eng. Progr. Symp. Ser.*, **49**, 42 (1953).

Doherty, M. F., and M. F. Malone, *Conceptual Design of Distillation Systems*, McGraw-Hill, New York (2001).

Fenske, M. R., *Ind. Eng. Chem.*, **24**, 482 (1932).

Gilliland, E. R., *Ind. Eng. Chem.*, **32**, 1220 (1940).

Hachmuth, K. H., *Chem. Eng. Prog.*, **47**, 523, 621 (1951).

Holman, J. P., *Heat Transfer*, 7th ed., McGraw-Hill, New York (1990).

Hwalek, J., J., University of Maine (2001).

"International Critical Tables," vol. V, p. 213, McGraw-Hill, New York (1929).

King, C. J., *Separation Processes*, 2nd ed., McGraw-Hill, New York (1980).

Kirkbride, C. G., *Petroleum Refiner.*, **23**, 87 (1944).

Kister, H. Z., *Distillation Design*, McGraw-Hill, New York (1992).

Krishna, R., et al., *Trans. I. Chem. E.*, **55**, 178 (1977).

Krishna, R., and G. L. Standardt, *Chem. Eng. Comm.*, **3**, 201 (1979).

Krishnamurthy, R., and R. Taylor, *AIChE J.*, **31**, 449, 456 (1985).

McCabe, W. L., and E. W. Thiele, *Ind. Eng. Chem.*, **17**, 605 (1925).

McWilliams, M. L., *Chem. Eng.*, **80**, 138 (1973).

Molokanov, Y. K., et al., *Int. Chem. Eng.*, **12**, 209 (1972).

Perry, R. H., and C. H. Chilton (eds.), *Chemical Engineers' Handbook,* 5th ed., McGraw-Hill, New York (1973).

Raal, J. D., and M. K. Khurani, *Can. J. Chem. Eng.*, **51**, 162 (1973).

Reid, R. C., J. M. Prausnitz, and B. E. Poling, *The Properties of Gases and Liquids*, 4th ed., McGraw-Hill, Boston (1987).

Seader, J. D., and E. J. Henley, *Separation Process Principles*, Wiley, New York (1998).

Shiras, R. N., et al., *Ind. Eng. Chem.*, **42**, 871 (1950).

Smith, J. M., H. C. Van Ness, and M. M. Abbott, *Introduction to Chemical Engineering Thermodynamics*, 5th ed., McGraw-Hill., New York (1996).

Taylor, R. and R. Krishna, *Multicomponent Mass Transfer*, Wiley, New York, (1993).

Taylor, R., et al., *Comput. Chem. Engng.*, **18**, 205 (1994).

Treybal, R. E., *Mass-Transfer Operations*, 3rd ed., McGraw-Hill, New York (1980).

Wang, J. C., and G. E. Henke, *Hydrocarbon Processing*, **45**, 155 (1966).

Wankat, P. C., *Equilibrium Staged Separations*, Elsevier, New York (1988).

7

Liquid–Liquid Extraction

7.1 INTRODUCTION

Liquid–liquid extraction, sometimes called *solvent extraction*, is the separation of the constituents of a liquid solution by contact with another insoluble liquid. If the substances constituting the original solution distribute themselves differently between the liquid phases, a certain degree of separation is achieved. This can be enhanced by the use of multiple contacts, or their equivalent, in the manner of gas absorption and distillation.

A simple example will indicate the scope of the operation and some of its characteristics. Acetic acid is produced by methanol carbonylation or oxidation of acetaldehyde, or as a byproduct of cellulose acetate manufacture. In all three cases, a mixture of acetic acid and water must be separated to give glacial acetic acid (99.8 wt%, min). Separation by distillation is expensive because of the need to vaporize large amounts of water with its very high heat of vaporization. Accordingly, an alternative, less expensive liquid–liquid extraction process is often used.

If a solution of acetic acid in water is agitated with a liquid such as ethyl acetate, some of the acid but relatively little water will enter the ester phase. Since at equilibrium the densities of the aqueous and ester layers are different, they will settle when agitation stops and can be decanted from each other. Since now the ratio of acid to water in the ester layer is different from that in the original solution and also different from that in the residual water solution, a certain degree of separation will have occurred. This is an example of stagewise contact, and it can be carried out either in batch or in continuous fashion. The residual water can be repeatedly extracted with more ester to reduce the acid content still further, or we can arrange a countercurrent cascade of stages. Another possibility is to use some sort of countercurrent continuous-contact device, where discrete stages are not involved. The use of reflux may enhance the ultimate separation still further.

384

There are some other applications of liquid–liquid extraction where it seems uniquely qualified as a separation technique. Many pharmaceutical products (e. g., penicillin) are produced in mixtures so complex that only liquid–liquid extraction is a feasible separation process (Seader and Henley, 1998).

In all such operations, the solution which is to be extracted is called the *feed,* and the liquid with which it is contacted is the *solvent*. The solvent-rich product of the operation is called the *extract*, and the residual liquid from which solute has been removed is the *raffinate* (Treybal, 1980).

7.2 LIQUID EQUILIBRIA

Your objectives in studying this section are to be able to:

1. Plot extraction equilibrium data on equilateral- and right-triangular diagrams.
2. Explain the difference between type I and type II extraction equilibrium behavior.
3. Find the saturated extract, saturated raffinate, and conjugate lines on a right-triangular diagram.

Extraction involves the use of systems composed of at least three substances, and although for the most part the insoluble phases are chemically very different, generally all three components appear at least to some extent in both phases. A number of different phase diagrams and computational techniques have been devised to determine the equilibrium compositions in ternary mixtures. In what follows, A and B are pure, substantially insoluble liquids, and C is the distributed solvent. Mixtures to be separated by extraction are composed of A and C, and B is the extracting solvent.

Equilateral-triangular diagrams are extensively used in the chemical literature to graphically describe the concentrations in ternary systems. It is the property of an equilateral triangle that the sum of the perpendicular distances from any point within the triangle to the three sides equals the altitude of the triangle. We can, therefore, let the altitude represent 100 percent composition and the distances to the three sides the percentages or fractions of the three components (refer to Figure 7.1). Each apex of the triangle represents a pure component, as marked. The perpendicular distance from any point such as K to the base AB represents the percentage of C in the mixture at K, the distance to the base AC the percentage of B, and that to the base CB the percentage of A. Thus the composition at point K in Figure 7.1 is 40% A, 20% B, and 40% C.

Any point on a side of the triangle represents a binary mixture. Point D, for example, is a binary mixture containing 80% A, 20% B. All points on the line DC represent mixtures containing the same ratio of A to B and can be considered as mixtures originally at D to which C has been added.

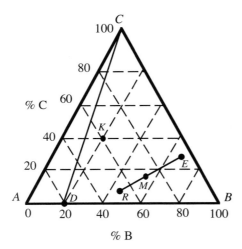

Figure 7.1 Equilateral-triangular diagram.

If R kg of a mixture at point R is added to E kg of a mixture at E, the new mixture is shown on the straight line RE at point M such that (see Prob. 7.1)

$$\frac{R}{E} = \frac{\text{line } ME}{\text{line } RM} = \frac{x_E - x_M}{x_M - x_R} \tag{7-1}$$

where x is the weight fraction of C in the solvent-lean, or raffinate liquids. This is known as *the lever-arm rule*. Similarly, if a mixture at M has removed from it a mixture of composition E, the new mixture is on the straight line EM extended in the direction opposite to E and located at R such that equation (7-1) applies.

Extraction systems are noted for the wide variety of equilibrium behavior that can occur in them. The most common type of system behavior is shown in Figure 7.2, often called a type I system, since there is one pair of immiscible binary compounds. The triangular coordinates are used as *isotherms*, or diagrams at constant temperature. Liquid C dissolves completely in A and B, but A and B dissolve only to a limited extent in each other to give rise to the saturated liquid solutions at L (A-rich) and at K (B-rich). A binary mixture J anywhere between L and K will separate into two insoluble liquid phases of compositions at L and K. The relative amounts of the phases depends upon the position of J, according to the principle of equation (7-1). A typical example of a system exhibiting type I behavior is water (A)–chloroform (B)–acetone (C).

Curve $LRPEK$ is the binodal solubility curve, indicating the change in solubility of the A- and B-rich phases upon addition of C. Any mixture outside this curve will be a homogeneous solution of one liquid phase. Any ternary mixture underneath the curve, such as M, will form two insoluble, saturated liquid phases of equilibrium compositions indicated by R (A-rich) and E (B-rich). The line RE joining these equilibrium compositions is a *tie line*, which must necessarily pass through M representing the mixture as a whole. There are an infinite number of tie lines in the two-phase region. The tie lines converge to point P, the *plait point*, where the two phases become one phase. The plait point is ordinarily not at the maximum value of C on the solubility curve.

The percentage of C in solution E is clearly greater than in solution R, and it is said that, in this case, the distribution of C favors the B-rich phase. This is conveniently shown on the distribution diagram xy in Figure 7.2. Here, y is the weight fraction of C in the solvent-rich (B-rich) liquid, or extract. The ratio $y*/x$, *the distribution coefficient,* is in this case greater than unity. The distribution curve lies above the diagonal $y = x$. Should the tie lines on Figure 7.2 slope in the opposite direction, the distribution curve will lie below the diagonal.

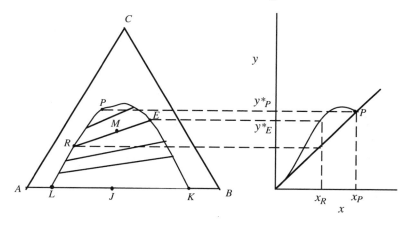

Figure 7.2 Type I equilibrium diagram: A and B partially miscible.

A less common type of system equilibrium behavior, type II, is illustrated in Figure 7.3, a typical isotherm. In this case, A and C are completely soluble, while the pairs A–B and B–C show only limited solubility. At the prevailing temperature, points J and K represent the mutual solubilities of A and B, and points H and L those of B and C. Curves KRH (A-rich) and JEL (B-rich) are the ternary solubility curves, and mixtures outside the band between these two curves form homogeneous single-phase liquid solutions. Mixtures such as M, inside the heterogeneous area, form two liquid phases at equilibrium at E and R, joined in the diagram by a tie line. Figure 7.3 also shows the corresponding distribution curve. Notice that type II systems do not have a plait point. This type is exemplified by the system chlorobenzene (A), water (B), and methyl ethyl ketone (C).

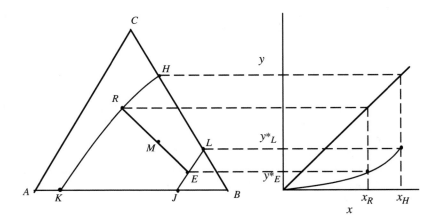

Figure 7.3 Type II equilibrium diagram: A–B and B–C partially miscible.

Whether a ternary system is type I or type II often depends on the temperature. For example, data for the system n-hexane (A), methylcyclopentane (C), and aniline (B) for temperatures of 298 K, 307.5 K, and 318 K are shown in Figure 7.4. At the lowest temperature, we have a type II system. As the temperature increases, the solubility of C in B increases more rapidly than the solubility of A in B until at 34.5 °C the system is at the border between type II and type I behavior. At 45 °C the system is clearly of type I.

The coordinate scales of equilateral triangles are necessarily always the same. In order to be able to expand one concentration scale relative to the other, right-triangular coordinates can be used. Here the concentrations in weight percent of any two of the three components are given, the concentration of the third obtained by difference from 100 wt%. Figure 7.5 is an example of a right-triangular diagram.

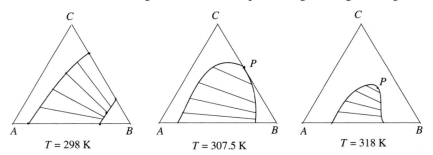

Figure 7.4 Effect of temperature on solubility for the system n-hexane (A), aniline (B), and methylcyclopentane (C).

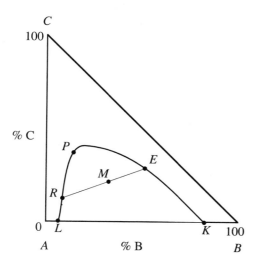

Figure 7.5 Right-triangular diagram.

We will use right-triangular diagrams exclusively in the remainder of this chapter because they are easy to read, they do not require special paper, the scales of the axes can be varied, and portions of the diagram can be enlarged. Although equilateral-triangular diagrams have none of these advantages, they are used extensively in the literature for reporting extraction data; therefore it is important to be able to read and use this type of extraction diagram (Wankat, 1988). Notice that the lever-arm rule also applies to right-triangular diagrams.

Example 7.1 Equilibrium: System Water–Chloroform–Acetone

Table 7.1 presents equilibrium extraction data obtained by Alders (1959) for the system water–chloroform–acetone at 298 K and 1 atm. Plot these data on a right-triangular diagram, including in the diagram a *conjugate* or *auxiliary line* that will allow graphical construction of tie lines.

Solution

Figure 7.6 is the right-triangular diagram for this system. We have chosen chloroform as solvent (B), water as diluent (A), and acetone as solute (C). This system exhibits a type I equilibrium behavior. The graphical construction to generate the conjugate line from the tie-line data given (for example, line *RE*) is shown, as well as the use of the conjugate line to generate other tie lines not included in the original data. Line *LRP* is the saturated raffinate line, while line *PEK* is the saturated extract line.

Table 7.1 Water–Chloroform–Acetone Equilibrium Data

Water-phase, wt %				Chloroform-phase, wt%		
Water	Chloroform	Acetone		Water	Chloroform	Acetone
82.97	1.23	15.80		1.30	70.00	28.70
73.11	1.29	25.60		2.20	55.70	42.10
62.29	1.71	36.00		4.40	42.90	52.70
45.60	5.10	49.30		10.30	28.40	61.30
34.50	9.80	55.70		18.60	20.40	61.00
23.50	16.90	59.60		23.50	16.90	59.60

Water and chloroform phases on the same line are in equilibrium with each other. Data were obtained by Alders (1959) at 298 K and 1 atm.

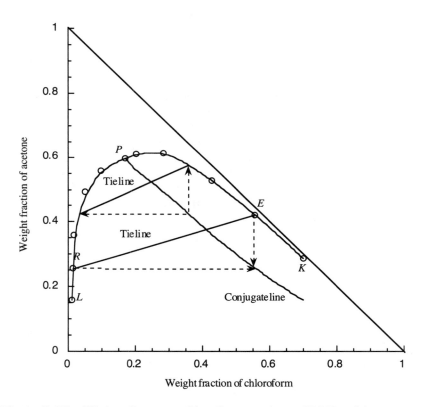

Figure 7.6 Equilibrium for water-chloroform-acetone at 298 K and 1 atm.

7.3 STAGEWISE LIQUID–LIQUID EXTRACTION

Your objectives in studying this section are to be able to:

1. Explain the concept and importance of dimensional analysis in correlating experimental data on convective mass-transfer coefficients.
2. Use the Buckingham method to determine the dimensionless groups significant to a given mass-transfer problem.

Extraction in equipment of the stage type can be carried on according to a variety of flowsheets, depending upon the nature of the system and the extent of separation desired. In the discussion which follows it is to be understood that each stage is a *theoretical* or *equilibrium* stage, such that the effluent extract and raffinate solutions are in equilibrium with each other. Each stage must include facilities for contacting the insoluble liquids and separating the product streams. A combination of a mixer and a settler may therefore constitute a stage, and in multistage operations these may be arranged in cascades as desired. In countercurrent multistage operation, it is also possible to use towers of the multistage type as described in previous chapters.

7.3.1 Single-Stage Extraction

This may be a batch or continuous operation (refer to Figure 7.7). Feed of mass F (if batch) of F mass/time (if continuous) contains substances A and C at $x_{C,F}$ weight fraction of C. This is contacted with mass S (or mass/time) of a solvent, principally B, containing $y_{C,S}$ weight fraction of C, to give an equilibrium extract E and raffinate R, each measured in mass or mass/time. Solvent recovery then involves separate removal of solvent B from each product stream (not shown).

The operation can be followed in the right-triangular diagram as shown. If the solvent is pure B ($y_{C,S} = 0$), it will be plotted at the B apex. If the solvent has been recovered from a previous extraction, it may contain small amounts of A and B as shown by the location of S on the diagram. Adding S to F produces in the extraction stage a mixture M which, on settling, separates into the equilibrium phases E and R joined by the tie line through M. A total material balance is

$$F + S = M = E + R \qquad (7\text{-}2)$$

and point M can be located on line FS by the lever arm rule of equation (7-1), but it is usually more satisfactory to locate M by computing its C concentration. Thus, a C balance provides

$$Fx_{C,F} + Sy_{C,S} = Mx_{C,M} \qquad (7\text{-}3)$$

from which $x_{C,M}$ can be computed. Alternatively, the amount of solvent required to provide a given location for M on the line FS can be computed:

$$\frac{S}{F} = \frac{x_{C,F} - x_{C,M}}{x_{C,M} - y_{C,S}} \qquad (7\text{-}4)$$

The quantities of extract and raffinate can be computed from the lever arm rule, or by the material balance for C:

$$Ey_{C,E} + Rx_{C,R} = Mx_{C,M} \qquad (7\text{-}5)$$

$$E = \frac{M\left(x_{C,M} - x_{C,R}\right)}{y_{C,E} - x_{C,R}} \qquad (7\text{-}6)$$

and R can be determined from equation (7-2).

Since two insoluble phases must form for an extraction operation, point M must lie within the heterogeneous liquid area, as shown. The minimum amount of solvent is thus found by locating M at D, which would then provide an infinitesimal

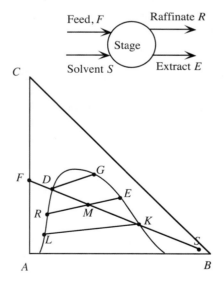

Figure 7.7 Single-stage extraction.

amount of extract at G, and the maximum amount of solvent is found by locating M_1 at K, which would then provide an infinitesimal amount of raffinate at L.

Example 7.2 Single-Stage Extraction.

Pure chloroform is used to extract acetone from a feed containing 60 wt% acetone and 50 wt% water. The feed rate is 50 kg/h, and the solvent rate is also 50 kg/h. Operation is at 298 K and 1 atm. Find the extract and raffinate flow rates and compositions when one equilibrium stage is used for the separation.

Solution

This problem can be solved graphically or algebraically. Next, we will illustrate both the graphical and algebraic solutions.

(a) Graphical solution:

The equilibrium data for this system can be obtained from Table 7.1 and Figure 7.6. Plot stream F ($x_{CF} = 0.6$, $x_{BF} = 0.0$) and S ($y_{CS} = 0.0$, $y_{BS} = 1.0$). After locating streams F and S, M is on the line FS; its exact location is found by calculating x_{CM} from

$$x_{CM} = \frac{Fx_{CF} + Sy_{CS}}{F + S} = \frac{(50)(0.6) + (50)(0)}{50 + 50} = 0.3$$

A tie line is then constructed through M by trial and error, using the conjugate line, and the extract (E) and raffinate (R) locations are obtained (see Figure 7.8). The concentrations are $x_{CR} = 0.189$, $x_{BR} = 0.013$, $y_{CE} = 0.334$, $y_{BE} = 0.648$. The flow rates of extract and raffinate can be obtained from equations (7.6) and (7.2):

$$E = \frac{M(x_{CM} - x_{CR})}{y_{CE} - x_{CR}} = \frac{(50 + 50)(0.300 - 0.189)}{(0.334 - 0.189)} = 76.71 \text{ kg / h}$$

$$R = F + S - E = 50 + 50 - 76.71 = 23.29 \text{ kg / h}$$

(b) Algebraic solution:

Using the data from Table 7.1, generate a cubic spline interpolation formula for the saturated raffinate curve, one for the saturated extract curve, and one for the tie-line data (see the Mathcad program in Appendix G-1). Equations in the forms

$$x_{B,R}(x_{C,R}) = f_{raf}(x_{C,R}) \qquad y_{C,E}(x_{C,R}) = f_{tie}(x_{C,R}) \qquad y_{B,E}(x_{C,R}) = f_{ext}(y_{C,E}(x_{C,R}))$$

are convenient because they reduce the number of unknown equilibrium concentrations from four $(x_{B,R}, x_{C,R}, y_{B,E}, y_{C,E})$ to only one $(x_{C,R})$. Then, the total material balance balance represented by equation (7-2) is solved simultaneously with the solute (C) material balance:

$$Fx_{C,F} + Sy_{C,S} = Rx_{C,R} + Ey_{C,E}(x_{C,R}) \qquad (7\text{-}7)$$

and the solvent (B) material balance:

$$Fx_{B,F} + Sy_{B,S} = Rx_{B,R}(x_{C,R}) + Ey_{B,E}(x_{C,R}) \qquad (7\text{-}8)$$

Equations (7-2), (7-7), and (7-8) constitute a system of three simultaneous equations in three unknowns, E, R, and $x_{C,R}$, which are easily solved using the "solve block" feature of Mathcad. From Appendix G-1, the results are: $x_{CR} = 0.189$, $E = 76.71$ kg/h, $R = 23.29$ kg/h, which agree with the values obtained graphically.

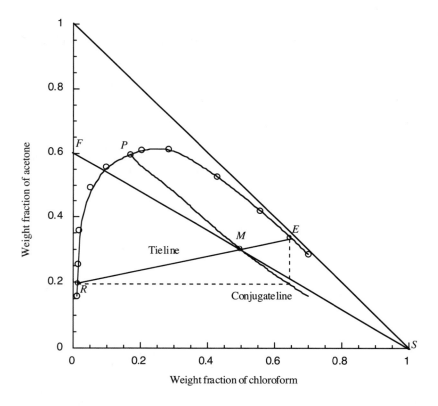

Figure 7.8 Graphical solution of Example 7.2: single-stage extraction.

7.3.2 Multistage Crosscurrent Extraction

This is an extension of single-stage extraction wherein the raffinate is successively contacted with fresh solvent, and may be done continuously or in batches. Figure 7.9 is a schematic diagram for a three-stage crosscurrent extraction process. A single final raffinate results, and the extracts can be combined to provide a composited extract, as shown. As many stages as necessary can be used.

Computations are shown in Figure 7.9 on a right-triangular diagram. All the material balances for a single stage now apply, of course, to the first stage. Subsequent stages are dealt with in the same manner, except that the feed to any stage is the raffinate from the previous stage. Unequal amounts of solvent can be used in the various stages. For a given final raffinate concentration, the greater the number of stages the less total solvent will be used.

Example 7.3 Multistage Crosscurrent Extraction

The feed of Example 7.2 is extracted three times with pure chloroform at 298 K, using 8 kg/h of solvent in each stage. Determine the flow rates and compositions of the various streams.

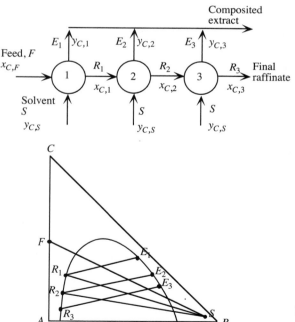

Figure 7.9 Multistage crosscurrent extraction.

Solution

This problem can be solved graphically, as shown schematically in Figure 7.9, or algebraically. Appendix G-2 presents a modification of the Mathcad program of Appendix G-1 that can handle multistage crosscurrent extraction algebraic calculations. The following table summarizes the results, stage by stage.

| | Flow Rates, kg/h | | Compositions, wt% | | | |
| | | | Raffinate | | Extract | |
Stage #	Raffinate	Extract	Acetone	Chloroform	Acetone	Chloroform
1	36.36	21.64	46.6	4.1	60.4	30.2
2	35.76	18.60	31.1	1.4	48.0	49.1
3	21.42	12.34	18.5	1.3	32.8	65.4

The composited extract flows at the rate of 52.58 kg/h, and its composition is 5.4 wt% water, 45.1 wt% chloroform, and 49.5 wt% acetone. Notice that the acetone content of the final raffinate (18.5 wt%) is slightly lower than the acetone content of the raffinate obtained in Example 7.2 in a single-stage extraction with 50 kg/h of chloroform. Remarkably, this is true even though the total amount of solvent used in the multistage crosscurrent extraction process is only $3 \times 8 = 24$ kg/h.

7.3.3 Countercurrent Extraction Cascades

A countercurrent cascade allows for more complete removal of the solute, and the solvent is reused so less is needed. Figure 7.10 is a schematic diagram of a countercurrent extraction cascade. Extract and raffinate streams flow from stage to stage in countercurrent fashion and yield two final products, raffinate R_N and extract E_1. For a given degree of separation, this type of operation requires fewer stages for a given amount of solvent, or less solvent for a fixed number of stages, than the crosscurrent extraction method described above.

In the usual design problem, the extraction column temperature and pressure, the flow rates and compositions of streams F and S, and the desired composition (or percent removal) of solute in the raffinate product are specified. The designer must

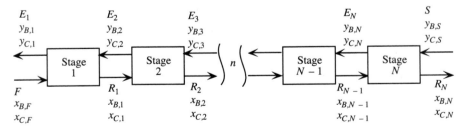

Figure 7.10 Schematic diagram of countercurrent multistage extraction.

determine the number of equilibrium stages needed and the flow rate and composition of the outlet extract stream.

The graphical approach to solution is developed in Figure 7.11 on right-triangular coordinates. A total material balance around the entire cascade is

$$F + S = E_1 + R_N = M \qquad (7-9)$$

Point M can be located on line FS through a material balance for substance C:

$$Fx_{C,F} + Sy_{C,S} = E_1 y_{C,1} + R_N x_{C,N} = Mx_{C,M} \qquad (7-10)$$

$$x_{C,M} = \frac{Fx_{C,F} + Sy_{C,S}}{F + S} \qquad (7-11)$$

Equation (7-9) indicates that point M must lie on line $R_N E_1$ as shown. Rearranging equation (7-9) provides

$$R_N - S = F - E_1 = \Delta_R \qquad (7-12)$$

where Δ_R, called a *difference point*, is the net flow outward at the last stage N. According to equation (7-12), the extended lines $E_1 F$ and SR_N must intersect at Δ_R, as shown in Figure 7.11. A material balance for stages n through N is

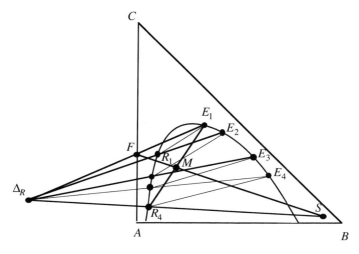

Figure 7.11 Countercurrent multistage extraction on a right-triangular diagram.

$$R_{n-1} + S = R_N + E_n \qquad \text{(7-13)}$$

or

$$R_{n-1} - E_n = R_N - S = \Delta_R \qquad \text{(7-14)}$$

so that the difference in flow rates at a location between any two adjacent stages is constant , Δ_R. Line $E_n R_{n-1}$ extended must therefore pass through Δ_R as shown on the figure.

The graphical construction is then as follows. After location of points F, S, M, E_1, R_N, and Δ_R, a tie line from E_1 provides R_1 since extract and raffinate from the first ideal stage are in equilibrium. A line from Δ_R through R_1 when extended provides E_2; a tie line from E_2 provides R_2, and so forth. The lowest possible value of $x_{C,N}$ is that given by the A-rich end of the tie line which passes through S.

As the amount of solvent is increased, point M representing the overall balance moves toward S on Figure 7.12 and point Δ_R moves further to the left. At an amount of solvent such that lines $E_1 F$ and SR_N are parallel, point Δ_R will be at an infinite distance. Greater amounts of solvent will cause these lines to intersect on the right-hand side of the diagram rather than as shown, with point Δ_R nearer B for increasing solvent quantities. The interpretation of the difference point is, however, still the same: a line from Δ_R intersects the two branches of the solubility curve at points representing extract and raffinate from adjacent stages.

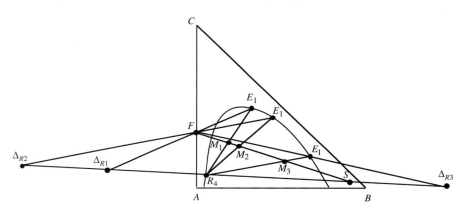

Figure 7.12 Effect of increasing solvent/feed ratio on the location of the difference point $[(S/F)_3 > (S/F)_2 > (S/F)_1]$.

If a line from point Δ_R should coincide with a tie line, an infinite number of stages will be required to reach this condition. The maximum amount of solvent for which this occurs corresponds to the *minimum solvent/feed ratio* which can be used for the specified products. The procedure for determining the minimum amount of solvent is indicated in Figure 7.13. All tie lines below that marked JK are extended to line SR_N to give intersections with line SR_N as shown. The intersection farthest from S (if on the left-hand side of the diagram) or nearest S (if on the right)

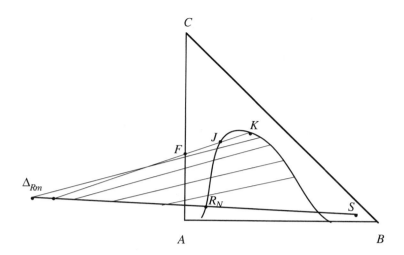

Figure 7.13 Minimum solvent for countercurrent extraction.

represents the difference point for the minimum solvent, Δ_{Rm}. The actual position of Δ_R must be farther from S (if on the left) or nearer to S (if on the right) for a finite number of stages. Usually, but not in the instance shown, the tie line which when extended passes through F, that is, tie line JK, will locate Δ_{Rm} for minimum solvent.

Stepping off a lot of stages on a triangular diagram can be difficult and inaccurate. More accurate calculations can be done with a McCabe-Thiele diagram. Therefore, the construction indicated in Figure 7.14 may be convenient. A few lines are drawn at random from point Δ_R to intersect the two branches of the solubility curve as shown, where the points of intersection do not now necessarily indicate streams between two actual adjacent stages. The concentrations $x_{C,op}$ and $y_{C,op}$ corresponding to these are plotted on x_C, y_C as shown to generate an operating curve. Tie-line data provide the equilibrium curve y_C^* vs. x_C, and the theoretical stages are stepped off in the resulting McCabe-Thiele diagram in the manner used for distillation and gas absorption.

Example 7.4 Multistage Countercurrent Extraction

A solution of acetic acid (C) in water (A) is to be extracted using isopropyl ether (C) as the solvent The feed is 1,000 kg/h of a solution containing 35 wt% acid and 65 wt% water. The solvent used is essentially pure isopropyl ether. The exiting raffinate stream should contain 2 wt% acetic acid. Operation is at 293 K and 1 atm. Determine (a) the minimum amount of solvent which can be used and (b) the number of theoretical stages if the solvent rate used is 60% above the minimum. Table 7.2 gives the equilibrium data.

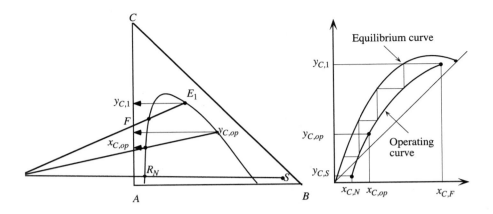

Figure 7.14 From right-triangular coordinates to McCabe-Thiele diagram.

Table 7.2 Water–Isopropyl Ether–Acetic Acid Equilibrium Data at 293 K and 1 atm.

Water phase, wt %			Isopropyl ether phase, wt%		
Water	Ether	Acetic acid	Water	Ether	Acetic acid
98.10	1.2	0.69	0.5	99.3	0.18
97.10	1.5	1.41	0.7	98.9	0.37
95.50	1.6	2.89	0.8	98.4	0.79
91.70	1.9	6.42	1.0	97.1	1.93
84.40	2.3	13.30	1.9	93.3	4.82
71.10	3.4	25.30	3.9	84.7	11.40
58.90	4.4	36.70	6.9	71.5	21.60
45.10	10.6	44.30	10.8	58.1	31.10
37.10	16.5	46.40	15.1	48.7	36.20

Water and isopropyl ether phases on the same line are in equilibrium with each other.
Source: Treybal (1980).

Solution
(a) Figure 7.15 is the right-triangular diagram for this system. Locate points F, S, and R_N. Extend the tie line that passes through point F until it meets the extension of line $R_N S$. The intersection of the two lines locates the difference point corresponding

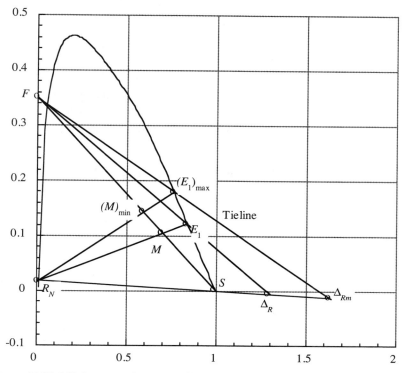

Figure 7.15 Minimum and actual difference points for Example 7.4.

to the minimum solvent rate, Δ_{Rm}. Line $F\Delta_{Rm}$ intercepts the extract branch of the solubility curve at point $(E_1)_{max}$, which gives the composition of the final extract if the minimum amount of solvent is used. Draw the line from $(E_1)_{max}$ to R_N. The intersection of this line with the line FS gives $(M)_{min}$. From Figure 7.15, $(x_{C,M})_{min} = 0.144$. Equation (7-11) can be written for the minimum solvent rate condition, S_{min}:

$$(x_{C,M})_{min} = \frac{Fx_{C,F} + S_{min}y_{C,S}}{F + S_{min}} \tag{7-15}$$

For this example, $x_{C,F} = 0.35$, $y_{C,S} = 0$, $(x_{C,M})_{min} = 0.144$, and $F = 1{,}000$ kg/h. Substituting in equation (7-15), $S_{min} = 1{,}431$ kg/h.

(b) For $S = 1.6S_{min} = 1{,}860$ kg/h, equation (7-11) gives $x_{C,M} = 0.106$. Point M is then located on line FS. Line $R_N E_1$ is located passing through point M and ending on the extract branch of the solubility curve. Extend the line FE_1 until it meets the extension of the line R_NS. The intersection of the two lines corresponds to the location of the actual difference point, Δ_R. A few lines are drawn at random from point Δ_R to intersect the two branches of the solubility curve. The concentrations $x_{C,op}$ and $y_{C,op}$ corresponding to these are given in Table 7.3, and plotted on x_C, y_C coordinates as shown in Figure 7.16 to generate an operating curve. Tie-line data provide the equilibrium curve y_C^* vs. x_C, and the theoretical stages are stepped off in the resulting McCabe-Thiele diagram. A total of 8 equilibrium stages are required for the specified separation.

Table 7.3 Operating Line Data for Example 7.4

$x_{C,op}$	0.02	0.05	0.10	0.15	0.20	0.25	0.30	0.35
$y_{C,op}$	0.000	0.009	0.023	0.037	0.054	0.074	0.096	0.121

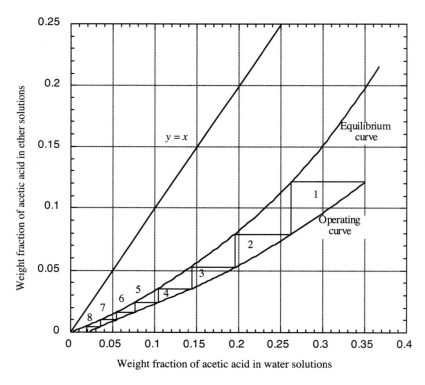

Figure 7.16 McCabe-Thiele diagram for Example 7.4.

7.3.4 Insoluble Liquids

When the liquids A and B are insoluble over the range of solute concentrations encountered, the stage computation is done more simply in terms of the mass flow rates of A (A mass/time) and B (B mass/time), and the mass ratios, defined here as:

x' = mass of C/mass of A in the raffinate liquids
y' = mass of C/mass of B in the extract liquids

An overall plant balance for substance C is

$$By'_S + Ax'_F = Ax'_N + By'_1 \tag{7-16}$$

or

$$\frac{A}{B} = \frac{y'_1 - y'_S}{x'_F - x'_N} \tag{7-17}$$

which is the equation of a straight line, the operating line, of slope A/B, through points (y'_1, x'_F) and (y'_S, x'_N).

For the special case where the equilibrium curve is of constant slope $m' = y'_i/x'_i$, a special form of Kremser equations applies:

$$N = \frac{\ln\left[\dfrac{x'_F - y'_S / m'}{x'_N - y'_S / m'}\left(1 - \dfrac{1}{EF}\right) + \dfrac{1}{EF}\right]}{\ln(EF)} \tag{7-18}$$

where the *extraction factor, EF*, is given by

$$EF = \frac{m'B}{A} \tag{7-19}$$

Equation (7-18) is valid for $EF \neq 1$. If the Murphree stage efficiency expressed in terms of extract compositions, $\mathbf{E}_{ME}$, is constant, then the overall efficiency for the cascade is given by

$$\mathbf{E}_O = \frac{\text{eqilibrium trays}}{\text{real trays}} = \frac{\ln\left[1 + \mathbf{E}_{ME}(EF - 1)\right]}{\ln(EF)} \tag{7-20}$$

Example 7.5 Multistage Extraction: Insoluble Liquids

A solution of nicotine (C) in water (A) is to be extracted at 293 K using pure kerosene (B) as the solvent. Water and kerosene are essentially insoluble at this temperature. The feed is 1,000 kg/h of a solution containing 1.0 wt% nicotine. The exiting raffinate stream should contain 0.1 wt% nicotine. Determine (a) the minimum amount of solvent which can be used, (b) the number of theoretical stages if the solvent rate used is 20% above the minimum, and (c) the number of real stages if the Murphree stage efficiency, expressed in terms of extract compositions, is $E_{ME} = 0.6$. The equilibrium curve in this range of compositions is basically a straight line with slope $m' = 0.926$ kg water/kg kerosene (Treybal, 1980).

Solution
(a) Because the solutions are so dilute, the concentrations expressed in terms of mass ratios are virtually equal to the mass fractions. Therefore, $x'_F = 0.01$ kg nicotine/kg water, $x'_N = 0.001$ kg nicotine/kg water, $y'_S = 0.0$ kg nicotine/kg kerosene. Because, in this case, both the equilibrium and operating lines are straight, if the minimum solvent flow rate (B_{min}) is used, the concentration of the exiting extract, $y'_1(max)$, will be in equilibrium with x'_F. Then,

$$y'_1(max) = m'x'_N = 0.926 \times 0.01 = 0.00926 \text{ kg nicotine / kg kerosene}$$

The feed to the process flows at the rate of 1,000 kg/h and is 99% water. Therefore, $A = 990$ kg water/h. From equation (7-17)

$$B_{min} = A\frac{x'_F - x'_N}{y'_1(max) - y'_S} = 990 \times \frac{0.01 - 0.001}{0.00926 - 0.0} = 962 \text{ kg kerosene/h}$$

(b) For a solvent flow rate that is 205 above the minimum,

$$B = 1.2B_{min} = 1,154 \text{ kg kerosene/h}$$

Then,

$$EF = m'B / A = 0.926 \times 1,154 / 990 = 1.079$$

Substituting in equation (7-18), $N = 6.66$ equilibrium stages.

(c) For $EF = 1.079$, and $E_{ME} = 0.6$ equation (7-20) yields $E_O = 0.609$. Then, the number of real stages required is $6.66/0.609 = 11$ stages.

7.3.5 Continuous Countercurrent Extraction with Reflux

Whereas in ordinary countercurrent operation the richest possible extract product leaving the plant is at best only in equilibrium with the feed solution, the use of reflux at the extract end of the cascade can provide a product even richer, as in the rectifying section of a distillation column. Reflux is not needed at the raffinate end of the cascade since, unlike distillation, where heat must be carried in from the reboiler by a vapor reflux, in extraction the solvent (the analog of heat) can enter without a carrier stream.

An arrangement for this is shown in Figure 7.17. The feed to be separated into its components is introduced at an appropriate place in the cascade, through which extract and raffinate liquids are flowing countercurrently. The concentration of solute C is increased in the extract-enriching section by countercurrent contact with a raffinate liquid rich in C. This is provided by removing the solvent from extract E_1 to produce the solvent-free stream E', part of which is removed as extract product P_E' and part returned as reflux R_0. The raffinate-stripping section of the cascade is the same as the countercurrent extractor of Figure 7.10, and C is stripped from the raffinate by countercurrent contact with the solvent.

Graphical determination of the number of stages required for such an operation is usually inconvenient to carry out in triangular coordinates because of crowding (Seader and Henley, 1998). Instead, a *Janecke diagram*, used in conjunction with a distribution diagram, is very useful for this purpose. In Janecke diagrams, which use convenient rectangular coordinates, solvent concentration on a solvent-free basis, N_R or N_E, is plotted as the ordinate against solute concentration on a solvent-free basis, X or Y, as the abscissa (Janecke, 1906). These are defined here as:

N_R = mass B/(mass of A + mass of C) in the raffinate liquids
N_E = mass B/(mass of A + mass of C) in the extract liquids
X = mass C/(mass of A + mass of C) in the raffinate liquids
Y = mass C/(mass of A + mass of C) in the extract liquids

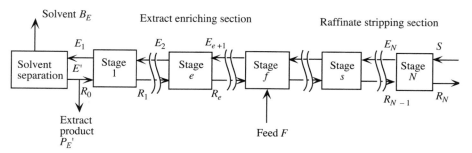

Figure 7.17 Countercurrent multistage extraction with extract reflux.

The application of Janecke diagrams to liquid–liquid extraction of a type II system with the use of reflux was considered in detail by Maloney and Schubert (1940), who used an auxiliary distribution diagram of the McCabe-Thiele type, but on a solvent-free basis to facilitate visualization of the stages. This method is also referred to as the *Ponchon-Savarit method for extraction*. Although the Janecke diagram can also be applied to type I systems, it becomes difficult to use when the liquids A and B are highly miscible because the resulting values of the ordinate can become very large.

For the development that follows, primed letters indicate solvent-free quantities. Thus, for example, E' = mass of B-free solution/time. Consider now the extract-enriching section in Figure 7.17. An (A + C) balance around the solvent separator is

$$E_1' = E' = P_E' + R_0' \qquad (7\text{-}21)$$

Let Δ_E represent the net rate of flow outward from this section. Then, for its (A + C) content

$$\Delta_E' = P_E' \qquad (7\text{-}22)$$

and for its C content

$$X_{\Delta E} = X_{PE} \qquad (7\text{-}23)$$

while for its B content

$$B_E = \Delta_E' N_{\Delta E} \qquad (7\text{-}24)$$

The point Δ_E is plotted on Figure 7.18, which is drawn for a system of two partly miscible component pairs. For all stages through e, an (A + C) balance is

$$E_{e+1}' = P_E' + R_e' = \Delta_E' + R_e' \qquad (7\text{-}25)$$

or

$$\Delta_E' = E_{e+1}' - R_e' \qquad (7\text{-}26)$$

A balance on component C is

$$\Delta_E' X_{\Delta E} = E_{e+1}' Y_{e+1} - R_e' X_e \qquad (7\text{-}27)$$

and a B balance is

$$\Delta_E' N_{\Delta E} = E_{e+1}' N_{E,e+1} - R_e' N_{R,e} \qquad (7\text{-}28)$$

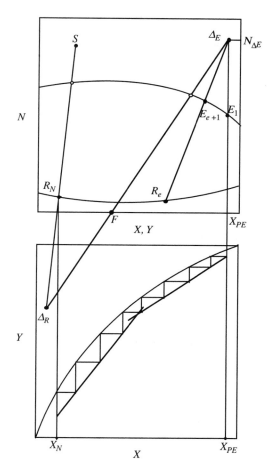

Figure 7.18 Use of Janecke diagram with auxiliary distribution curve for countercurrent extraction with reflux.

Since e is any stage in the extract enriching section, lines radiating from point Δ_E cut the solubility curves of Figure 7.18 at points representing extract and raffinate flowing between any two adjacent stages. Δ_E is therefore a difference point, constant for all stages in this section. Then, alternating tie lines and lines from Δ_E establish the equilibrium stages in this section, starting with stage 1 and continuing to the feed stage.

Combining equations (7-25), (7-27), and (7-28) we get an expression for the internal reflux ratio at any stage,

$$\frac{R'_e}{E'_{e+1}} = \frac{N_{\Delta E} - N_{E,e+1}}{N_{\Delta E} - N_{R,e}} \qquad (7\text{-}29)$$

The external reflux ratio can be calculated from equation (7-29)

$$\frac{R'_0}{P'_E} = \frac{R_0}{P'_E} = \frac{N_{\Delta E} - N_{E,1}}{N_{E,1}} \tag{7-30}$$

which can be used to locate Δ_E for any specified reflux ratio.

Consider now the raffinate stripping section of Figure 7.17. The balance for stages s through N is

$$R'_N - S' = R'_{s-1} - E'_s = \Delta'_R \tag{7-31}$$

where Δ'_R is the difference in solvent-free flow, out minus in, at stage N. It is also the constant difference of solvent-free flows of the streams between any two adjacent stages in this section of the cascade. Δ_R is therefore a difference point, constant for all stages in this section. Then, alternating tie lines and lines from Δ_R establish the equilibrium stages in this section, starting with the feed stage and continuing to the last stage N.

Material balances may also be written for the whole cascade. A balance for (A + C) is of the form

$$F' + S' = P'_E + R'_N \tag{7-32}$$

Combining equation (7-32) with (7-22) and (7-31)

$$F' = \Delta'_E + \Delta'_R \tag{7-33}$$

where normally $F' = F$. Point F must therefore lie on the line joining the two difference points as shown in Figure 7.18.

The higher the location of Δ_E (and the lower Δ_R), the larger the reflux ratio and the smaller the number of ideal stages. At total reflux, $N_{\Delta E} = \infty$ and the minimum number of stages results. The capacity of the plant falls to zero, feed must be stopped, and solvent B_E is recirculated to become S. The minimum number of stages is easily determined on the XY equilibrium distribution diagram using the 45° diagonal as operating lines for both sections of the cascade.

An infinite number of stages is required if a line radiating from either Δ_E or Δ_R coincides with a tie line, and the greatest reflux ratio for which this occurs is the *minimum reflux ratio*. Frequently the tie line that when extended passes through point F, representing the feed condition, will establish the minimum reflux ratio. This will always be the case if the XY equilibrium distribution curve is everywhere concave downward

Example 7.6 Countercurrent Extraction with Extract Reflux

A solution containing 60% styrene (C) and 40% ethylbenzene (A) is to be separated at 298 K at the rate of 1,000 kg/h into products containing 10% and 90% styrene respectively, with diethylene glycol (B) as solvent. Determine (a) the minimum number of theoretical stages, (b) the minimum extract reflux ratio, and (c) the number of theoretical stages and the important flow quantities at an extract reflux ratio of 1.5 times the minimum value.

Solution

Equilibrium data (Boobar, et al., 1951) have been converted to a solvent-free basis and are given in Table 7.4 (treybal, 1980). These are plotted in the form of a Janecke diagram in Figure 7.19. The corresponding equilibrium distribution diagram is Figure 7.20. From the problem statement, $F' = F = 1,000$ kg/h, $X_F = 0.6$ wt fraction of styrene, $X_{PE'} = 0.9, X_N = 0.1$, all on a solvent-free basis.

(a) Minimum theoretical stages are determined on the XY equilibrium distribution diagram, stepping them off from the diagonal line to the equilibrium curve, beginning at $X_{PE'} = 0.9$ and ending at $X_N = 0.1$, as shown in Figure 7.20. The answer is $N_{min} = 9$ ideal stages.

Table 7.4 Equilibrium Data for Ethylbenzene (A)–Diethylene Glycol (B)–Styrene (C) at 298 K

Raffinate Solutions		Extract Solutions	
X	N_R	Y	N_E
kg C/kg (A + B)	kg B/kg (A + B)	kg C/kg (A + B)	kg B/kg (A + B)
0.0000	0.00675	0.0000	8.62
0.0870	0.00617	0.1429	7.71
0.1883	0.00938	0.2730	6.81
0.2880	0.01010	0.3860	6.04
0.3840	0.01101	0.4800	5.44
0.4580	0.01215	0.5570	5.02
0.4640	0.01215	0.5650	4.95
0.5610	0.01410	0.6550	4.46
0.5730	0.01405	0.6740	4.37
0.7810	0.01833	0.8630	3.47
0.9000	0.02300	0.9500	3.10
1.0000	0.02560	1.0000	2.69

Sources: Boobar, et al., (1951) and Treybal (1980).

(b) Since the equilibrium distribution curve is everywhere concave downward, the tie line which when extended passes through F provides the minimum reflux ratio. From Figure 7.19, $N_{\Delta Em} = 11.04$, and $N_{E1} = 3.1$. From equation (7-30)

$$\left(\frac{R_0'}{P_E'}\right)_{min} = \frac{11.04 - 3.1}{3.1} = 2.561 \text{ kg reflux/kg extract product}$$

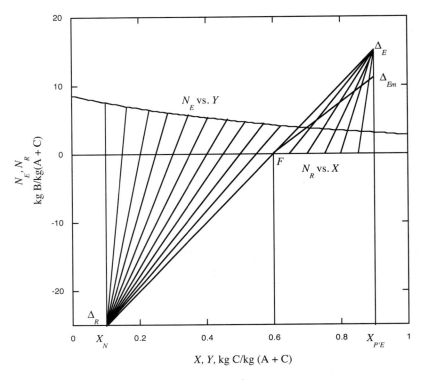

Figure 7.19 Janecke diagram for Example 7.6.

(c) For $R_0/P'_E = (1.5)(2.561) = 3.842$ kg reflux/kg extract, equation (7.30) gives $N_{\Delta E}$ = 15.01. Point Δ_E is plotted as shown. A straight line from Δ_E through F intersects line $X = X_N = 0.10$ at Δ_R; from the diagram, $N_{\Delta R} = -24.90$. Random lines are drawn from Δ_E for concentrations to the right of F, and from Δ_R; for those to the left. Intersections of these with the solubility curves provide the coordinates of the operating lines. Figure 7.21 shows the McCabe-Thiele type construction to determine the number of equilibrium stages required. A total of 17.5 equilibrium stages is required, and the feed is to be introduced into the fifth stage from the extract-product of the cascade.

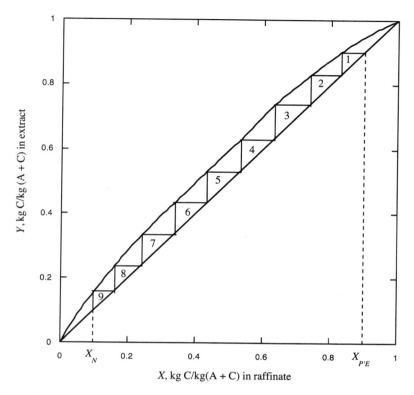

Figure 7.20 Equilibrium distribution diagram, and minimum number of stages for Example 7.6.

From Figure 7.19, for $X_N = 0.01$, $N_{R,N} = 0.0083$. On the basis of 1 hour, an overall plant balance is

$$F = 1,000 = P'_E + R'_N$$

A balance on C is

$$FX_F = 600 = P'_E \times 0.9 + R'_N \times 0.01$$

Solving simultaneously,

$$P'_E = 625 \text{ kg/h}, \quad R'_N = 375 \text{ kg/h}$$
$$R' = R'_0 = 3.842 P'_E = 2,401 \text{ kg/h}$$
$$E'_1 = R'_0 + P'_E = 3,026 \text{ kg/h}$$

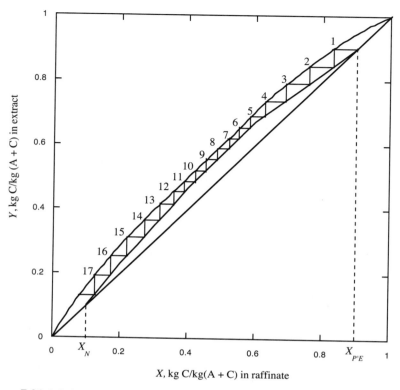

Figure 7.21 McCabe-Thiele construction: number of ideal stages for Example 7.6.

$$B_E = E'_1 N_{E,1} = 3{,}026(3.1) = 9{,}381 \text{ kg/h}$$
$$E_1 = B_E + E'_1 = 12{,}407 \text{ kg/h}$$
$$R_N = R'_N (1 + N_{R,N}) = 375(1 + 0.0083) = 378 \text{ kg/h}$$
$$S = B_E + R'_N N_{R,N} = 9{,}384 \text{ kg/h}$$

7.4 EQUIPMENT FOR LIQUID–LIQUID EXTRACTION

Your objectives in studying this section are to be able to:

1. Identify the most common types of equipment used for liquid–liquid extraction.
2. Make preliminary estimates of the dimensions of the equipment, and of their power requirements.

Given the wide diversity of applications of liquid–liquid extraction, one might suspect a correspondingly large variety of extraction devices. Indeed, such is the case. Equipment such as that used for absorption, stripping, and distillation is sometimes used. However, such devices are inefficient unless liquid viscosities are low and the difference in phase density is high. For that reason, centrifugal and mechanically agitated devices are often preferred. Often, the choice of suitable extraction equipment is between a cascade of *mixer-settler* units or a multicompartment column-type extractor with mechanical agitation. Methods for estimating size and power requirements for these two general types of extractors are presented next. Column devices with no mechanical agitation are also considered.

7.4.1 Mixer-Settler Cascades

In mixer-settlers, the two liquids are first mixed and then separated by settling. Any number of mixer-settlers units may be connected together to form a multistage, countercurrent cascade. During mixing one of the liquid is dispersed in the form of small droplets into the other liquid phase. The dispersed phase may be either the lighter or the heavier of the two phases. The mixing step is commonly conducted in an agitated vessel, with sufficient agitation and residence time so that a reasonable approach to equilibrium (e. g., 80% to 90% stage efficiency) is attained. The mixer is usually agitated by impellers. The settling step is by gravity in a second vessel called a settler or *decanter*.

Sizing of mixer-settler units is done most accurately by scale-up from batch or continuous runs in laboratory or pilot-plant equipment. However, preliminary sizing calculations can be done using available theory and empirical correlations. Experimental data by Flynn and Treybal (1955) show that when liquid viscosities are less than 5 cP and the specific gravity difference between the two liquid phases is greater than about 0.1, the average residence time required of the two liquid phases in the mixing vessel to achieve at least 90% stage efficiency may be as low as 30 s, and is usually not more than 5 min, when an agitator-power input per mixer volume of 0.788 kW/m^3 is used. For a vertical, cylindrical mixer of height H and diameter D_T, the economic ratio of H to D_T is approximately 1. The mixing vessel is closed with the two liquid phases entering at the bottom and the effluent, in the form of a two-phase emulsion, leaving at the top.

Based on experiments reported by Ryan, et al. (1959), the capacity of a settler vessel can be estimated as 0.2 m^3/min of combined extract and raffinate flow per square meter of phase disengaging area. For a horizontal, cylindrical vessel of length L and diameter D_T, the economic ratio of L to D_T is approximately 4. Thus, if the interphase is located at the middle of the vessel, the disengaging area is $D_T L$, or, using the economic L/D_T ratio, $4D_T^2$. Frequently, the settling vessel will be bigger than the mixing vessel, as is the case in the following example.

Example 7.7 Design of a Mixer-Settler Extractor

Benzoic acid is to be continuously extracted from a dilute solution in water using toluene as solvent in a series of mixer-settler vessels operated in countercurrent flow. The flow rates of the feed and solvent are 1.89 and 2.84 m^3/min, respectively. The residence time in each mixer is 2 min. Estimate:

(a) Diameter and height of each mixing vessel.

(b) Agitator power for each mixer.

(c) Diameter and length of a settling vessel.

(d) Residence time in settling vessel, in minutes.

Solution
(a) Let Q = total flow rate = 1.89 + 2.84 = 4.73 m^3/min. For a residence time, t_{res} = 2 min, the vessel volume is $V_T = Qt_{res}$ = 4.73 × 2 = 9.46 m^3. For a cylindrical vessel with $H = D_T$

$$V_T = \pi D_T^3 / 4$$

$$H = D_T = \left(4V_T / \pi\right)^{1/3} = 2.3 \text{ m}$$

(b) Based on the recommendation by Flynn and Treybal (1955),

$$\text{Mixer power} = 0.788 \text{ kW} / \text{m}^3 \times 9.46 \text{ m}^3 = 7.45 \text{ kW}$$

(c) Based on the recommendation by Ryan, et al. (1959), the disengaging area in the settler is

$$D_T L = 4.73 \text{ m}^3/\text{min}/(0.2 \text{ m}^3/\text{min-m}^2) = 23.65 \text{ m}^2$$

For $L/D_T = 4$,

$$D_T = (23.65 / 4)^{0.5} = 2.43 \text{ m}$$
$$L = 4 \times 2.43 = 9.72 \text{ m}$$

(d) Total volume of the settler, $V_T = \pi D_T^2 L/4 = 45.1$ m^3

$$t_{res} = V_T / Q = 45.1/4.73 = 9.5 \text{ min}$$

Mixing is accomplished by an appropriate, centrally located impeller, as shown in Figure 7.22. Vertical side baffles are usually installed to prevent vortex formation in open tanks, and to minimize swirling and improve circulation patterns in closed tanks. Although no standards exist for vessel and turbine geometry, the values recommended in Figure 7.22 give good dispersion performance in liquid-liquid agitation (Seader and Henley, 1998).

To achieve a high stage efficiency (between 90 and 100%) it is necessary to provide fairly vigorous agitation. Based on the work of Skelland and Ramsay (1987), a minimum impeller rate of rotation Ω_{min} is required for complete and uniform dispersion of one liquid into another. For a flat-blade impeller in a baffled vessel of the type discussed above, this minimum rotation rate can be estimated from

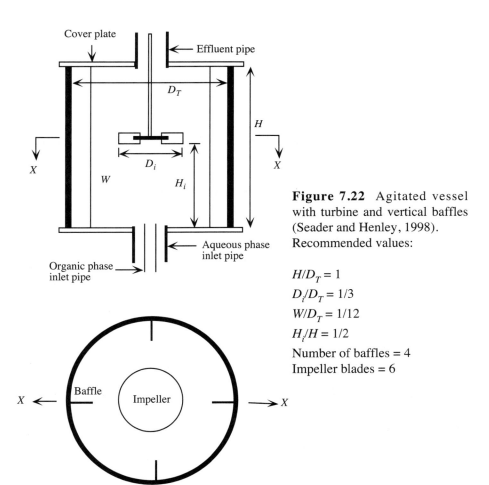

Figure 7.22 Agitated vessel with turbine and vertical baffles (Seader and Henley, 1998). Recommended values:

$H/D_T = 1$

$D_i/D_T = 1/3$

$W/D_T = 1/12$

$H_i/H = 1/2$

Number of baffles = 4
Impeller blades = 6

$$\frac{\Omega_{min}^2 \rho_M D_i}{g \Delta \rho} = 1.03 \phi_D^{0.106} \left(\frac{D_T}{D_i} \right)^{2.76} \left(\frac{\mu_M^2 \sigma}{D_i^5 \rho_M g^2 (\Delta \rho)^2} \right)^{0.084} \tag{7-34}$$

where:

$\Delta \rho$	= absolute value of the difference in density between the liquids
σ	= interfacial tension between the liquid phases
ϕ_D	= fractional holdup in the tank of the dispersed liquid phase
ρ_M	= two-phase mixture density, given by

$$\rho_M = \rho_C \phi_C + \rho_D \phi_D \tag{7-35}$$

ϕ_C	= fractional holdup in the tank of the continuous liquid phase
ρ_C	= density of the continuous liquid phase
ρ_D	= density of the dispersed liquid phase
μ_M	= two-phase mixture viscosity, given by

$$\mu_M = \frac{\mu_C}{\phi_C} \left(1 + \frac{5 \mu_D \phi_D}{\mu_C + \mu_D} \right) \tag{7-36}$$

The viscosity of the mixture given by equation (7-36) can exceed that of either constituent (Treybal, 1980).

The agitator power, P, can be estimated from an empirical correlation in terms of a *power number*, Po, which depends on an *impeller Reynolds number*, Re, where

$$\text{Po} = \frac{P}{\Omega^3 D_i^5 \rho_M} \qquad \text{Re} = \frac{\Omega D_i^2 \rho_M}{\mu_M} \tag{7-37}$$

A correlation of experimental data on Po vs. Re for baffled vessels with six-bladed flat-blade impellers was developed by Laity and Treybal (1957). They found that for Re > 10,000, the power number tends to an asymptotic value of Po = 5.7.

Example 7.8 Power Requirements of a Mixer-Settler Extractor

Furfural is to be continuously extracted from a dilute solution in water using pure toluene as solvent in a series of mixer-settler vessels operated in countercurrent flow. The flow rates of the feed and solvent are 2.57 and 1.41 kg/s, respectively. The residence time in each mixer is 2 min. The density of the feed is 998 kg/m^3, and its viscosity is 0.89 cP; the density of the solvent is 868 kg/m^3, and its viscosity is 0.59

cP. The interfacial tension is 0.025 N/m. Estimate for the raffinate as the dispersed phase:

(a) Dimensions of the mixing vessel and the diameter of the flat-blade impeller.

(b) The minimum rate of rotation of the impeller for complete and uniform dispersion.

(c) The power requirement of the agitator at 1.20 times the minimum rotation rate.

Solution

Calculate the feed volumetric flow rate, $Q_F = 2.57 \times 60/998 = 0.155$ m³/min, and the solvent volumetric flow rate, $Q_S = 1.41 \times 60/868 = 0.097$ m³/min. Because of the dilute concentration of solute in the feed and sufficient agitation to achieve complete and uniform dispersion, assume that the fractional volumetric holdups of raffinate and extract in the vessel are equal to the corresponding volume fractions in the combined feed and solvent entering the mixer:

$$\phi_R = 0.155/(0.155 + 0.097) = 0.615 \qquad \phi_E = 1 - 0.615 = 0.385$$

(a) Let Q = total flow rate = $0.155 + 0.097 = 0.252$ m³/min. For a residence time of 2 min, the vessel volume is $V_T = Qt_{res} = 0.252 \times 2 = 0.504$ m³. For a cylindrical vessel with $H = D_T$,

$$V_T = \pi D_T^3 / 4$$
$$H = D_T = \left(4V_T / \pi\right)^{1/3} = 0.863 \text{ l}$$
$$D_i = D_T/3 = 0.288 \text{ m}$$

(b) For the raffinate phase dispersed:

$$\phi_D = \phi_R = 0.615 \qquad \phi_C = \phi_E = 0.385$$
$$\Delta\rho = 998 - 868 = 130 \text{ kg/m}^3$$
$$\rho_M = 0.615 \times 998 + 0.385 \times 868 = 948 \text{ kg/m}^3$$

$$\mu_M = \frac{0.59}{0.385}\left[1 + \frac{1.5 \times 0.89 \times 0.615}{0.59 \times 0.89}\right] = 2.38 \text{ cP}$$

Substituting in equation (7-34), $\Omega_{min} = 165$ rpm.

(c) For $\Omega = 1.2\Omega_{min} = 198$ rpm, from equation (7-37), Re $= 1.09 \times 10^5$. Then, according to Laity and Treybal (1957), the power number, Po $= 5.7$. From equation (7-37), $P = 0.385$ kW. The power density is $P/V_T = 0.385/0.504 = 0.764$ kW/m³, close to the recommended value of 0.788 kW/m³ (Flynn and Treybal, 1955).

When dispersion is complete and uniform, the contents of the vessel are perfectly mixed with respect to both phases. In that case, the concentration of the solute in each of the two phases in the vessel is uniform and equal to the concentrations in the two-phase emulsion leaving the mixing tank. This is the so-called ideal CFSTR (*continuous-flow-stirred-tank-reactor*) model, sometimes called the perfectly mixed model. Next we develop an equation to estimate the Murphree-stage efficiency for liquid–liquid extraction *in a* perfectly mixed vessel.

The Murphree dispersed-phase efficiency for liquid–liquid extraction, based on the raffinate as the dispersed phase, can be expressed as the fractional approach to equilibrium. In terms of bulk molar concentrations of the solute,

$$E_{MD} = \frac{c_{D,in} - c_{D,out}}{c_{D,in} - c_D^*} \qquad (7\text{-}38)$$

where c_D^* is the solute concentration in the dispersed phase in equilibrium with the solute concentration in the exiting continuous phase, $c_{C,out}$. The rate of mass transfer of solute from the dispersed phase to the continuous phase, n, can be expressed as

$$n = K_{OD}a\left(c_{D,out} - c_D^*\right)V_T \qquad (7\text{-}39)$$

where the concentration driving force for mass transfer is uniform throughout the well-mixed vessel and is equal to the driving force based on the the exit concentrations, a is the interfacial area for mass transfer per unit volume of the liquid phases, V_T is the total volume of the liquid phases in the vessel, and K_{OD} is the overall mass-transfer coefficient based on the dispersed phase. The overall mass-transfer coefficient is given in terms of the separate resistances of the dispersed and continuous phases by

$$\frac{1}{K_{OD}} = \frac{1}{k_D} + \frac{1}{mk_C} \qquad (7\text{-}40)$$

where equilibrium is assumed at the interphase between the two liquids, and m is the slope of the equilibrium curve for the solute expressed as $c_C = f(c_D)$. For dilute solutions, changes in volumetric flow rate of the raffinate and extract are small, and thus the rate of mass transfer based on the change in solute concentration in the dispersed phase is given by the material balance:

$$n = Q_D\left(c_{D,in} - c_{D,out}\right) \qquad (7\text{-}41)$$

where Q_D is the volumetric flow rate of the dispersed phase.

To obtain an expression for E_{MD} in terms of $K_{OD}a$, equations (7-38), (7-39), and (7-41) are combined as follows. From equation (7-38),

$$\frac{\mathbf{E}_{MD}}{1-\mathbf{E}_{MD}} = \frac{c_{D,in} - c_{D,out}}{c_{D,out} - c_D^*} \qquad (7\text{-}42)$$

On the other hand, the number of dispersed-phase overall transfer units, N_{tOD}, for this case is given by [compare to equation (5-23) for dilute solutions]

$$N_{tOD} \cong \int_{c_{D,out}}^{c_{D,in}} \frac{dc_D}{c_D - c_D^*} \qquad (7\text{-}43)$$

It has been established that the liquids in a well-agitated vessel are thoroughly back-mixed, so that everywhere the concentrations are constant at the effluent values (Schindler and Treybal, 1968). Then,

$$N_{tOD} \cong \int_{c_{D,out}}^{c_{D,in}} \frac{dc_D}{c_D - c_D^*} \cong \frac{c_{D,in} - c_{D,out}}{c_{D,out} - c_D^*} \qquad (7\text{-}44)$$

Combining equations (7-39) and (7-41),

$$\frac{c_{D,in} - c_{D,out}}{c_{D,out} - c_D^*} = \frac{K_{OD}aV_T}{Q_D} \cong N_{tOD} \qquad (7\text{-}45)$$

Combining equations (7-42) and (7-45),

$$\mathbf{E}_{MD} = \frac{N_{tOD}}{1 + N_{tOD}} = \frac{K_{OD}aV_T/Q}{1 + K_{OD}aV_T/Q} \qquad (7\text{-}46)$$

From equations (7-40) and (7-46), it is evident that an estimate of $\mathbf{E}_{MD}$ requires generalized correlations of experimental data for the interfacial area for mass transfer and for the dispersed- and continuous-phase mass-transfer coefficients. The population of dispersed-phase droplets in an agitated vessel will cover a range of sizes and shapes. For each droplet, it is useful to define d_e, the equivalent diameter of a spherical drop, using the method of Lewis, et al. (1951):

$$d_e = \left(d_1^2 d_2\right)^{1/3} \qquad (7\text{-}47)$$

where d_1 and d_2 are the major and minor axis, respectively, of an ellipsoidal drop image. For a spherical drop, d_e is simply the diameter of the drop. For the population of drops, it is useful to define an average or mean diameter. for mass-transfer calculations, the surface-mean diameter, d_{vs} (also called the *Sauter mean diameter*) is the most appropriate. It is usually determined from experimental drop-size distribution data. The interfacial area for mass transfer per unit volume of a two-phase mixture in terms of the Sauter mean diameter is given by (Seader and Henley, 1998)

$$a = \frac{6\phi_D}{d_{vs}}$$ (7-48)

Early experimental investigations, such as those of Vermeulen, et al. (1955), found that d_{vs} is dependent on the Weber number, We, defined as

$$We = \frac{D_i^3 \Omega^2 \rho_C}{\sigma}$$ (7-49)

high Weber numbers give small droplets and high interfacial areas. Gnanasundaram, et al. (1979) recommended the following correlations:

$$\frac{d_{vs}}{D_i} = 0.052[We]^{-0.6}\exp(4\phi_D), \qquad We \leq 10,000$$ (7-50)

$$\frac{d_{vs}}{D_i} = 0.39[We]^{-0.6}, \qquad We > 10,000$$ (7-51)

Typical values of We for industrial extractors are less than 10,000, so equation (7-50) usually applies.

Example 7.9 Drop Size and Interfacial Area in an Extractor

For the conditions and results of Example 7.8, estimate the Sauter mean drop diameter and the interfacial area.

Solution
From equation (7-49), We = 9,032. Equation (7-50) yields d_{vs} = 0.742 mm. Substituting in equation (7-48), a = 4,973 m²/m³.

Experimental studies conducted since the early 1940s show that mass transfer in mechanically agitated liquid–liquid systems is very comple. The reasons for this camplexity are many. for example, the magnitude of k_D depends on drop diameter, solute diffusivity, and fluid motion within the drop. When drop diameter is less than 1 mm, interfacial tension is higher than 15 dyne/cm, and trace amounts of surface-active ingredients are present, droplets are rigid and behave like solid particles (Davies, 1978). Under those conditions, k_D can be estimated from (Treybal, 1963)

$$\text{Sh}_D = \frac{k_D d_{vs}}{D_D} = 6.6 \tag{7-52}$$

For the continuous-phase mass-transfer coefficient, k_C, Skelland and Moeti (1990) proposed the following,

$$\text{Sh}_C = \frac{k_D d_{vs}}{D_D} = 1.27 \times 10^{-5}\,\text{Sc}_C^{1/3}\text{Fr}^{5/12}\text{Eo}^{5/4}\phi_D^{-1/2}\left(\frac{D_i}{d_{vs}}\right)^2\left(\frac{d_{vs}}{D_T}\right)^{1/2} \tag{7-53}$$

where:

$$
\begin{aligned}
\text{Re}_C &= D_i{}^2\Omega\rho_C/\mu_C \\
\text{Sh}_C &= k_C d_{vs}/D_C \\
\text{Sc}_C &= \mu_C/\rho_C D_C \\
\text{Fr} &= D_i\Omega^2/g \\
\text{Eo} &= \rho_D d_{vs}{}^2 g/\sigma \ \text{(known as Eotvos number)}
\end{aligned}
$$

Example 7.10 Mass-Transfer Coefficients in Agitated Extractor

For the conditions and results of Example 7.8 and 7.9, estimate:

(a) The dispersed-phase mass-transfer coefficient.

(b) The continuous-phase mass-transfer coefficient.

(c) The Murphree dispersed-phase efficiency.

(d) The fractional extraction of furfural.

The molecular diffusivities of furfural in water (dispersed) and toluene are, respectively, $D_D = 1.15 \times 10^{-9}$ m^2/s; $D_C = 2.15 \times 10^{-9}$ m^2/s. The equilibrium distribution coefficient $m = 10.15$ m^3 raffinate/m^3 extract.

Solution
(a) From equation (7-52), $k_D = 6.6 \times 1.15 \times 10^{-9}/(7.42 \times 10^{-4}) = 1.02 \times 10^{-5}$ m/s.
(b) To apply equation (7-53), compute each of the following dimensionless groups:

$$\text{Sc}_C = \mu_C/\rho_C D_C = 316.6 \qquad \text{Re}_C = D_i{}^2\Omega\rho_C/\mu_C = 402{,}700$$

$$\mathrm{Fr} = D_i\Omega^2/g = 0.3198 \qquad\qquad \mathrm{Eo} = \rho_D d_{vs}^{\ 2}g/\sigma = 0.216$$

Substituting in equation (7-53), $\mathrm{Sh}_C = 236.5$. Then,

$$k_C = \mathrm{Sh}_C D_C/d_{vs} = 6.84 \times 10^{-4}\ \mathrm{m/s}.$$

(c) From equation (7-40),

$$K_{OD} = [(1.02 \times 10^{-5})^{-1} + (10.15 \times 6.84 \times 10^{-4})^{-1}]^{-1} = 1.01 \times 10^{-5}\ \mathrm{m/s}$$

From equation (7-45),

$$N_{tOD} = K_{OD}aV_T/Q_D = 1.01 \times 10^{-5} \times 4{,}973 \times 0.504 \times 60\ /\ 0.155 = 9.8$$

From equation (7-46),

$$\mathbf{E}_{MD} = N_{tOD}/(1 + N_{tOD}) = 9.8\ /\ 10.8 = 0.907$$

(d) For dilute solutions, where the total volume of the raffinate and extract liquids remain fairly constant, the fractional extraction of furfural, f_{ext}, is defined as

$$f_{ext} = \frac{c_{D,in} - c_{D,out}}{c_{D,in}} = 1 - \frac{c_{D,out}}{c_{D,in}} \qquad (7\text{-}54)$$

By a material balance on furfural,

$$Q_D(c_{D,in} - c_{D,out}) = Q_C c_{C,out} \qquad (7\text{-}55)$$

From the definition of the equilibrium distribution coefficient,

$$c_D^* = \frac{c_{C,out}}{m} \qquad (7\text{-}56)$$

Combining equations (7-38), (7-54), (7-55), and (7-56),

$$f_{ext} = \frac{\mathbf{E}_{MD}}{1 + \mathbf{E}_{MD}\dfrac{Q_D}{mQ_C}} \qquad (7\text{-}57)$$

Substituting in equation (7-57), $f_{ext} = 0.794$.

7.4.2 Multicompartment Columns

Sizing extraction columns, which may or may not include mechanical agitation, involves the determination of the column diameter and column height. The diameter must be sufficiently large to permit the two phases to flow countercurrently through the column without flooding. The column height must be sufficient to achieve the number of equilibrium stages corresponding to the desired degree of separation.

Sieve-tray towers are very effective, both with respect to liquid-handling capacity and extraction efficiency, particularly for systems of low interfacial tension which do not require mechanical agitation for good dispersion. The general assembly of plates and downspouts is much the same as for gas–liquid contact except that no weir is required. Towers packed with the same random packing used for gas–liquid contact have also been used for liquid extractors; however, mass-transfer rates are poor. It is recommended instead that sieve-tray towers be used for systems of low interfacial tension and mechanically agitated extractors for those of high interfacial tension (Treybal, 1980).

The most important mechanically agitated columns are those that employ rotating agitators, driven by a shaft that extends axially through the column. The agitators create shear mixing zones, which alternate with settling zones in the column. Differences among the various agitated columns lie primarily in the mixers and settling chambers used.

Perhaps the first mechanically agitated extractor of importance was the *Scheibel column* (Scheibel, 1948). The liquid phases are contacted at fixed intervals by unbaffled, flat-bladed, turbine-type agitators mounted on a vertical shaft (see Figure 7.23). In the unbaffled separation or calming zones, knitted wire-mesh packing is installed to prevent backmixing between mixing zones and to induce coalescence and settling of drops. For more economical designs for larger-diameter installations (> 1 m), Scheibel added outer and inner horizontal annular baffles to divert the vertical flow of the phases in the mixing zone and to ensure complete mixing. For systems with high interfacial surface tension and viscosities, the wire mesh is removed.

Another type of column with rotating agitators is the *rotating disk contactor* (RDC). On a worldwide basis, it is probably the most extensively used liquid–liquid extraction device (Seader and Henley, 1998). Horizontal disks, mounted on a centrally located shaft, are the agitation elements. Mounted at the column wall are annular stator rings with an opening larger than the agitator disk diameter. Thus, the agitator shaft assembly is easily removed from the column. Because the rotational speed of the rotor controls the drop size, the rotor speed can be continuously varied over a wide range.

Rather than provide agitation by rotating impellers on a vertical shaft, Karr and Lo (1976) devised a reciprocating perforated-plate extractor, called the *Karr column*, in which the plates move up and down approximately two times per second with a stroke length of about 20 mm (see Figure 7.24). Annular baffle plates are provided periodically in the plate stack to minimize axial mixing. The perforated plates use large holes (typically 14 mm) and a high hole area fraction (typically 58%). The central shaft, which supports both sets of plates, is reciprocated by a drive mechanism located at the top of the column.

A number of industrial centrifugal extractors have been available since 1944 when the *Podbielniak (POD) extractor*, with its short residence time, was successfully applied to penicillin extraction (Barson and Beyer, 1953). In the POD, several concentric sieve trays are arranged around a horizontal axis through which the two liquids flow countercurrently. Liquid inlet pressures of 4 to 7 atm are required to overcome pressure drop and centrifugal force. As many as five theoretical stages can be achieved in one unit.

Because of the large number of important variables, an accurate estimation of column diameter for liquid–liquid contacting devices is far more complex and lesss

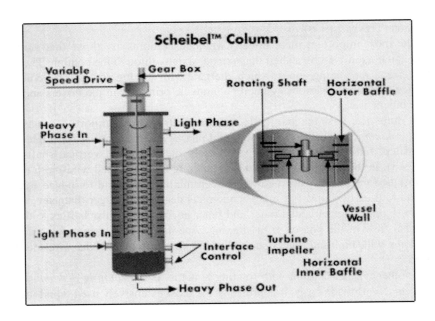

Figure 7.23 Scheibel column (courtesy of Koch Modular Process Systems, Paramus, NJ).

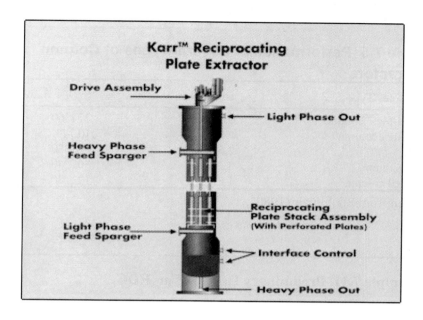

Figure 7.24 Karr reciprocating column (courtesy of Koch Modular Process Systems, Paramus, NJ).

certain than for vapor–liquid contactors. Column diameter is best determined by scale-up from tests run in standard laboratory or pilot plant test units. The sum of the measured superficial velocities of the two liquid phases ($v_C + v_D$) in the test unit can then be assumed to hold for larger commercial units.

Despite their compartmentalization, mechanically assisted liquid–liquid extraction columns, such as the RDC and Karr columns, operate more nearly like differential contacting devices than like staged contactors. Therefore, it is more common to consider stage efficiency for these columns in terms of HETS (*height equivalent to a theoretical stage*) or as some function of mass-transfer parameters, such as HTU (height of a transfer unit). Although it is not as sound on a theoretical basis as the HTU, the HETS is preferred here because it can be applied directly to determine column height from the number of equilibrium stages.

Because of the great complexity of liquid–liquid systems and the large number of variables that influence contacting efficiency, general correlations for HETS have been difficult to develop. For small diameter columns, rough estimates of the diameter and height can be made using the results of a study by Stichlmair (1980). Typical results from that study are summarized in Table 7.5. It is preferred to obtain values of HETS by conducting small-scale laboratory experiments with systems of interest. These values are scaled to commercial-size units by assuming that HETS increases with column diameter raised to an exponent, which varies from 0.2 to 0.4.

Table 7.5 Performance of Several Types of Column Extractors

Extractor Type	1/HETS, m^{-1}	$(v_D + v_C)$, m/h
Sieve-plate column	0.8 – 1.2	27 – 60
Scheibel column	5 – 9	10 – 14
RDC	2.5 – 3.5	15 – 30
Karr column	3.7 – 7.0	30 – 40
Packed column	1.5 – 2.5	12 – 30

Taken from Stichlmair (1980).

Example 7.11 Preliminary Design of an RDC

Estimate the diameter and height of an RDC to extract acetone from a dilute toluene–acetone solution into water at 293 K. The flow rates for the dispersed organic and continuous aqueous phases are 12,250 kg/h and 11,340 kg/h, respectively. The density of the organic phase is 858 kg/m^3 and that of the aqueous phase is 998 kg/m^3. For the desired degree of separation, 12 equilibrium stages are required.

Solution
Calculate the total volume flow rate of the two liquids:

$$Q_D = 12{,}250/858 = 14.28 \text{ m}^3/\text{h} \qquad Q_C = 11{,}340/998 = 11.36 \text{ m}^3/\text{h}$$

(a) Assume that, based on the information on Table 7.5, $(v_D + v_C) = 22$ m/h. Then, the column cross-section area $A_c = (Q_D + Q_C)/(v_D + v_C) = 1.165$ m^2. The column diameter $D_T = (4A_c/\pi)^{0.5} = 1.22$ m.

(b) Assume that, based on the information on Table 7.5, HETS = 0.333 m/theoretical stage. Then, the column height is $Z = 12 \times 0.333 = 4.0$ m.

PROBLEMS

The problems at the end of each chapter have been grouped into four classes (designated by a superscript after the problem number).

Class a: Illustrates direct numerical application of the formulas in the text.
Class b: Requires elementary analysis of physical situations, based on the subject material in the chapter.
Class c: Requires somewhat more mature analysis.
Class d: Requires computer solution.

7.1[a]. Lever-arm rule.

Derive equation (7-1).

7.2[b]. Single-stage extraction.

Repeat Example 7.2 using a solvent rate of 80 kg/h.

$$\text{Answer: } x_{C,R} = 0.136$$

7.3[a]. Single-stage extraction: insoluble liquids.

A feed of 13,500 kg/h consists of 8 wt% acetic acid in water. Acetic acid will be removed from the solution by extraction with pure methyl isobutyl ketone at 298 K. If the raffinate is to contain only 1 wt% of acetic acid, estimate the kilograms/hour of solvent required if a single equilibrium stage is used. Assume that water and methyl isobutyl ketone are insoliuble. For this system, $m' = 0.657$ kg water/kg methyl isobutyl ketone (Perry and Chilton, 1993).

$$\text{Answer: } 145{,}000 \text{ kg/h}$$

7.4[a]. Single-stage extraction: insoluble liquids.

Repeat Problem 7.3 using methyl acetate as solvent, for which $m' = 1.273$ kg water/kg methyl acetate.

7.5[b]. Multistage crosscurrent extraction: insoluble liquids.

Consider the extraction process of Problem 7.3. Calculate the total amount of solvent required if the extraction is done in a crosscurrent cascade consisting of 5 ideal stages. Use equation (3-60) from Problem (3.19).

Answer: 50,900 kg/h

7.6[a]. Multistage crosscurrent extraction: insoluble liquids.

Repeat Example 7.3 using 10 kg/h of solvent in each stage.

7.7[a]. Multistage countercurrent extraction.

Determine the number of ideal stages required in Example 7.4 if the solvent rate used is twice the minimum.

7.7[a]. Multistage countercurrent extraction: insoluble liquids.

A water solution containing 0.005 mole fraction of benzoic acid is to be extracted using pure benzene as the solvent. If the feed rate is 100 moles/h and the solvent rate is 10 moles/h, find the number of equilibrium stages required to reduce the concentration of benzoic acid in the aqueous solution to 0.0001 mole fraction. Operation is isothermal at 280 K where the equilibrium data can be represented as (Wankat, 1988):

Mole fraction of benzoic acid in water = 0.0446 × (mole fraction of benzoic acid in benzene)

Problems 7.9 to 7.12 refer to the system water (A)–chlorobenzene (B)–pyridine (C) at 298 K. Equilibrium tie-line data taken from Treybal (1980) in weight percent are given in Table 7.6.

7.9[a]. Right-triangular diagrams.

Plot the equilibrium data for the system water–chlorobenzene–pyridine in right-triangular coordinates.

7.10[b]. Single-stage extraction.

It is desired to reduce the pyridine concentration of 5,000 kg/h of an aqueous solution from 50 to 5 wt% in a single batch extraction with pure chlorobenzene. What amount of solvent is required? Solve on right-triangular coordinates.

Table 7.6 Water–Chlorobenzene–Pyridine Equilibrium Data

Water	Chlorobenzene-phase, wt % Chlorobenzene	Pyridine	Water	Water-phase, wt% Chlorobenzene	Pyridine
0.05	99.95	0.00	99.92	0.08	0.00
0.67	88.28	11.05	94.82	0.16	5.02
1.15	79.90	18.95	88.71	0.24	11.05
1.62	74.28	24.10	80.72	0.38	18.90
2.25	69.15	28.60	73.92	0.58	25.50
2.87	65.58	31.55	62.05	1.85	36.10
3.95	61.00	35.05	50.87	4.18	44.95
6.40	53.00	40.60	37.90	8.90	53.20
13.20	37.80	49.00	13.20	37.80	49.00

Water and chlorobenzene phases on the same line are in equilibrium with each other. Data were obtained from Treybal (1980) at 298 K and 1 atm.

7.11[b]. Multistage crosscurrent extraction.

A 5,000-kg batch of pyridine-water solution, 50 wt% pyridine, is to be extracted with an equal weight of pure chlorobenzene in each of the stages of a crosscurrent cascade. How many ideal stages and what total solvent is required to reduce the concentration of pyridine to 5 wt% in the final raffinate? Solve on right-triangular coordinates.

7.12[b]. Multistage countercurrent extraction.

A pyridine-water solution, 50 wt% pyridine, is to be continuously and countercurrently extracted at the rate of 2.25 kg/s with pure chlorobenzene to reduce the pyridine concentration to 2 wt%.

(a) Determine the minimum solvent rate required.

(b) If 2.3 kg/s of solvent is used, calculate the number of theoretical stages required, and the flow rates of final extract and raffinate.

7.13[b]. Multistage countercurrent extraction.

We wish to remove acetic acid from water using pure isopropyl ether as solvent. The operation is at 293 K and 1 atm (see Table 7.2). The feed is 45 wt% acetic acid and 55 wt% water. Feed flow rate is 2,000 kg/h. A multistage countercurrent extraction cascade is used to produce a final extract that is 20 wt% acetic acid and a final raffinate that is also 20 wt% acetic acid. Calculate how much solvent and how many equilibrium stages are required.

7.14[b]. Multistage countercurrent extraction.

We are extracting acetic acid from water with isopropyl ether at 293 K and 1 atm (see Table 7.2). The column has three equilibrium stages. The entering feed rate is 1,000 kg/h, and is 40 wt% acid and 60 wt% water. The final extract stream flows at the rate of 2,500 kg/h and is 20 wt% acetic acid. The entering solvent (which is *not* pure isopropyl ether) contains no water. Calculate:

(a) The final raffinate concentration and flow rate.

(b) The entering solvent concentration and flow rate.

7.15[b]. Single-stage and multistage crosscurrent extraction.

Water-dioxane solutions form a minimum-boiling point azeotrope at atmospheric pressure and cannot be separated by ordinary distillation methods. Benzene forms no azeotrope with dioxane and can be used as an extraction solvent. At 298 K the equilibrium distribution of dioxane between water and benzene is as follows (Treybal, 1980):

Wt% dioxane in water	5.1	18.9	25.2
Wt% dioxane in benzene	5.2	22.5	32.0

At these concentrations water and benzene are substantially insoluble. A 1,000-kg batch of a 25 wt% dioxane–75 wt% water solution is to be extracted with pure benzene to remove 955 of the dioxane.

(a) Calculate the solvent requirement for a single batch operation.

(b) If the extraction were done with equal amounts of solvent in five crosscurrent ideal stages, how much solvent would be required?

7.16[b]. Multistage countercurrent extraction.

A 25 wt% solution of dioxane in water is to be extracted continuously at the rate of 1,000 kg/h in countercurrent fashion with pure benzene to remove 95% of the dioxane. Equilibrium data are given in Problem 7.15.

(a) What is the minimum solvent requirement?

(b) If 900 kg/h of solvent is used, how many theoretical stages are required?

Problems 7.17 to 7.19 refer to the system cottonseed oil (A)–liquid propane (B)–oleic acid (C) at 372 K and 42.5 atm, Equilibrium tie-line data taken from Treybal (1980) in weight percent are:

Oil	Popane	Acid	Oil	Propane	Acid
63.5	36.5	0.0	2.30	97.7	0.00
57.2	37.3	5.5	1.95	97.3	0.76
52.0	39.0	9.0	1.78	97.0	1.22
42.7	39.5	13.8	1.50	96.6	1.90
39.8	41.5	18.7	1.36	95.9	2.74
31.0	42.7	26.3	1.20	95.0	3.80
26.9	43.7	29.4	1.10	94.5	4.40
21.0	46.6	32.4	1.00	93.9	5.10
14.2	48.4	37.4	0.80	93.1	6.10
8.3	52.2	39.5	0.70	92.1	7.20
4.5	54.4	41.1	0.40	93.5	6.10
0.8	55.5	43.7	0.20	94.3	5.50

7.17[b]. Janecke diagram for liquid–liquid extraction.

Generate the Janecke diagram and the distribution curve for the system cottonseed oil (A)–propane (B)–oleic acid (C) at 372 K and 42.5 atm.

7.18[b]. Crosscurrent extraction on the Janecke diagram.

If 100 kg of a cottonseed oil–oleic acid solution containing 25 wt% acid is extracted twice in crosscurrent fashion, each time with 1,000 kg of pure propane, determine the compositions (wt%) and the weights of the mixed extract and the final raffinate. Make the computations on a solvent-free basis using the Janecke diagram generated in Problem 7.17.

7.19[c]. Countercurrent extraction with extract reflux.

If 1,000 kg of a cottonseed oil–oleic acid solution containing 25 wt%

acid is to be continuously separated into two products containing 2 and 90 wt% acid (solvent-free compositions) by countercurrent extraction with propane, calculate:

(a) Minimum number of theoretical stages required.

(b) Minimum external extract-reflux ratio required.

(c) For an external extract-reflux ratio of 4.5, the number of theoretical stages, the position of the feed stage, and the quantities, in kg/h, of the important streams.

7.20^c. Countercurrent extraction with extract reflux.

A feed mixture containing 50 wt% n-heptane and 50 wt% methylcyclohexane (MCH) at 298 K and 1 atm is to be separated by liquid–liquid extraction into one product containing 92.5 wt% MCH and another containing 7.5 wt% MCH. Aniline will be used as the solvent. Using the equilibrium data given in Table 7.7 to construct a Janecke diagram, calculate:

(a) Minimum number of theoretical stages required.

(b) Minimum external extract-reflux ratio required.

(c) For an external extract-reflux ratio of 7.0, the number of theoretical stages, the position of the feed stage, and the quantities, in kg/h, of the important streams.

7.21^b. Design of a mixer-settler extraction unit.

Acetic acid is to be extracted from a dilute aqueous solution with isopropyl ether at 298 K in a countercurrent cascade of mixer-settler units. In one of the units, the following conditions apply:

	Raffinate	Extract
Flow rate, kg/s	2.646	6.552
Density, kg/m^3	1,017	726
Viscosity, cP	3.0	1.0

Interfacial tension = 0.0135 N/m

If the raffinate is the dispersed phase, and the mixer residence time is 2.5 min, estimate for the mixer:

(a) The dimensions of a closed, baffled vessel.

(b) The diameter of a flat-bladed impeller.

(c) The minimum rate of rotation of the impeller.

(d) The power requirement of the agitator at minimum rate of rotation.

Table 7.7 Equilibrium Data for System n-Heptane (A)–Aniline (B)–MCH (C) at 298 K (solvent-free basis)

Hydrocarbon Layer		Solvent Layer	
wt% MCH	kg B/kg (A + B)	wt% MCH	kg B/kg (A + B)
0.0	0.0799	0.0	15.12
9.9	0.0836	11.8	13.72
20.2	0.0870	33.8	11.50
23.9	0.0894	37.0	11.34
36.9	0.0940	50.6	9.98
44.5	0.0952	60.0	9.00
50.5	0.0989	67.3	8.09
66.0	0.1062	76.7	6.83
74.6	0.1111	84.3	6.45
79.7	0.1135	88.8	6.00
82.1	0.1160	90.4	5.90
93.9	0.1272	96.2	5.17
100.0	0.1350	100.0	4.92

Source: Seader and Henley (1998).

7.22[c]. Mass transfer in mixer unit for liquid–liquid extraction.

For the conditions and results of Example 7.7, involving the extraction of benzoic acid from a dilute aqueous solution using toluene as solvent, determine the following when using a six-flat-blade impeller in a closed vessel with baffles and with the extract phase dispersed:

(a) The minimum rate of rotation of the impeller for complete and uniform dispersion.

(b) The power requirement of the agitator at the minimum rate of rotation.

(c) The Sauter mean droplet diameter, and the interfacial area.

(d) The overall mass-transfer coefficient, K_{OD}.

(e) The fractional extraction of benzoic acid.

Liquid properties are:

	Raffinate	Extract
Density, g/cm^3	0.995	0.860
Viscosity, cP	0.95	0.59
Diffusivity, cm^2/s	2.2×10^{-5}	1.5×10^{-5}

Interfacial tension = 0.022 N/m

Distribution coefficient of benzoic acid = $c_D/c_C = 21$

7.23^a. Sizing of an RDC extraction column.

Estimate the diameter and the HETS of an RDC column to extract acetic acid from water using isopropyl ether for the conditions and data of Problem 7.21.

7.24^a. Sizing of a Karr extraction column.

Estimate the diameter and the HETS of a Karr column to extract aceti-cacid from water using isopropyl ether for the conditions and data of Problem 7.21.

REFERENCES

Alders, L., *Liquid–Liquid Extraction,* 2nd ed., Elsevier, Amsterdam (1959).

Barson, N., and G. H. Beyer, *Chem. Eng. Prog.,* **49**, 243 (1953).

Boobar, N., et al., *Ind. Eng. Chem.,* **43**, 2922 (1951).

Davies, J. T., *Turbulence Phenomena,* Academic Press, New York (1978).

Flynn, A. W., and R. E. Teybal, *AIChE J.,* **1**, 324 (1955).

Gnanasundaram, S., et al., *Can. J. Chem. Eng.,* **57**, 141 (1979).

Janecke, E., *Z. Anorg. Allg. Chem.,* **51**, 132 (1906).

Karr, A. E., and T. C. Lo, *Chem. Eng. Prog.,* **72**, 243 (1976).

Laity, D. S., and R. E. Treybal, *AIChE J.,* **3**, 176 (1957).

Lewis, J. B., et al., *Trans. Inst. Chem. Eng.,* **29,** 126 (1951).

Maloney, J. O., and A. E. Schubert, *Trans. AIChE,* **36**, 741 (1940).

Perry, R. H., and C. H. Chilton (eds.), *Chemical Engineers' Handbook,* 5 th ed., McGraw-Hill, New York (1973).

Ryan , A. D., et al., *Chem. Eng. Prog.,* **55**, 70 (1959).

Scheibel, E. G., *Chem. Eng. Prog.,* **44**, 681 (1948).

Schindler, H. D., and R. E. Treybal, *AIChE J.,* **14**, 790 (1968).

Seader, J. D., and E. J. Henley, *Separation Process Principles,* Wiley, New York (1998).

Skelland, A. H. P., and L. T. Moeti, *Ind. Eng. Chem. Res.,* **29**, 2258 (1990).

Skelland, A. H. P., and G. G. Ramsey, *Ind. Eng. Chem. Res.,* **26**, 77 (1987).

Stichlmair, J., *Chemie–Ingeniur–Technik,* **52**, 253 (1980).

Treybal, R. E., *Liquid Extraction,* 2nd ed., McGraw-Hill, New York (1963).

Treybal, R. E., *Mass-Transfer Operations,* 3rd ed., McGraw-Hill, New York (1980).

Vermeulen, T., et al., *Chem. Eng. Prog.,* **51**, 85F (1955).

Wankat, P. C., *Equilibrium Staged Separations,* Elsevier, New York (1988).

Appendix A

Binary Diffusion Coefficients

Table A.1 Mass Diffusivities in Gases

System	T, K	$D_{AB}P$, m²-Pa/s	System	T, K	$D_{AB}P$, m²-Pa/s
Air			Carbon dioxide		
Ammonia	273	2.006	Ethanol	273	0.702
Aniline	298	0.735	Ethyl ether	273	0.548
Benzene	298	0.974	Hydrogen	273	5.572
Bromine	293	0.923	Methane	273	1.550
Carbon dioxide	273	1.378	Methanol	298.6	1.064
Carbon disulfide	273	0.894	Nitrogen	298	1.672
Chlorine	273	1.256	Nitrous oxide	298	1.185
Diphenyl	491	1.621	Propane	298	0.874
Ethyl acetate	273	0.718	Water	298	1.661
Ethanol	298	1.337	Hydrogen		
Ethyl ether	293	0.908	Ammonia	293	8.600
Iodine	298	0.845	Argon	293	7.800
Methanol	298	1.641	Benzene	273	3.211
Mercury	614	4.791	Ethane	273	4.447
Naphthalene	303	0.870	Methane	273	6.331
Nitrobenzene	298	0.879	Oxygen	273	7.061
n-Octane	298	0.610	Pyridine	318	4.427
Oxygen	273	1.773	Nitrogen		
Propyl acetate	315	0.932	Ammonia	293	2.441
Sulfur dioxide	273	1.236	Ethylene	298	1.651
Toluene	298	0.855	Hydrogen	288	7.527
Water	298	2.634	Iodine	273	0.709
Ammonia			Oxygen	273	1.834
Ethylene	293	1.793	Oxygen		
Carbon dioxide			Ammonia	293	2.563
Benzene	318	0.724	Benzene	296	0.951
Carbon disulfide	318	0.724	Ethylene	293	1.844
Ethyl acetate	319	0.675	Water	308.1	2.857

Source: Welty, J. R., C. E. Wicks, and R. E. Wilson, *Fundamentals of Momentum, Heat, and Mass Transfer,* 3rd ed., Wiley, New York (1984).

Table A.2 Mass Diffusivities in Liquids at Infinite Dilution

System	T, K	$D^o_{AB} \times 10^5$, cm^2/s	System	T, K	$D^o_{AB} \times 10^5$, cm^2/s
Chloroform (solvent B)			Ethyl acetate (solvent B)		
Acetone	298	2.35	Acetic acid	293	2.18
Benzene	288	2.51	Acetone	293	3.18
Ethanol	288	2.20	Ethyl benzoate	293	1.85
Ethyl ether	298	2.13	MEK	303	2.93
Ethyl acetate	298	2.02	Nitrobenzene	293	2.25
MEK	298	2.13	Water	298	3.20
Benzene (solvent B)			Water (solvent B)		
Acetic acid	298	2.09	Methane	298	1.49
Aniline	298	1.96	Air	298	2.00
Benzoic acid	298	1.38	Carbon dioxide	298	1.92
Bromobenzene	281	1.45	Chlorine	298	1.25
Ciclohexane	298	2.09	Argon	298	2.00
Ethanol	288	2.25	Benzene	298	1.02
Formic acid	298	2.28	Ethanol	288	1.00
n-Heptane	353	4.25	Ethane	298	1.20
MEK	303	2.09	Oxygen	298	2.10
Naphthalene	281	1.19	Pyridine	288	0.58
Toluene	298	1.85	Aniline	293	0.92
Vinyl chloride	281	1.77	Ammonia	298	1.64
Acetone (solvent B)			Ethylene	298	1.87
Acetic acid	288	2.92	Allyl alcohol	288	0.90
Acetic acid	313	4.04	Acetic acid	293	1.19
Benzoic acid	298	2.62	Benzoic acid	298	1.00
Formic acid	298	3.77	Propionic acid	298	1.06
Water	298	4.56	Vinyl chloride	298	1.34
Ethanol (solvent B)			Ethylbenzene	293	0.81
Benzene	298	1.81	Sulfuric acid	298	1.73
Water	298	1.24	Nitric acid	298	2.60

Sources: Reid, R. C., J. M. Prausnitz, and B. E. Poling, *The Properties of Gases and Liquids,* 4th ed., McGraw-Hill, New York (1987); Cussler, E. L., *Diffusion, Mass Transfer in Fluid Systems,* 2nd ed., Cambridge University Press, Cambridge, UK (1997).

Table A.3 Mass Diffusivities in the Solid State

System	T, K	Diffusivity, cm^2/s
Hydrogen in iron	283	1.66×10^{-9}
	323	11.4×10^{-9}
	373	124×10^{-9}
Hydrogen in nickel	358	11.6×10^{-9}
	438	10.5×10^{-8}
Carbon monoxide in nickel	1223	4.00×10^{-8}
	1323	14.0×10^{-8}
Aluminum in copper	1123	2.20×10^{-9}
	293	1.30×10^{-30}
Uranium in tungsten	2000	1.30×10^{-11}
Cerium in tungsten	2000	95.0×10^{-11}
Yttrium in tungsten	2000	1820×10^{-11}
Tin in lead	558	1.60×10^{-10}
Gold in lead	558	4.60×10^{-10}
Gold in silver	1033	3.60×10^{-10}
Antimony in silver	293	3.60×10^{-10}
Zinc in aluminum	773	2.00×10^{-9}
Silver in aluminum	323	1.20×10^{-9}
Bismuth in lead	293	1.10×10^{-16}
Cadmium in copper	293	2.70×10^{-15}
Carbon in iron	1073	1.50×10^{-8}
	1373	45.0×10^{-8}
Helium in silica	293	4.00×10^{-10}
	773	7.80×10^{-8}
Hydrogen in silica	473	6.50×10^{-10}
	773	1.30×10^{-8}
Helium in Pyrex	293	4.50×10^{-11}
	773	2.00×10^{-8}

Sources: Barrer, R. M., *Diffusion in and through Solids,* Macmillan, New York (1941); American Society for Metals, *Diffusion,* ASM (1973); Cussler, E. L., *Diffusion, Mass Transfer in Fluid Systems,* 2nd ed., Cambridge University Press, Cambridge, UK (1997).

Appendix B
Lennard–Jones Constants

Determined from Viscosity Data

Compound		σ, Å	ε/k, K
Ar	Argon	3.542	93.3
He	Helium	2.551	10.22
Kr	Krypton	3.655	178.9
Ne	Neon	2.820	32.8
Xe	Xenon	4.047	231.0
Air	Air	3.620	97.0
AsH_3	Arsine	4.145	259.8
BCl_3	Boron chloride	5.127	337.7
BF_3	Boron fluoride	4.198	186.3
$B(OCH_3)_3$	Methyl borate	5.503	396.7
Br_2	Bromine	4.296	507.9
CCl_4	Carbon tetrachloride	5.947	322.7
CF_4	Carbon tetrafluoride	4.662	134.0
$CHCl_3$	Chloroform	5.389	340.2
CH_2Cl_2	Methylene chloride	4.898	356.3
CH_3Br	Methyl bromide	4.118	449.2
CH_3Cl	Methyl chloride	4.182	350.0
CH_3OH	Methanol	3.626	481.8
CH_4	Methane	3.758	148.6
CO	Carbon monoxide	3.690	91.7
COS	Carbonyl sulfide	4.130	336.0
CO_2	Carbon dioxide	3.941	195.2
CS_2	Carbon disulfide	4.483	467.0
C_2H_2	Acetylene	4.033	231.8
C_2H_4	Ethylene	4.163	224.7
C_2H_6	Ethane	4.443	215.7
C_2H_5Cl	Ethyl chloride	4.898	300.0
C_2H_5OH	Ethanol	4.530	362.6
C_2N_2	Cyanogen	4.361	348.6
CH_3OCH_3	Methyl ether	4.307	395.0
CH_2CHCH_3	Propylene	4.678	298.9

Lennard–Jones Constants Determined from Viscosity Data

Compound		σ, Å	ε/k, K
CH_3CHH	Methylacetylene	4.761	251.8
C_3H_6	Cyclopropane	4.807	248.9
C_3H_8	Propane	5.118	237.1
$n\text{-}C_3H_7OH$	n-Propyl alcohol	4.549	576.7
CH_3COCH_3	Acetone	4.600	560.2
CH_3COOCH_3	Methyl acetate	4.936	469.8
$n\text{-}C_4H_{10}$	n-Butane	4.687	531.4
$iso\text{-}C_4H_{10}$	Isobutane	5.278	330.1
$C_2H_5OC_2H_5$	Ethyl ether	5.678	313.8
$CH_3COOC_2H_5$	Ethyl acetate	5.205	521.3
$n\text{-}C_5H_{12}$	n-Pentane	5.784	341.1
$C(CH_3)_4$	2,2-Dimethylpropanone	6.464	193.4
C_6H_6	Benzene	5.349	412.3
C_6H_{12}	Cyclohexane	6.182	297.1
$n\text{-}C_6H_{14}$	n-Hexane	5.949	399.3
Cl_2	Chlorine	4.217	316.0
F_2	Fluorine	3.357	112.6
HBr	Hydrogen bromide	3.353	449.0
HCN	Hydrogen cyanide	3.630	569.1
HCl	Hydrogen chloride	3.339	344.7
HF	Hydrogen fluoride	3.148	330.0
HI	Hydrogen iodide	4.211	288.7
H_2	Hydrogen	2.827	59.7
H_2O	Water	2.641	809.1
H_2O_2	Hydrogen peroxide	4.196	289.3
H_2S	Hydrogen sulfide	3.623	301.1
Hg	Mercury	2.969	750.0
$HgBr_2$	Mercuric bromide	5.080	686.2
$HgCl_2$	Mercuric chloride	4.550	750.0
HgI_2	Mercuric iodide	5.625	695.6
I_2	Iodine	5.160	474.2
NH_3	Ammonia	2.900	558.3
NO	Nitric oxide	3.492	116.7
N_2	Nitrogen	3.798	71.4
N_2O	Nitrous oxide	3.828	232.4
O_2	Oxygen	3.467	106.7
SF_6	Sulfur hexafluoride	5.128	222.1
SO_2	Sulfur dioxide	4.112	335.4
UF_6	Uranium hexafluoride	5.967	236.8

Source: Reid, R. C., J. M. Praunitz, and B. E. Poling, *The Properties of Gases and Liquids,* 4th ed., McGraw-Hill, New York (1987).

Appendix C-1

Solution of Maxwell-Stefan Equations by Orthogonal Collocation ($NC = 3$)

Numerical data correspond to Example 1.15

Orthogonal collocation matrices

$$A := \begin{bmatrix} -9.990 & 13.514 & -5.352 & 2.827 & -1.00 \\ -2.472 & -0.016 & 3.500 & -1.524 & 0.517 \\ 0.747 & -2.672 & 0.00 & 2.672 & -0.747 \\ -0.517 & 1.529 & -3.500 & 0.016 & 2.472 \\ 1.00 & -2.827 & 5.352 & -13.514 & 9.990 \end{bmatrix} \qquad \eta := \begin{bmatrix} 0.000 \\ 0.173 \\ 0.500 \\ 0.827 \\ 1.000 \end{bmatrix}$$

The pressure and temperature in the vapor phase are $\qquad \qquad$ ORIGIN $\equiv 1$

$$P := 99.4 \cdot 10^3 \cdot Pa \qquad T := 328.5 \cdot K$$

The Maxwell-Stefan diffusion coefficients are

$$D12 := 8.48 \cdot \frac{mm^2}{sec} \qquad D13 := 13.72 \cdot \frac{mm^2}{sec} \qquad D23 := 19.91 \cdot \frac{mm^2}{sec}$$

The length of the diffusion path is

$$\delta := 0.24 \cdot m$$

The density of the gas phase follows from the ideal gas law:

$$R := 8.314 \cdot \frac{Pa \cdot m^3}{mole \cdot K}$$

$$c := \frac{P}{R \cdot T} \qquad c = 36.395 \cdot m^{-3} \cdot mole$$

$$F_{12} := c \cdot \frac{D12}{\delta} \qquad F_{12} = 1.286 \cdot 10^{-3} \cdot m^{-2} \cdot sec^{-1} \cdot mole$$

$$F_{13} := \frac{c \cdot D13}{\delta} \qquad F_{13} = 2.081 \cdot 10^{-3} \cdot m^{-2} \cdot sec^{-1} \cdot mole$$

$$F_{23} := c \cdot \frac{D23}{\delta} \qquad F_{23} = 3.019 \cdot 10^{-3} \cdot m^{-2} \cdot sec^{-1} \cdot mole$$

441

Initial estimates of the fluxes

$$\begin{bmatrix} N1 \\ N2 \\ N3 \end{bmatrix} := \begin{bmatrix} F_{12} \\ F_{23} \\ 0.0 \cdot \dfrac{mole}{m^2 \cdot sec} \end{bmatrix}$$

Initial estimates of the concentrations

$$\begin{bmatrix} y1_1 \\ y2_1 \\ y3_1 \end{bmatrix} := \begin{bmatrix} 0.319 \\ 0.528 \\ 1 - 0.319 - 0.528 \end{bmatrix}$$

$y1_2 := 0.3 \qquad y1_3 := 0.2 \qquad y1_4 := 0.0 \qquad y1_5 := 0.0$

$y2_2 := 0.4 \qquad y2_3 := 0.25 \qquad y2_4 := 0.0 \qquad y2_5 := 0.0$

Given

$$A_{2,1} \cdot y1_1 + A_{2,2} \cdot y1_2 + A_{2,3} \cdot y1_3 + A_{2,4} \cdot y1_4 \ldots + A_{2,5} \cdot y1_5 = \frac{y1_2 \cdot N2 - y2_2 \cdot N1}{F_{12}} \ldots + \frac{y1_2 \cdot N3 - \left(1 - y1_2 - y2_2\right) \cdot N1}{F_{13}}$$

$$\left(A_{2,1} \cdot y2_1 + A_{2,2} \cdot y2_2 + A_{2,3} \cdot y2_3 + A_{2,4} \cdot y2_4\right) \ldots + A_{2,5} \cdot y2_5 = \frac{y2_2 \cdot N1 - y1_2 \cdot N2}{F_{12}} \ldots + \frac{y2_2 \cdot N3 - \left(1 - y1_2 - y2_2\right) \cdot N2}{F_{23}}$$

$$\left(A_{3,1} \cdot y1_1 + A_{3,2} \cdot y1_2 + A_{3,3} \cdot y1_3 + A_{3,4} \cdot y1_4\right) \ldots + A_{3,5} \cdot y1_5 = \frac{y1_3 \cdot N2 - y2_3 \cdot N1}{F_{12}} \ldots + \frac{y1_3 \cdot N3 - \left(1 - y1_3 - y2_3\right) \cdot N1}{F_{13}}$$

$$\left(A_{3,1} \cdot y2_1 + A_{3,2} \cdot y2_2 + A_{3,3} \cdot y2_3 + A_{3,4} \cdot y2_4\right) \ldots + A_{3,5} \cdot y2_5 = \frac{y2_3 \cdot N1 - y1_3 \cdot N2}{F_{12}} \ldots + \frac{y2_3 \cdot N3 - \left(1 - y1_3 - y2_3\right) \cdot N2}{F_{23}}$$

$$\left(A_{4,1} \cdot y1_1 + A_{4,2} \cdot y1_2 + A_{4,3} \cdot y1_3 + A_{4,4} \cdot y1_4 \atop + A_{4,5} \cdot y1_5\right) \dots = \frac{y1_4 \cdot N2 - y2_4 \cdot N1}{F_{12}} \dots$$
$$+ \frac{y1_4 \cdot N3 - \left(1 - y1_4 - y2_4\right) \cdot N}{F_{13}}$$

$$\left(A_{4,1} \cdot y2_1 + A_{4,2} \cdot y2_2 + A_{4,3} \cdot y2_3 + A_{4,4} \cdot y2_4 \atop + A_{4,5} \cdot y2_5\right) \dots = \frac{y2_4 \cdot N1 - y1_4 \cdot N2}{F_{12}} \dots$$
$$+ \frac{y2_4 \cdot N3 - \left(1 - y1_4 - y2_4\right) \cdot}{F_{23}}$$

$$\left(A_{5,1} \cdot y1_1 + A_{5,2} \cdot y1_2 + A_{5,3} \cdot y1_3 + A_{5,4} \cdot y1_4 \atop + A_{5,5} \cdot y1_5\right) \dots = \frac{y1_5 \cdot N2 - y2_5 \cdot N1}{F_{12}} \dots$$
$$+ \frac{y1_5 \cdot N3 - \left(1 - y1_5 - y2_5\right) \cdot N1}{F_{13}}$$

$$\left(A_{5,1} \cdot y2_1 + A_{5,2} \cdot y2_2 + A_{5,3} \cdot y2_3 + A_{5,4} \cdot y2_4 \atop + A_{5,5} \cdot y2_5\right) \dots = \frac{y2_5 \cdot N1 - y1_5 \cdot N2}{F_{12}} \dots$$
$$+ \frac{y2_5 \cdot N3 - \left(1 - y1_5 - y2_5\right) \cdot N}{F_{23}}$$

$$\begin{bmatrix} N1 \\ N2 \\ y1_2 \\ y1_3 \\ y1_4 \\ y2_2 \\ y2_3 \\ y2_4 \end{bmatrix} := \text{Find}\left(N1, N2, y1_2, y1_3, y1_4, y2_2, y2_3, y2_4\right)$$

$$N1 = 1.791 \cdot 10^{-3} \cdot m^{-2} \cdot sec^{-1} \cdot mole \qquad N2 = 3.095 \cdot 10^{-3} \cdot m^{-2} \cdot sec^{-1} \cdot mole$$

$$y1 := \begin{bmatrix} y1_1 \\ y1_2 \\ y1_3 \\ y1_4 \\ y1_5 \end{bmatrix} \qquad y2 := \begin{bmatrix} y2_1 \\ y2_2 \\ y2_3 \\ y2_4 \\ y2_5 \end{bmatrix} \qquad y1 = \begin{bmatrix} 0.319 \\ 0.305 \\ 0.244 \\ 0.12 \\ 0 \end{bmatrix} \qquad y2 = \begin{bmatrix} 0.528 \\ 0.485 \\ 0.366 \\ 0.158 \\ 0 \end{bmatrix}$$

$$y3 := \overrightarrow{(1 - y1 - y2)}$$

$$z := \overrightarrow{(\eta \cdot \delta)} \qquad\qquad y3 = \begin{bmatrix} 0.153 \\ 0.21 \\ 0.39 \\ 0.721 \\ 1 \end{bmatrix} \qquad z = \begin{bmatrix} 0 \\ 0.042 \\ 0.12 \\ 0.198 \\ 0.24 \end{bmatrix} \cdot m$$

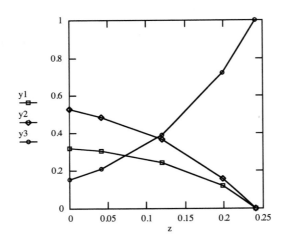

Appendix C-2

Solution of Maxwell-Stefan Equations by Orthogonal Collocation ($NC = 3$)

Numerical data correspond to Example 1.16

Orthogonal collocation matrices

$$A := \begin{bmatrix} -9.990 & 13.514 & -5.352 & 2.827 & -1.00 \\ -2.472 & -0.016 & 3.500 & -1.524 & 0.517 \\ 0.747 & -2.672 & 0.00 & 2.672 & -0.747 \\ -0.517 & 1.529 & -3.500 & 0.016 & 2.472 \\ 1.00 & -2.827 & 5.352 & -13.514 & 9.990 \end{bmatrix} \qquad \eta := \begin{bmatrix} 0.000 \\ 0.173 \\ 0.500 \\ 0.827 \\ 1.000 \end{bmatrix}$$

The pressure and temperature in the vapor phase are $\qquad$ ORIGIN $\equiv 1$

$$P := 101.3 \cdot 10^3 \cdot Pa \qquad T := 548 \cdot K$$

The Maxwell-Stefan diffusion coefficients are

$$D12 := 72 \cdot \frac{mm^2}{sec} \qquad D13 := 230 \cdot \frac{mm^2}{sec} \qquad D23 := 230 \cdot \frac{mm^2}{sec}$$

The length of the diffusion path is $\qquad\qquad$ Reaction rate constant

$$\delta := 0.001 \cdot m \qquad\qquad\qquad k_r := 10 \cdot \frac{mole}{m^2 \cdot sec}$$

The density of the gas phase follows from the ideal gas law:

$$R := 8.314 \cdot \frac{Pa \cdot m^3}{mole \cdot K}$$

$$c := \frac{P}{R \cdot T} \qquad c = 22.234 \cdot m^{-3} \cdot mole$$

$$F_{12} := c \cdot \frac{D12}{\delta} \qquad F_{12} = 1.601 \cdot m^{-2} \cdot sec^{-1} \cdot mole$$

$$F_{13} := \frac{c \cdot D13}{\delta} \qquad F_{13} = 5.114 \cdot m^{-2} \cdot sec^{-1} \cdot mole$$

$$F_{23} := c \cdot \frac{D23}{\delta} \qquad F_{23} = 5.114 \cdot m^{-2} \cdot sec^{-1} \cdot mole$$

445

Initial estimates of the fluxes

$$\begin{bmatrix} N1 \\ N2 \\ N3 \end{bmatrix} := \begin{bmatrix} F_{12} \\ -F_{12} \\ -F_{12} \end{bmatrix}$$

Initial estimates of the concentrations

$$\begin{bmatrix} y1_1 \\ y2_1 \\ y3_1 \end{bmatrix} := \begin{bmatrix} 0.6 \\ 0.2 \\ 0.2 \end{bmatrix}$$

$y1_2 := 0.5$	$y1_3 := 0.4$	$y1_4 := 0.3$	$y1_5 := 0.2$
$y2_2 := 0.3$	$y2_3 := 0.4$	$y2_4 := 0.5$	$y2_5 := 0.6$

Given

$$A_{2,1} \cdot y1_1 + A_{2,2} \cdot y1_2 + A_{2,3} \cdot y1_3 + A_{2,4} \cdot y1_4 \dots = \frac{y1_2 \cdot N2 - y2_2 \cdot N1}{F_{12}} \dots$$
$$+ A_{2,5} \cdot y1_5 \qquad\qquad + \frac{y1_2 \cdot N3 - \left(1 - y1_2 - y2_2\right) \cdot N1}{F_{13}}$$

$$\left(A_{2,1} \cdot y2_1 + A_{2,2} \cdot y2_2 + A_{2,3} \cdot y2_3 + A_{2,4} \cdot y2_4\right) \dots = \frac{y2_2 \cdot N1 - y1_2 \cdot N2}{F_{12}} \dots$$
$$+ A_{2,5} \cdot y2_5 \qquad\qquad + \frac{y2_2 \cdot N3 - \left(1 - y1_2 - y2_2\right) \cdot N2}{F_{23}}$$

$$\left(A_{3,1} \cdot y1_1 + A_{3,2} \cdot y1_2 + A_{3,3} \cdot y1_3 + A_{3,4} \cdot y1_4\right) \dots = \frac{y1_3 \cdot N2 - y2_3 \cdot N1}{F_{12}} \dots$$
$$+ A_{3,5} \cdot y1_5 \qquad\qquad + \frac{y1_3 \cdot N3 - \left(1 - y1_3 - y2_3\right) \cdot N1}{F_{13}}$$

$$\left(A_{3,1} \cdot y2_1 + A_{3,2} \cdot y2_2 + A_{3,3} \cdot y2_3 + A_{3,4} \cdot y2_4\right) \dots = \frac{y2_3 \cdot N1 - y1_3 \cdot N2}{F_{12}} \dots$$
$$+ A_{3,5} \cdot y2_5 \qquad\qquad + \frac{y2_3 \cdot N3 - \left(1 - y1_3 - y2_3\right) \cdot N2}{F_{23}}$$

$$\left(A_{4,1} \cdot y1_1 + A_{4,2} \cdot y1_2 + A_{4,3} \cdot y1_3 + A_{4,4} \cdot y1_4 \atop + A_{4,5} \cdot y1_5 \right) \dots = \frac{y1_4 \cdot N2 - y2_4 \cdot N1}{F_{12}} \dots$$
$$+ \frac{y1_4 \cdot N3 - \left(1 - y1_4 - y2_4\right) \cdot N1}{F_{13}}$$

$$\left(A_{4,1} \cdot y2_1 + A_{4,2} \cdot y2_2 + A_{4,3} \cdot y2_3 + A_{4,4} \cdot y2_4 \atop + A_{4,5} \cdot y2_5 \right) \dots = \frac{y2_4 \cdot N1 - y1_4 \cdot N2}{F_{12}} \dots$$
$$+ \frac{y2_4 \cdot N3 - \left(1 - y1_4 - y2_4\right) \cdot N2}{F_{23}}$$

$$\left(A_{5,1} \cdot y1_1 + A_{5,2} \cdot y1_2 + A_{5,3} \cdot y1_3 + A_{5,4} \cdot y1_4 \atop + A_{5,5} \cdot y1_5 \right) \dots = \frac{y1_5 \cdot N2 - y2_5 \cdot N1}{F_{12}} \dots$$
$$+ \frac{y1_5 \cdot N3 - \left(1 - y1_5 - y2_5\right) \cdot N1}{F_{13}}$$

$$\left(A_{5,1} \cdot y2_1 + A_{5,2} \cdot y2_2 + A_{5,3} \cdot y2_3 + A_{5,4} \cdot y2_4 \atop + A_{5,5} \cdot y2_5 \right) \dots = \frac{y2_5 \cdot N1 - y1_5 \cdot N2}{F_{12}} \dots$$
$$+ \frac{y2_5 \cdot N3 - \left(1 - y1_5 - y2_5\right) \cdot N2}{F_{23}}$$

$$N1 = k_r \cdot y1_5 \qquad\qquad N2 = -N1 \qquad\qquad N3 = -N1$$

$$\begin{bmatrix} N1 \\ N2 \\ N3 \\ y1_2 \\ y1_3 \\ y1_4 \\ y1_5 \\ y2_2 \\ y2_3 \\ y2_4 \\ y2_5 \end{bmatrix} := \text{Find}\left(N1, N2, N3, y1_2, y1_3, y1_4, y1_5, y2_2, y2_3, y2_4, y2_5 \right)$$

$$N1 = 0.888 \cdot m^{-2} \cdot sec^{-1} \cdot mole \qquad N2 = -0.888 \cdot m^{-2} \cdot sec^{-1} \cdot mole$$

$$y1 := \begin{bmatrix} y1_1 \\ y1_2 \\ y1_3 \\ y1_4 \\ y1_5 \end{bmatrix} \qquad y2 := \begin{bmatrix} y2_1 \\ y2_2 \\ y2_3 \\ y2_4 \\ y2_5 \end{bmatrix} \qquad y1 = \begin{bmatrix} 0.6 \\ 0.501 \\ 0.327 \\ 0.168 \\ 0.089 \end{bmatrix} \qquad y2 = \begin{bmatrix} 0.2 \\ 0.274 \\ 0.405 \\ 0.524 \\ 0.583 \end{bmatrix}$$

$$y3 := \overrightarrow{(1 - y1 - y2)} \qquad y3 = \begin{bmatrix} 0.2 \\ 0.225 \\ 0.268 \\ 0.308 \\ 0.329 \end{bmatrix} \qquad z = \begin{bmatrix} 0 \\ 1.73 \cdot 10^{-4} \\ 5 \cdot 10^{-4} \\ 8.27 \cdot 10^{-4} \\ 1 \cdot 10^{-3} \end{bmatrix} \cdot m$$

$$z := \overrightarrow{(\eta \cdot \delta)}$$

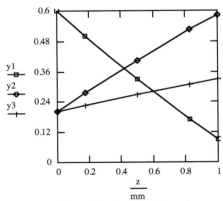

Distance from the bulk of the gas-phase

Appendix D

Packed Column Design Program

This program calculates the diameter of a packed column to satisfy a given gas-pressure drop criterion, and estimates the volumetric mass-transfer coefficients. Data presented are from Example 4.4.

Enter data related to the gas and liquid streams

Enter liquid flow rate, mL, in kg/s

$$mL := 0.804 \cdot kg \cdot sec^{-1}$$

Enter liquid density, in kg/m3

$$\rho L := 986 \cdot kg \cdot m^{-3}$$

Enter liquid viscosity, Pa-s

$$\mu L := 0.000631 \cdot Pa \cdot sec$$

Enter temperature, T, in K

$$T := 303 \cdot K$$

Enter data related to the packing

Enter packing factor, Fp, in ft2/ft3

$$Fp := 16$$

Introduce a units conversion factor in Fp

$$Fpu := Fp \cdot 1.0 \cdot sec^{1.9} \cdot m^{-2} \cdot Pa^{-0.1}$$

Enter porosity, fraction

$$\varepsilon := 0.977$$

Enter pressure drop constant, Cp

$$Cp := 0.421$$

Calculate flow parameter, X

$$X := \frac{mL}{mG} \cdot \sqrt{\frac{\rho G}{\rho L}} \qquad\qquad X = 0.016$$

Calculate Y at flooding conditions

$$Yflood := \exp\left[-\left(3.5021 + 1.028 \cdot \ln\left(X\right) + 0.11093 \cdot \ln\left(X\right)^{2}\right)\right]$$

Enter gas flow rate, mG, in kg/s

$$mG := 2.202 \cdot kg \cdot sec^{-1}$$

Enter gas density, kg/m3

$$\rho G := 1.923 \cdot kg \cdot m^{-3}$$

Enter gas viscosity, Pa-s

$$\mu G := 1.45 \cdot 10^{-5} \cdot Pa \cdot sec$$

Enter total pressure, P, in Pa

$$P := 110000 \cdot Pa$$

$$R := 8.314510 \cdot \frac{joule}{mole \cdot K}$$

Enter specific area, a, m2/m3

$$a := 92.3 \cdot m^{-1}$$

$$g := 9.8 \cdot m \cdot sec^{-2}$$

Enter loading constant, Ch

$$Ch := 0.876$$

Enter allowed pressure drop, in Pa/m

$$\Delta P := 300 \cdot Pa \cdot m^{-1}$$

449

Yflood = 0.317

Calculate gas velocity at flooding, vGf

$$Csf := \sqrt{\frac{Yflood}{Fpu \cdot \mu L^{0.1}}}$$

$Csf = 0.203 \cdot m \cdot sec^{-1}$

$$vGf := \frac{Csf}{\sqrt{\frac{\rho G}{\rho L - \rho G}}}$$

$vGf = 4.602 \cdot m \cdot sec^{-1}$

As a first estimate of the column diameter, D, design for 70% of flooding

$f := 0.7$

Calculate gas volume flow rate, QG, in m3/s

$$QG := \frac{mG}{\rho G}$$

$QG = 1.145 \cdot m^3 \cdot sec^{-1}$

$$D := \sqrt{\frac{4 \cdot QG}{f \cdot vGf \cdot \pi}}$$

$D = 0.673 \cdot m$

Calculate liquid volume flow rate, QL, in m3/s

$$QL := \frac{mL}{\rho L}$$

$QL = 8.154 \cdot 10^{-4} \cdot m^3 \cdot sec^{-1}$

Calculate effective particle size, dp, in m

$$dp := \frac{6 \cdot \left(1 - \varepsilon\right)}{a}$$

$dp = 1.495 \cdot 10^{-3} \cdot m$

$$vL\left(D\right) := \frac{4 \cdot QL}{\pi \cdot D^2}$$

$$ReL\left(D\right) := \frac{vL\left(D\right) \cdot \rho L}{a \cdot \mu L}$$

$$FrL\left(D\right) := \frac{vL\left(D\right)^2 \cdot a}{g}$$

$$ah\left(D\right) := \begin{vmatrix} a \cdot 0.85 \cdot Ch \cdot ReL\left(D\right)^{0.25} \cdot FrL\left(D\right)^{0.1} & \text{if} \quad ReL\left(D\right) \geq 5 \\ a \cdot Ch \cdot ReL\left(D\right)^{0.25} \cdot FrL\left(D\right)^{0.1} & \text{if} \quad ReL\left(D\right) < 5 \end{vmatrix}$$

$$hL\left(D\right) := \left(\frac{12 \cdot FrL\left(D\right)}{ReL\left(D\right)}\right)^{\frac{1}{3}} \cdot \left(\frac{ah\left(D\right)}{a}\right)^{\frac{2}{3}} \qquad KW\left(D\right) := \left[1 + \frac{2 \cdot dp}{3 \cdot D \cdot \left(1 - \varepsilon\right)}\right]^{-1}$$

$$vG\left(D\right) := \frac{4 \cdot QG}{\pi \cdot D^2} \qquad\qquad ReG\left(D\right) := \frac{vG\left(D\right) \cdot dp \cdot \rho G \cdot KW\left(D\right)}{\left(1 - \varepsilon\right) \cdot \mu G}$$

$$\psi0\left(D\right) := Cp \cdot \left(\frac{64}{ReG\left(D\right)} + \frac{1.8}{ReG\left(D\right)^{0.08}}\right)$$

$$\Delta P0\left(D\right) := \frac{\psi0\left(D\right) \cdot a \cdot \rho G \cdot vG\left(D\right)^2}{2 \cdot \varepsilon^3 \cdot KW\left(D\right)}$$

Iterate to find the tower diameter for the given pressure drop

Given

$$\Delta P = \Delta P0\left(D\right) \cdot \left[\frac{\varepsilon}{\varepsilon - hL\left(D\right)}\right]^{1.5} \cdot \exp\left(\frac{ReL\left(D\right)}{200}\right)$$

$$D := Find\left(D\right)$$

Column diameter, in meters

$$D = 0.737 \cdot m$$

Fractional approach to flooding

$$f := \frac{vG\left(D\right)}{vGf} \qquad\qquad f = 0.583$$

$$ah\left(D\right) = 58.664 \cdot m^{-1} \qquad vG\left(D\right) = 2.685 \cdot m \cdot sec^{-1} \qquad hL\left(D\right) = 0.017$$

$$ReG\left(D\right) = 2.186 \cdot 10^4 \qquad ReL\left(D\right) = 32.372 \qquad vL\left(D\right) = 1.912 \cdot 10^{-3} \cdot m \cdot sec^{-1}$$

$$\Delta P0 \left(\tilde{D}\right) = 248.429 \cdot \frac{Pa}{m} \qquad\qquad FrL \left(\tilde{D}\right) = 3.444 \cdot 10^{-5}$$

Calculate the volumetric mass-transfer coefficents

Enter the gas-phase diffusion coefficient, DG, m2/s, and the liquid-phase diffusion coefficient, DL, m2/s

$$DG := 8.5 \cdot 10^{-6} \cdot m^2 \cdot sec^{-1} \qquad\qquad DL := 1.91 \cdot 10^{-9} \cdot m^2 \cdot sec^{-1}$$

Enter the packing empirical coefficients for the mass-transfer correlations

$$CL := 1.168 \qquad\qquad CV := 0.408$$

$$kL \left(\tilde{D}\right) := 0.757 \cdot CL \cdot \sqrt{\frac{DL \cdot a \cdot vL \left(\tilde{D}\right)}{\varepsilon \cdot hL \left(\tilde{D}\right)}} \qquad\qquad kL \left(\tilde{D}\right) = 1.25 \cdot 10^{-4} \cdot m \cdot sec^{-1}$$

Volumetric liquid mass-transfer coefficient

$$kL \left(\tilde{D}\right) \cdot ah \left(\tilde{D}\right) = 7.33 \cdot 10^{-3} \cdot sec^{-1}$$

$$ScG := \frac{\mu G}{\rho G \cdot DG} \qquad\qquad ScG = 0.887$$

$$ky \left(\tilde{D}\right) := \frac{0.1304 \cdot CV \cdot DG \cdot P \cdot a \cdot \left(\dfrac{ReG \left(\tilde{D}\right)}{KW \left(\tilde{D}\right)}\right)^{0.75} \cdot ScG^{\frac{2}{3}}}{R \cdot T \cdot \left[\varepsilon \cdot \left(\varepsilon - hL \left(\tilde{D}\right)\right)\right]^{0.5}}$$

$$ky \left(\tilde{D}\right) = 3.261 \cdot m^{-2} \cdot sec^{-1} \cdot mole$$

Volumetric gas mass-transfer coefficient

$$ky \left(\tilde{D}\right) \cdot ah \left(\tilde{D}\right) = 191.314 \cdot m^{-3} \cdot sec^{-1} \cdot mole$$

Appendix E

Sieve-Tray Design Program

This program calculates the diameter of a sieve-tray tower to satisfy an approach to flooding criterion, and estimates the tray efficiency. Data presented are from Example 4.10.

Enter data related to the gas and liquid streams

Enter liquid flow rate, mL, in kg/s

$mL := 0.347 \cdot kg \cdot sec^{-1}$

Enter liquid density, in kg/m 3

$\rho L := 791 \cdot kg \cdot m^{-3}$

Enter temperature, T, in K

$T := 353 \cdot K$

Enter liquid surface tension, in dyne/cm

$\sigma := 21 \cdot \dfrac{dyne}{cm}$

Enter foaming factor, dimensionless

$FF := 0.9$

Enter water density at T, kg/m3

$\rho W := 970 \cdot kg \cdot m^{-3}$

Enter molecular weights of gas and liquid

$ML := 32 \qquad\qquad MG := 34.2$

Enter data related to the tray design

Enter hole diameter and pitch

$do := 4.5 \cdot mm$

$p := 12 \cdot mm$

Enter fractional approach to flooding

$f := 0.8$

Enter gas flow rate, mG, in kg/s

$mG := 0.494 \cdot kg \cdot sec^{-1}$

Enter gas density, kg/m3

$\rho G := 1.18 \cdot kg \cdot m^{-3}$

Enter gas viscosity, Pa-s

$\mu G := 1.05 \cdot 10^{-5} \cdot Pa \cdot sec$

Enter total pressure, P, in Pa

$P := 101300 \cdot Pa$

$R := 8.314510 \cdot \dfrac{joule}{mole \cdot K}$

$qL := \dfrac{mL}{\rho L} \qquad QG := \dfrac{mG}{\rho G}$

$g = 9.8066 \cdot m \cdot sec^{-2}$

Enter local slope of equilibrium curve

$me := 0.42$

Enter diffusivities of gas and liquid

$DG := 0.158 \cdot \dfrac{cm^2}{sec}$

$DL := 2.07 \cdot 10^{-5} \cdot \dfrac{cm^2}{sec}$

Enter plate thickness

$l := 2 \cdot mm$

Enter weir height

$hw := 5 \cdot cm$

453

Calculate flow parameter, X

$$X := \frac{mL}{mG} \cdot \sqrt{\frac{\rho G}{\rho L}} \qquad\qquad X = 0.0271$$

Specify the ratio of downcomer area to total area, AdAt

$$AdAt := \begin{vmatrix} 0.1 & \text{if} & X \le 0.1 \\ 0.2 & \text{if} & X \ge 1 \\ 0.1 + \dfrac{X - 0.1}{9} & & \text{otherwise} \end{vmatrix} \qquad AdAt = 0.1$$

If X is smaller than 0.1, use X = 0.1 in equation (4-31)

$$X := \begin{vmatrix} X & \text{if} & X \ge 0.1 \\ 0.1 & \text{otherwise} \end{vmatrix}$$

$$X = 0.1$$

Calculate the ratio of hole to active area, Ah/Aa

$$AhAa := 0.907 \cdot \left(\frac{do}{p}\right)^2 \qquad\qquad AhAa = 0.1275$$

Calculate FHA

$$FHA := \begin{vmatrix} \left(5 \cdot AhAa + 0.5\right) & \text{if} & AhAa < 0.1 \\ 1 & \text{otherwise} \end{vmatrix}$$

$$FHA = 1$$

Calculate FST

$$FST := \left[\frac{\sigma}{20 \cdot \dfrac{dyne}{cm}}\right]^{0.2} \qquad\qquad FST = 1.0098$$

$$C1 := FST \cdot FHA \cdot FF \qquad\qquad C1 = 0.9088$$

Enterparameters

$$\alpha 1 := 0.0744 \cdot m^{-1} \qquad\qquad \alpha 2 := 0.01173$$

$$\beta 1 := 0.0304 \cdot m^{-1} \qquad\qquad \beta 2 := 0.015$$

Iterate to find diameter and tray spacing

$$\alpha (t) := \alpha 1 \cdot t + \alpha 2$$

$$\beta (t) := \beta 1 \cdot t + \beta 2$$

$$CF (t) := \alpha (t) \cdot \log \left(X^{-1} \right) + \beta (t) \qquad\qquad C (t) := C1 \cdot CF (t) \cdot \frac{m}{sec}$$

$$vGF (t) := C (t) \cdot \sqrt{\frac{\rho L - \rho G}{\rho G}}$$

Initial estimates of tray spacing and diameter $D := 2 \cdot m$ $t := 0.5 \cdot m$

Given

$$D = \sqrt{\frac{4 \cdot QG}{\sqrt{f \cdot vGF (t) \cdot (1 - AdAt) \cdot \pi}}}$$

Use is made here of the step function $\Phi(x)$ to define the recommended values of tray spacing according to Table 4.3.

$$t = 0.5 \cdot m \cdot \Phi (1 \cdot m - D) + 0.6 \cdot m \cdot \left(\Phi (3 \cdot m - D) - \Phi (1 \cdot m - D) \right) \dots$$
$$+ 0.75 \cdot m \cdot \left(\Phi (4 \cdot m - D) - \Phi (3 \cdot m - D) \right) \dots$$
$$+ 0.9 \cdot m \cdot \left(\Phi (20 \cdot m - D) - \Phi (4 \cdot m - D) \right)$$

$$\begin{bmatrix} D \\ t \end{bmatrix} := find (D, t)$$

$$D = 0.6308 \cdot m \qquad\qquad t = 0.5 \cdot m$$

Calculate some further details of the tray design

$$At := \frac{\pi \cdot D^2}{4}$$ $At = 0.3125 \cdot m^2$ total area

$$Ad := AdAt \cdot At$$ $Ad = 0.0313 \cdot m^2$ downcomer area

$$Aa := At - 2 \cdot Ad \qquad\qquad Aa = 0.25 \cdot m^2 \qquad\qquad \text{active area}$$

$$Ah := AhAa \cdot Aa \qquad\qquad Ah = 0.0319 \cdot m^2 \qquad\qquad \text{hole area}$$

$$\theta 1 := 1.2 \cdot rad \qquad\qquad \text{first estimate}$$

Given

$$AdAt = \frac{\theta 1 - \sin\left(\theta 1\right)}{2 \cdot \pi} \qquad\qquad \text{equation (4-36)}$$

$$\theta 1 := find\left(\theta 1\right) \qquad\qquad \theta 1 = 1.6268$$

$$Lw := D \cdot \sin\left(\frac{\theta 1}{2}\right) \qquad\qquad Lw = 0.4583 \cdot m \qquad\qquad \text{weir length}$$

$$rw := \frac{D}{2} \cdot \left(\cos\left(\frac{\theta 1}{2}\right)\right) \qquad\qquad rw = 0.2167 \cdot m \qquad\qquad \begin{array}{l}\text{distance from tower center}\\\text{to weir}\end{array}$$

Estimate the gas-pressure drop through the tray

Dry tray head loss, hd

Calculate orifice gas velocity, vo

$$vo := \frac{QG}{Ah} \qquad\qquad vo = 13.1286 \cdot m \cdot sec^{-1}$$

Calculate orifice coefficient, Co, fom Eq. (4-39)

$$Co := 0.85032 - 0.04231 \cdot \frac{do}{l} + 0.0017954 \cdot \left(\frac{do}{l}\right)^2 \qquad\qquad Co = 0.7642$$

$$hd := 0.0051 \cdot \left(cm \cdot m \cdot \frac{sec^2}{kg}\right) \cdot \left(\frac{vo}{Co}\right)^2 \cdot \rho G \cdot \frac{\rho W}{\rho L} \cdot \left(1 - AhAa^2\right)$$

$$hd = 0.0214 \cdot m$$

Equivalent head of clear liquid, hl

Calculate gas velocity based on active area, va

$$va := \frac{QG}{Aa} \qquad\qquad va = 1.6745 \cdot m \cdot sec^{-1}$$

Calculate capacity parameter, Ks $\qquad\qquad qL = 4.3869 \cdot 10^{-4} \cdot m^3 \cdot sec^{-1}$

$$Ks := va \cdot \sqrt{\frac{\rho G}{\rho L - \rho G}} \qquad\qquad Ks = 0.0647 \cdot m \cdot sec^{-1}$$

Calculate froth density ϕe

$$\phi e := exp\left[-12.55 \cdot \left(\frac{Ks}{1 \cdot m \cdot sec^{-1}}\right)^{0.91}\right] \qquad \phi e = 0.3537$$

$$CL1 := 50.12 \cdot cm \cdot sec^{\frac{2}{3}} \cdot m^{\frac{-4}{3}} \qquad\qquad CL2 := 43.89 \cdot cm \cdot sec^{\frac{2}{3}} \cdot m^{\frac{-4}{3}}$$

$$CL := CL1 + CL2 \cdot exp\left(-1.378 \cdot cm^{-1} \cdot hw\right)$$

$$hl := \phi e \cdot \left[hw + CL \cdot \left(\frac{qL}{Lw \cdot \phi e}\right)^{\frac{2}{3}}\right] \qquad\qquad hl = 0.0211 \cdot m$$

Head loss due to surface tension, $h\ \sigma$

$$h\sigma := \frac{6 \cdot \sigma}{g \cdot \rho L \cdot do} \qquad\qquad h\sigma = 3.6096 \cdot 10^{-3} \cdot m$$

Total head loss, ht

$$ht := hd + hl + h\sigma \qquad\qquad ht = 0.0462 \cdot m$$

Convert head loss to pressure drop, ΔP

$$\Delta P := ht \cdot \rho L \cdot g \qquad\qquad \Delta P = 358.1205 \cdot Pa$$

Check tray design for excessive weeping; calculate orifice Froude number, Fro
If Fro > 0.5, there is no weeping problem.

$$Fro := \sqrt{\frac{\rho G \cdot vo^2}{\rho L \cdot g \cdot hl}} \qquad\qquad Fro = 1.1139$$

Calculate fractional entrainment, E

$$\kappa := 0.5 \cdot \left(1 - \tanh\left(1.3 \cdot \ln\left(\frac{hl}{do}\right) - 0.15\right)\right)$$

$$\kappa = 0.0236$$

$$h2\phi := \frac{hl}{\phi e} + 7.79 \cdot \left[1 + 6.9 \cdot \left(\frac{do}{hl}\right)^{1.85}\right] \cdot \frac{Ks^2}{\phi e \cdot g \cdot AhAa}$$

$$h2\phi = 0.1626 \cdot m$$

$$E := 0.00335 \cdot \left(\frac{h2\phi}{t}\right)^{1.1} \cdot \left[\frac{\rho L}{\rho G}\right]^{0.5} \cdot \left(\frac{hl}{h2\phi}\right)^{\kappa}$$

$$E = 0.024$$

Calculate point efficiency, EOG

$$ReFe := \frac{\rho G \cdot vo \cdot hl}{\mu G \cdot \phi e}$$

$$ReFe = 8.8142 \cdot 10^4$$

$$cG := \frac{\rho G}{MG}$$

$$cG = 0.0345 \cdot kg \cdot m^{-3}$$

$$cL := \frac{\rho L}{ML}$$

$$cL = 24.7188 \cdot kg \cdot m^{-3}$$

$$a1 := 0.4136 \qquad a2 := 0.6074 \qquad a3 := -0.3195$$

$$EOG := 1 - \exp\left[\frac{-0.0029}{1 + me \cdot \frac{cG}{cL}\sqrt{DG \cdot \frac{(1 - \phi e)}{DL \cdot AhAa}}} \cdot ReFe^{a1} \cdot \left(\frac{hl}{do}\right)^{a2} \cdot AhAa^{a3}\right]$$

$$EOG = 0.7597$$

Calculate Murphree tray efficiency, EMG

Check the degree of vapor mixing; calculate PeG.
If PeG > 50, or if t − h2 ϕ < 0, vapor is unmixed.

$$t - h2\phi = 0.3374 \cdot m \qquad\qquad DEG := 0.01 \cdot \frac{m^2}{sec}$$

$$PeG := \frac{4 \cdot QG \cdot rw^2}{DEG \cdot Aa \cdot \left(t - h2\phi\right)} \qquad\qquad PeG = 93.2179$$

Calculate PeL

$$DEL := 0.1 \cdot \sqrt{g \cdot h2\phi^3} \qquad\qquad DEL = 0.0205 \cdot m^2 \cdot sec^{-1}$$

$$PeL := \frac{4 \cdot qL \cdot rw^2}{Aa \cdot hl \cdot DEL} \qquad\qquad PeL = 0.7595$$

$$N := \frac{PeL + 2}{2} \qquad\qquad N = 1.3798$$

$$\lambda := me \cdot \frac{mG}{mL} \cdot \frac{ML}{MG} \qquad\qquad \lambda = 0.5595$$

For mixed vapor

$$EMGmixed := \frac{\left(1 + \frac{\lambda \cdot EOG}{N}\right)^N - 1}{\lambda} \qquad\qquad EMGmixed = 0.8016$$

For unmixed vapor

$$EMGunmixed := EMGmixed \cdot \left(1 - 0.0335 \cdot \lambda^{1.07272} \cdot EOG^{2.51844} \cdot PeL^{0.17524}\right)$$

$$EMGunmixed = 0.7947$$

$$EMG := \begin{vmatrix} EMGmixed & \text{if} & 0 < PeG < 50 \\ EMGunmixed & \text{otherwise} \end{vmatrix}$$

$$EMG = 0.7947$$

Correct efficiency for entrainment

$$EMGE := EMG \cdot \left(1 - 0.8 \cdot EOG \cdot \lambda^{1.543} \cdot \frac{E}{me}\right)$$

$$EMGE = 0.7835$$

Appendix F-1

McCabe–Thiele: Liquid Feed

McCabe–Thiele Metod: This MathCAD document creates a McCabe–Thiele Diagram for a binary distillation.

MathCAD File Name: McCabe2001-liquid-feed.MCD

MathCAD Version: 2001

Author: John Hwalek

E-mail: hwalek@maine.edu

Date: 8/22/01

Revisions:

Abstract: This document creates a McCabe–Thiele Diagram for a binary distillation using the constant molar overflow assumptions. It assumes:

1. Total condenser
2. Saturated liquid feed
3. Ideal stages

It uses activity coefficient methods to calculate vapor-liquid equilibrium. It will estimate a minimum reflux ratio (Rmin) assuming no pinch point in the rectifying section. The actual reflux ratio (RR) is set as a multiple of Rmin (R = C*Rmin).
The user must enter the feed molar flow rate (F) and mole fraction of light component, the mole fraction of the distillate (XD) and residue (XW) and the multiplier for the reflux ratio (C).

Note: subscripts on some variables are literal subscripts, not array subscripts.

460

$$\text{mol} := \text{mole} \qquad \text{kmol} \equiv 1000 \cdot \text{mole}$$

$$\text{kJ} := 10^3 \cdot \text{joule} \qquad \text{lbmol} \equiv \text{mole} \cdot \frac{\text{lb}}{\text{gm}}$$

-------------- Insert VLE thermodynamic method here --------------
see Thermodynamics folder for other VLE methods and BIP regression routines

$$\text{kPa} := 10^3 \cdot \text{Pa} \qquad R := 1.987 \cdot \frac{\text{cal}}{\text{mol} \cdot K}$$

Define saturation pressure as a function of temperature. In this case, Antoine's equation is used. (T must be in K)

(1) Methanol $\qquad A_1 := 16.5938 \qquad B_1 := 3644.3 \cdot K \qquad C_1 := 239.76 \cdot K$

(2) Water $\qquad A_2 := 16.2620 \qquad B_2 := 3799.89 \cdot K \qquad C_2 := 226.35 \cdot K$

$$\text{sat}(T) := \exp\left[A_1 - \frac{B_1}{(T - 273.15 \cdot K) + C_1}\right] \cdot \text{kPa} \qquad \text{P2sat}(T) := \exp\left[A_2 - \frac{B_2}{(T - 273.15 \cdot K) + C_2}\right] \cdot \text{kPa}$$

Define the activity coefficient as a function of composition (and temperature).

NRTL Equations

Define binary interaction parameters for Methanol(1)/Water(2).

$$b_{12} := -253.88 \cdot \frac{\text{cal}}{\text{mol}} \qquad\qquad b_{21} := 845.21 \cdot \frac{\text{cal}}{\text{mol}} \qquad\qquad \alpha := 0.2994$$

NRTL Equations

$$\tau_{12}(T) := \frac{b_{12}}{R \cdot T} \qquad\qquad \tau_{21}(T) := \frac{b_{21}}{R \cdot T}$$

$$G_{12}(T) := \exp\left(-\alpha \cdot \tau_{12}(T)\right) \qquad\qquad G_{21}(T) := \exp\left(-\alpha \cdot \tau_{21}(T)\right)$$

$$\gamma_1(x_1, x_2, T) := \exp\left[x_2^2 \cdot \left[\tau_{21}(T) \cdot \left[\frac{G_{21}(T)}{x_1 + x_2 \cdot G_{21}(T)}\right]^2 + \frac{G_{12}(T) \cdot \tau_{12}(T)}{\left(x_2 + x_1 \cdot G_{12}(T)\right)^2}\right]\right]$$

$$\gamma_2(x_1, x_2, T) := \exp\left[x_1^2 \cdot \left[\tau_{12}(T) \cdot \left[\frac{G_{12}(T)}{x_2 + x_1 \cdot G_{12}(T)}\right]^2 + \frac{G_{21}(T) \cdot \tau_{21}(T)}{\left(x_1 + x_2 \cdot G_{21}(T)\right)^2}\right]\right]$$

Define equilibrium relationship between liquid and vapor mole fractions at operating pressure

Operating pressure: $P := 1 \cdot atm$

Guess T: $T := \left(90 + 273.15\right) \cdot K$

$Teq\left(x1\right) := root\left[P - \gamma_1\left[x1, \left(1 - x1\right), T\right] \cdot x1 \cdot P1sat\left(T\right) - \gamma_2\left[x1, \left(1 - x1\right), T\right] \cdot \left(1 - x1\right) \cdot P2sat\left(T\right), T\right]$

Initial guesses for x and y. These may need to be changed if xeq(y) or yeq(x) are not converging.

$y := 0.1$ $x := 0.0$

$yeq\left(x\right) := \dfrac{\gamma_1\left[x, \left(1 - x\right), Teq\left(x\right)\right] \cdot x \cdot P1sat\left(Teq\left(x\right)\right)}{P}$

$xeq\left(y\right) := root\left(yeq\left(x\right) - y, x\right)$

Plot Vapor-Liquid Equilibrium diagram: N = number of points -1

$N := 20$ $k := 0 .. N$ $xe_k := \dfrac{k}{N}$

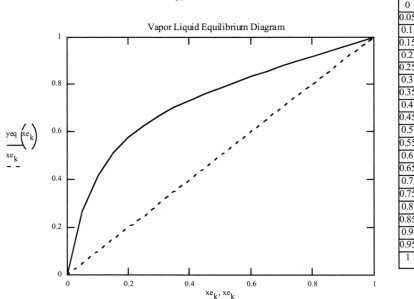

Vapor Liquid Equilibrium Diagram

xe_k	$yeq\left(xe_k\right)$
0	0
0.05	0.268
0.1	0.416
0.15	0.51
0.2	0.576
0.25	0.626
0.3	0.667
0.35	0.701
0.4	0.731
0.45	0.759
0.5	0.784
0.55	0.809
0.6	0.832
0.65	0.854
0.7	0.876
0.75	0.898
0.8	0.919
0.85	0.94
0.9	0.96
0.95	0.98
1	1

McCabe-Thiele Method for Binary Distillation

Assumes: Feed is saturated liquid
Total condenser

F = molar flow rate of feed (moles/tim
D = molar flow rate of distillate (moles
W = molar flow rate of residue (moles
x_F = mole fraction in feed
x_D = mole fraction in distillate
x_W = mole fraaction in residue

Set column conditions

Feed Conditions (x_F set globally below)

$$F = 277.778 \cdot \sec^{-1} \qquad x_F = 0.2$$

Set distillate and residue compositions (mole fractions) (set globally below)

Distillate: $\quad x_D = 0.995$ Residue: $\quad x_W = 1 \cdot 10^{-3}$

Calculate Rmin for intersection of top operating line and vertical feed line. This assumes there is r point or intersection between the top operating line and equilibrium curve.

Guess: $\qquad$ Rmin $:= 1$

Given

$$\frac{Rmin}{Rmin + 1} = \frac{yeq\left(x_F\right) - \dfrac{x_D}{Rmin + 1}}{x_F}$$

Rmin $:=$ Find $\left(Rmin\right)$

Rmin $= 1.116$

Set Reflux Ratio as multiple of Rmin (C set globally below)

RR $:= C \, Rmin$

Solve Overall Balances on Column

Guess: $\qquad D := \dfrac{F}{2} \qquad W := D$

Given

$$F = D + W$$

$$x_F \, F = x_D \, D + x_W \, W$$

$$D \geq 0 \, \frac{kmol}{hr} \qquad W \geq 0 \, \frac{kmol}{hr}$$

$$\begin{bmatrix} D \\ W \end{bmatrix} := \text{Find}\left(D, W\right)$$

$$D = 200.201 \cdot \frac{kmol}{hr} \qquad\qquad W = 799.799 \cdot \frac{kmol}{hr}$$

Stage balances in rectifying and stripping section s

L,V = liquid and vapor molar flow rates in rectifying section
l,v = liquid and vapor molar flow rates in stripping section

From the constant molal overflow assumption, L and V are constant as well as l and v

$$L := RR \cdot D$$

$$V := D + L$$

For a saturated liquid feed:

$$l := L + F$$

From a balance on the stripping section:

$$v := l - W$$

Stage balances in rectifying section for all stages including stage i and condenser

$$\text{yrop}\left(x\right) := \frac{RR}{RR + 1} x + \frac{x_D}{RR + 1}$$ Rectifying section operating line:

Calculate number of stages in rectifying section up to feed line

$$\text{rectify}\left(xD, xF\right) := \begin{vmatrix} i \leftarrow 0 \\ X_0 \leftarrow xD \\ Y_1 \leftarrow xD \\ \text{while} \quad X_i > xF \\ \quad \begin{vmatrix} i \leftarrow i + 1 \\ X_i \leftarrow \text{xeq}\left(Y_i\right) \\ Y_{i+1} \leftarrow \text{yrop}\left(X_i\right) \end{vmatrix} \\ \begin{bmatrix} X \\ Y \\ i \end{bmatrix} \end{vmatrix}$$

$$\begin{bmatrix} x \\ y \\ f \end{bmatrix} := \text{rectify}\left(x_D, x_F\right)$$

Stage balances in stripping section for all stages including stage j and reboile r

$$\text{ysop}\left(x\right) := \frac{1}{v} x + x_W \cdot \left[1 - \frac{1}{v}\right]$$ Stripping section operating line

Calculating number of stages in stripping section starting from feed stage

M = total number of stages including reboiler

$$\text{strip}\left(j, xj, yjp1, xW\right) := \begin{vmatrix} i \leftarrow j \\ X_j \leftarrow xj \\ Y_{j+1} \leftarrow \text{ysop}\left(X_j\right) \\ \text{while} \quad X_i \geq xW \\ \quad \begin{vmatrix} i \leftarrow i+1 \\ X_i \leftarrow \text{xeq}\left(Y_i\right) \\ Y_{i+1} \leftarrow \text{ysop}\left(X_i\right) \end{vmatrix} \\ \begin{bmatrix} X \\ Y \\ i \end{bmatrix} \end{vmatrix}$$

$$\begin{bmatrix} xs \\ ys \\ M \end{bmatrix} := \text{strip}\left(f, x_f, y_{f+1}, x\,W\right)$$

Combine rectifying and stripping section x,y values

$$i := 0 .. M \qquad x_i := \text{if}\left(i \leq f, x_i, xs_i\right) \qquad i := 1 .. M+1 \qquad y_i := \text{if}\left(i \leq f, y_i, ys_i\right)$$

Create points for equilibrium curve: $N := 50$ $k := 0 .. N$ $xe_k := \dfrac{k}{N}$

Create operating lines for rectifying and stripping sections

$j := 0 .. f + 1$ rectifying section stages

$h := (f - 1) .. M$ stripping section stages

$i := 0 .. M$ $y_{M+1} := ysop(x_M)$

Design Parameters (set globally here)

Feed molar flow rate: $F \equiv 1000 \cdot \dfrac{kmol}{hr}$

Results

Feed mole fraction: $x_F \equiv 0.2$ Ideal Stages: $M = 20$

Distillate mole fraction: $x_D \equiv 0.995$ Feed Stage: $f = 13$

Residue mole fraction: $x_W \equiv 0.001$ Reflux Ratio: $RR = 1.339$

Minimum Reflux Ratio: $Rmin = 1.116$

Set Reflux Ratio at C*Rmin $C \equiv 1.2$

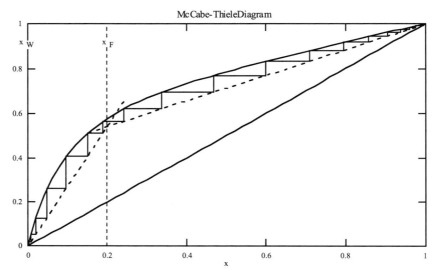

McCabe-ThieleDiagram

PRINT Page below for McCabe-Thiele Diagram only

Design Parameters   **Results**

Feed mole fraction: $x_F = 0.2$ Ideal Stages: $M = 20$ $F = 1000 \cdot \dfrac{kmol}{hr}$

Distillate mole fraction: $x_D = 0.995$ Feed Stage: $f = 13$

Residue mole fraction: $x_W = 1 \cdot 10^{-3}$ $D = 200.201 \cdot \dfrac{kmol}{hr}$

Minimum Reflux Ratio: $Rmin = 1.116$ Reflux Ratio: $RR = 1.339$ $W = 799.799 \cdot \dfrac{kmol}{hr}$

Set Reflux Ratio at C*Rmin $C = 1.2$

Appendix F-2

McCabe–Thiele: Vapor Feed

McCabe–Thiele Metod: This MathCAD document creates a McCabe–Thiele Diagram for a binary distillation.

MathCAD File Name: McCabe2001-vapor-feed.MCD

MathCAD Version: 2001

Author: John Hwalek

E-mail: hwalek@maine.edu

Date: 8/22/01

Revisions:

Abstract: This document creates a McCabe–Thiele Diagram for a binary distillation using the constant molar overflow assumptions. It assumes:

    1. Total condenser
    2. Saturated vapor feed
    3. Ideal stages

It uses activity coefficient methods to calculate vapor-liquid equilibrium. It will estimate a minimum reflux ratio (Rmin) assuming no pinch point in the rectifying section. The actual reflux ratio (RR) is set as a multiple of Rmin (R = C*Rmin).
The user must enter the feed molar flow rate (F) and mole fraction of light component, the mole fraction of the distillate (XD) and residue (XW) and the multiplier for the reflux ratio (C).

Note: subscripts on some variables are literal subscripts, not array subscripts.

470

$$mol := mole \qquad kmol \equiv 1000 \cdot mole$$

$$kJ := 10^3 \cdot joule \qquad lbmol \equiv mole \cdot \frac{lb}{gm}$$

-------------- Insert VLE thermodynamic method here ----------------
see Thermodynamics folder for other VLE methods and BIP regression routines

$$kPa := 10^3 \cdot Pa \qquad R := 1.987 \cdot \frac{cal}{mol \cdot K}$$

Define saturation pressure as a function of temperature. In this case, Antoine's equation is used. (T must be in K)

(1) Methanol $\qquad A_1 := 16.5938 \qquad B_1 := 3644.3 \cdot K \qquad C_1 := 239.76 \cdot K$

(2) Water $\qquad A_2 := 16.2620 \qquad B_2 := 3799.89 \cdot K \qquad C_2 := 226.35 \cdot K$

$$sat(T) := exp\left[A_1 - \frac{B_1}{(T - 273.15 \cdot K) + C_1}\right] \cdot kPa \qquad P2sat(T) := exp\left[A_2 - \frac{B_2}{(T - 273.15 \cdot K) + C_2}\right] \cdot kPa$$

Define the activity coefficient as a function of composition (and temperature).

NRTL Equations

Define binary interaction parameters for Methanol(1)/Water(2).

$$b_{12} := -253.88 \cdot \frac{cal}{mol} \qquad\qquad b_{21} := 845.21 \cdot \frac{cal}{mol} \qquad\qquad \alpha := 0.2994$$

NRTL Equations

$$\tau_{12}(T) := \frac{b_{12}}{R \cdot T} \qquad\qquad \tau_{21}(T) := \frac{b_{21}}{R \cdot T}$$

$$G_{12}(T) := exp\left(-\alpha \cdot \tau_{12}(T)\right) \qquad\qquad G_{21}(T) := exp\left(-\alpha \cdot \tau_{21}(T)\right)$$

$$\gamma_1(x_1, x_2, T) := exp\left[x_2^2 \cdot \left[\tau_{21}(T) \cdot \left[\frac{G_{21}(T)}{x_1 + x_2 \cdot G_{21}(T)}\right]^2 + \frac{G_{12}(T) \cdot \tau_{12}(T)}{\left(x_2 + x_1 \cdot G_{12}(T)\right)^2}\right]\right]$$

$$\gamma_2(x_1, x_2, T) := exp\left[x_1^2 \cdot \left[\tau_{12}(T) \cdot \left[\frac{G_{12}(T)}{x_2 + x_1 \cdot G_{12}(T)}\right]^2 + \frac{G_{21}(T) \cdot \tau_{21}(T)}{\left(x_1 + x_2 \cdot G_{21}(T)\right)^2}\right]\right]$$

Define equilibrium relationship between liquid and vapor mole fractions at operating pressure

Operating pressure: $P := 1 \cdot atm$

Guess T: $T := \left(90 + 273.15\right) \cdot K$

$Teq\left(x1\right) := root\left[P - x1 \cdot P1sat\left(T\right) - \left(1 - x1\right) \cdot P2sat\left(T\right), T\right]$

Initial guesses for x and y. These may need to be changed if xeq(y) or yeq(x) are not converging.

$y := 0.1$ $x := 0.0$

$yeq\left(x\right) := \dfrac{x \cdot P1sat\left(Teq\left(x\right)\right)}{P}$

$xeq\left(y\right) := root\left(yeq\left(x\right) - y, x\right)$

Plot Vapor-Liquid Equilibrium diagram: N = number of points -1

$N := 20$ $k := 0..N$ $xe_k := \dfrac{k}{N}$

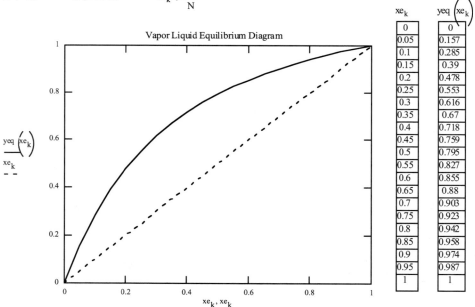

xe_k	$yeq\left(xe_k\right)$
0	0
0.05	0.157
0.1	0.285
0.15	0.39
0.2	0.478
0.25	0.553
0.3	0.616
0.35	0.67
0.4	0.718
0.45	0.759
0.5	0.795
0.55	0.827
0.6	0.855
0.65	0.88
0.7	0.903
0.75	0.923
0.8	0.942
0.85	0.958
0.9	0.974
0.95	0.987
1	1

Vapor Liquid Equilibrium Diagram

McCabe-Thiele Method for Binary Distillation

Assumes: Feed is saturated vapor
            Total condenser

F = molar flow rate of feed (moles/time)
D = molar flow rate of distillate (moles/time)
W = molar flow rate of residue (moles/time)
x_F = mole fraction in feed
x_D = mole fraction in distillate
x_W = mole fraaction in residue

Set column conditions

Feed Conditions (x_F set globally below)

$$F = 138.889 \cdot sec^{-1}$$

Set distillate and residue compositions (mole fractions) (set globally below)

Distillate: $x_D = 0.99$ Residue: $x_W = 0.02$

Calculate Rmin for intersection of top operating line and vertical feed line. This assumes there is no pinch point or intersection between the top operating line and equilibrium curve.

Guess: $Rmin := 1$

Given

$$\frac{Rmin}{Rmin + 1} = \frac{y_F - \dfrac{x_D}{Rmin + 1}}{xeq\left(y_F\right)}$$

$Rmin := Find\left(Rmin\right)$ $Rmin = 0.908$

Set Reflux Ratio as multiple of Rmin (C set globally below)

$RR := C\, Rmin$

Solve Overall Balances on Column

Guess: $D := \dfrac{F}{2}$ $W := D$

Given

$$F = D + W$$

$$y_F\, F = x_D\, D + x_W\, W$$

$$D \geq 0\, \frac{kmol}{hr} \qquad W \geq 0\, \frac{kmol}{hr}$$

$$\begin{bmatrix} D \\ W \end{bmatrix} := Find\left(D, W\right)$$

$$D = 350.515 \cdot \frac{kmol}{hr} \qquad\qquad W = 149.485 \cdot \frac{kmol}{hr}$$

Stage balances in rectifying and stripping section s

L,V = liquid and vapor molar flow rates in rectifying section
l,v = liquid and vapor molar flow rates in stripping section

From the constant molal overflow assumption, L and V are constant as well as l and v

$L := RR \cdot D$

$V := D + L$

For a saturated vapor feed:

$l := L$

From a balance on the stripping section:

$v := l - W$
$$x_F := \left[y_F - \frac{x_D}{RR + 1} \right] \cdot \left[\frac{RR + 1}{RR} \right]$$

Stage balances in rectifying section for all stages including stage i and condenser

$$\text{yrop}\left(x\right) := \frac{RR}{RR+1} x + \frac{x_D}{RR+1}$$ Rectifying section operating line:

Calculate number of stages in rectifying section up to feed line

$$\text{rectify}\left(xD, xF\right) := \begin{vmatrix} i \leftarrow 0 \\ X_0 \leftarrow xD \\ Y_1 \leftarrow xD \\ \text{while} \quad X_i > xF \\ \quad \begin{vmatrix} i \leftarrow i+1 \\ X_i \leftarrow xeq\left(Y_i\right) \\ Y_{i+1} \leftarrow yrop\left(X_i\right) \end{vmatrix} \\ \begin{bmatrix} X \\ Y \\ i \end{bmatrix} \end{vmatrix}$$

$$\begin{bmatrix} x \\ y \\ f \end{bmatrix} := \text{rectify}\left(x_D, x_F\right)$$

Stage balances in stripping section for all stages including stage j and reboile r

$$\text{ysop}\left(x\right) := \frac{1}{v} x + x_W \cdot \left[1 - \frac{1}{v}\right]$$ Stripping section operating line

Calculating number of stages in stripping section starting from feed stage

M = total number of stages including reboiler

$$
\text{strip}\Big(j, xj, yjp1, xW\Big) := \begin{array}{|l} i \leftarrow j \\ X_j \leftarrow xj \\ Y_{j+1} \leftarrow \text{ysop}\Big(X_j\Big) \\ \text{while} \quad X_i \geq xW \\ \qquad \begin{array}{|l} i \leftarrow i+1 \\ X_i \leftarrow \text{xeq}\Big(Y_i\Big) \\ Y_{i+1} \leftarrow \text{ysop}\Big(X_i\Big) \end{array} \\ \begin{bmatrix} X \\ Y \\ i \end{bmatrix} \end{array}
$$

$$
\begin{bmatrix} xs \\ ys \\ M \end{bmatrix} := \text{strip}\Big(f, x_f, y_{f+1}, x\,W\Big)
$$

Combine rectifying and stripping section x,y values

$$
i := 0..M \qquad x_i := \text{if}\Big(i \leq f, x_i, xs_i\Big) \qquad i := 1..M+1 \quad y_i := \text{if}\Big(i \leq f, y_i, ys_i\Big)
$$

Create points for equilibrium curve: $N := 50$ $k := 0..N$ $xe_k := \dfrac{k}{N}$

Create operating lines for rectifying and stripping sections

$j := 0..f+1$ rectifying section stages

$h := (f-1)..M$ stripping section stages

$i := 0..M$ $y_{M+1} := ysop\left(x_M\right)$

Design Parameters (set globally here)

Feed molar flow rate: $F \equiv 500 \cdot \dfrac{kmol}{hr}$

Feed mole fraction: $y_F \equiv 0.7$ **Results**

Distillate mole fraction: $x_D \equiv 0.99$ Ideal Stages: $M = 13$

Residue mole fraction: $x_W \equiv 0.02$ Feed Stage: $f = 7$

Minimum Reflux Ratio: $Rmin = 0.9077$ Reflux Ratio: $RR = 1.089$

Set Reflux Ratio at C*Rmin $C \equiv 1.2$

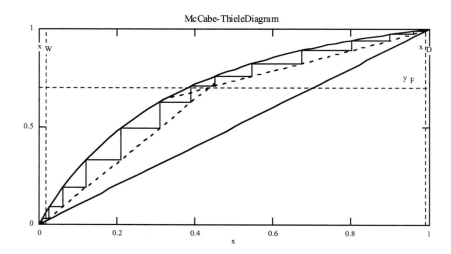

McCabe-ThieleDiagram

Design Parameters

Feed mole fraction:	$y_F = 0.7$	
Distillate mole fraction:	$x_D = 0.99$	
Residue mole fraction:	$x_W = 0.02$	
Minimum Reflux Ratio:	$Rmin = 0.908$	
Set Reflux Ratio at C*Rmin	$C = 1.2$	

Results

Ideal Stages:	$M = 13$	
Feed Stage:	$f = 7$	
Reflux Ratio:	$RR = 1.089$	

$$F = 500 \cdot \frac{kmol}{hr}$$

$$D = 350.515 \cdot \frac{kmol}{hr}$$

$$W = 149.485 \cdot \frac{kmol}{hr}$$

Appendix G-1

Single-Stage Extraction

Data presented are from Example 7-2: acetone-water-chloroform system at 298 K.

$$\text{vx}_C := \begin{bmatrix} .158 \\ .256 \\ .36 \\ .493 \\ .557 \\ .596 \end{bmatrix} \qquad \text{vx}_B := \begin{bmatrix} .0123 \\ .0129 \\ .0171 \\ .051 \\ .098 \\ .169 \end{bmatrix} \qquad \text{vy}_C := \begin{bmatrix} .287 \\ .421 \\ .527 \\ .613 \end{bmatrix} \qquad \text{vy}_B := \begin{bmatrix} .7 \\ .557 \\ .429 \\ .284 \end{bmatrix}$$

$$\text{vsraf} := \text{cspline}\left(\text{vx}_C, \text{vx}_B\right) \qquad\qquad \text{vsext} := \text{cspline}\left(\text{vy}_C, \text{vy}_B\right)$$

$$x_B\left(x_C\right) := \text{interp}\left(\text{vsraf}, \text{vx}_C, \text{vx}_B, x_C\right)$$

$$\text{vxtie}_C := \begin{bmatrix} .158 \\ .256 \\ .36 \\ .493 \\ .557 \\ .596 \end{bmatrix} \qquad\qquad \text{vytie}_C := \begin{bmatrix} .287 \\ .421 \\ .527 \\ .613 \\ .610 \\ .596 \end{bmatrix}$$

$$\text{vstie} := \text{cspline}\left(\text{vxtie}_C, \text{vytie}_C\right) \qquad y_C\left(x_C\right) := \text{interp}\left(\text{vstie}, \text{vxtie}_C, \text{vytie}_C, x_C\right)$$

$$y_B\left(x_C\right) := \text{interp}\left(\text{vsext}, \text{vy}_C, \text{vy}_B, y_C\left(x_C\right)\right)$$

$$F := 50 \qquad\qquad S := 50 \qquad\qquad x_{CF} := 0.60 \qquad\qquad y_{CS} := 0.0$$

$$x_{BF} := 0.0 \qquad\qquad y_{BS} := 1.0$$

Initial estimates: $\qquad E := \dfrac{F + S}{2} \qquad\qquad R := E \qquad x_C := .1$

Given

$$F + S = E + R$$

$$F \cdot x_{CF} + S \cdot y_{CS} = E \cdot y_C\left(x_C\right) + R \cdot x_C$$

$$F \cdot x_{BF} + S \cdot y_{BS} = E \cdot y_B\left(x_C\right) + R \cdot x_B\left(x_C\right)$$

$$\begin{bmatrix} E \\ R \\ x_C \end{bmatrix} := \text{Find}\left(E, R, x_C\right)$$

$$E = 76.71 \qquad\qquad R = 23.29$$

$$x_C = 0.189 \qquad\qquad x_B\left(x_C\right) = 0.013$$

$$y_C\left(x_C\right) = 0.334 \qquad\qquad y_B\left(x_C\right) = 0.648$$

Appendix G-2

Multistage Crosscurrent Extraction

Data presented are from Example 7-3: acetone-water-chloroform system at 298 K.

$$vx_C := \begin{bmatrix} .158 \\ .256 \\ .36 \\ .493 \\ .557 \\ .596 \end{bmatrix} \qquad vx_B := \begin{bmatrix} .0123 \\ .0129 \\ .0171 \\ .051 \\ .098 \\ .169 \end{bmatrix} \qquad vy_C := \begin{bmatrix} .287 \\ .421 \\ .527 \\ .613 \end{bmatrix} \qquad vy_B := \begin{bmatrix} .7 \\ .557 \\ .429 \\ .284 \end{bmatrix}$$

$$vsraf := cspline\left(vx_C, vx_B\right) \qquad\qquad vsext := cspline\left(vy_C, vy_B\right)$$

$$x_B\left(x_C\right) := interp\left(vsraf, vx_C, vx_B, x_C\right)$$

$$vxtie_C := \begin{bmatrix} .158 \\ .256 \\ .36 \\ .493 \\ .557 \\ .596 \end{bmatrix} \qquad\qquad vytie_C := \begin{bmatrix} .287 \\ .421 \\ .527 \\ .613 \\ .610 \\ .596 \end{bmatrix}$$

$$vstie := cspline\left(vxtie_C, vytie_C\right) \qquad y_C\left(x_C\right) := interp\left(vstie, vxtie_C, vytie_C, x_C\right)$$

$$y_B\left(x_C\right) := interp\left(vsext, vy_C, vy_B, y_C\left(x_C\right)\right)$$

$$F1 := 50 \qquad\qquad S1 := 8 \qquad\qquad x_{CF1} := 0.60 \qquad y_{CS1} := 0.0$$

$$x_{BF1} := 0.0 \qquad y_{BS1} := 1.0$$

Initial estimates:

$$E1 := \frac{F1 + S1}{2} \qquad R1 := E1 \qquad x_{C1} := .1$$

Given

$$F1 + S1 = E1 + R1$$

$$F1\, x_{CF1} + S1\, y_{CS1} = E1\, y_C\!\left(x_{C1}\right) + R1\, x_{C1}$$

$$F1\, x_{BF1} + S1\, y_{BS1} = E1\, y_B\!\left(x_{C1}\right) + R1\, x_B\!\left(x_{C1}\right)$$

$$\begin{bmatrix} E1 \\ R1 \\ x_{C1} \end{bmatrix} := \text{Find}\left(E1, R1, x_{C1}\right)$$

$$y_{C1} := y_C\!\left(x_{C1}\right) \quad y_{B1} := y_B\!\left(x_{C1}\right) \quad x_{B1} := x_B\!\left(x_{C1}\right)$$

$$E1 = 21.643 \qquad R1 = 36.357$$

$$x_{C1} = 0.466 \qquad x_{B1} = 0.041$$

$$y_{C1} = 0.604 \qquad y_{B1} = 0.302$$

$$F2 := R1 \qquad S2 := 8 \qquad x_{CF2} := x_{C1} \qquad y_{CS2} := 0.0$$

$$F2 = 36.357 \qquad x_{BF2} := x_B\!\left(x_{C1}\right) \qquad y_{BS2} := 1.0$$

$$x_{BF2} = 0.041 \qquad x_{CF2} = 0.466$$

Initial guesses:

$$x_{C2} := .18 \qquad R2 := 20 \qquad E2 := S2 + F2 - R2$$

Given

$$F2 + S2 = E2 + R2$$

$$F2\, x_{CF2} + S2\, y_{CS2} = E2\, y_C\!\left(x_{C2}\right) + R2\, x_{C2}$$

$$F2\, x_{BF2} + S2\, y_{BS2} = E2\, y_B\!\left(x_{C2}\right) + R2\, x_B\!\left(x_{C2}\right)$$

$$\begin{bmatrix} E2 \\ R2 \\ x_{C2} \end{bmatrix} := \text{Find}\left(E2, R2, x_{C2}\right)$$

$$y_{C2} := y_C\left(x_{C2}\right) \qquad y_{B2} := y_B\left(x_{C2}\right) \qquad x_{B2} := x_B\left(x_{C2}\right)$$

$$E2 = 18.594 \qquad\qquad R2 = 25.764$$

$$x_{C2} = 0.311 \qquad\qquad x_{B2} = 0.014$$

$$y_{C2} = 0.48 \qquad\qquad y_{B2} = 0.491$$

$$F3 := R2 \qquad\qquad S3 := 8 \qquad\qquad x_{CF3} := x_{C2} \qquad y_{CS3} := 0.0$$

$$x_{BF3} := x_B\left(x_{C2}\right) \qquad y_{BS3} := 1.0$$

$$x_{BF3} = 0.014 \qquad\qquad x_{CF3} = 0.311$$

Initial guesses:
$$x_{C3} := .18 \qquad\qquad R3 := 20 \qquad\qquad E3 := 60$$

Given

$$F3 + S3 = E3 + R3$$

$$F3 \, x_{CF3} + S3 \, y_{CS3} = E3 \, y_C\left(x_{C3}\right) + R3 \, x_{C3}$$

$$F3 \, x_{BF3} + S3 \, y_{BS3} = E3 \, y_B\left(x_{C3}\right) + R3 \, x_B\left(x_{C3}\right)$$

$$\begin{bmatrix} E3 \\ R3 \\ x_{C3} \end{bmatrix} := \text{Find}\left(E3, R3, x_{C3}\right)$$

$$y_{C3} := y_C\left(x_{C3}\right) \qquad\qquad y_{B3} := y_B\left(x_{C3}\right)$$

$$x_{B3} := x_B\left(x_{C3}\right)$$

$$E3 = 12.342 \qquad R3 = 21.422$$

$$x_{C3} = 0.185 \qquad x_{B3} = 0.013$$

$$y_{C3} = 0.328 \qquad y_{B3} = 0.654$$

Composited extract

$$E := E1 + E2 + E3$$

$$E = 52.578$$

$$y_C := \frac{E1 \cdot y_{C1} + E2 \cdot y_{C2} + E3\, y_{C3}}{E}$$

$$y_C = 0.495$$

$$y_B := \frac{E1 \cdot y_{B1} + E2 \cdot y_{B2} + E3\, y_{B3}}{E}$$

$$y_B = 0.451$$

Check material balances

Acetone
$$F1 \cdot x_{CF1} - E \cdot y_C - R3 \cdot x_{C3} = -3.997 \cdot 10^{-15}$$

Chloroform
$$F1 \cdot x_{BF1} + S1 + S2 + S3 - E \cdot y_B - R3\, x_{B3} = 0$$

Water
$$F1 \cdot \left(1 - x_{CF1} - x_{BF1}\right) - E \cdot \left(1 - y_B - y_C\right) - R3 \cdot \left(1 - x_{C3} - x_{B3}\right) = -3.553 \cdot 10^{-15}$$

Appendix H

Constants and Unit Conversions

Table H.1 Atomic Mass of Selected Elements.

Element	Symbol	Mass	Element	Symbol	Mass
Aluminum	Al	26.98	Lead	Pb	207.19
Arsenic	As	74.92	Lithium	Li	6.94
Barium	Ba	137.34	Magnesium	Mg	24.31
Berillium	Be	9.01	Manganese	Mn	54.94
Boron	B	10.81	Mercury	Hg	200.59
Bromine	Br	79.90	Nickel	Ni	58.71
Cadmium	Cd	112.40	Nitrogen	N	14.01
Calcium	Ca	40.88	Oxygen	O	16.00
Carbon	C	12.01	Phosphorus	P	30.97
Chlorine	Cl	35.45	Potassium	K	39.10
Chromium	Cr	52.00	Silicon	Si	28.09
Cobalt	Co	58.93	Silver	Ag	107.87
Copper	Cu	63.55	Sodium	Na	22.99
Fluorine	F	19.00	Sulfur	S	32.06
Germanium	Ge	72.59	Tin	Sn	118.69
Gold	Au	196.97	Titanium	Ti	47.90
Helium	He	4.00	Uranium	U	238.03
Hydrogen	H	1.01	Vanadium	V	50.94
Iodine	I	126.90	Zinc	Zn	65.37
Iron	Fe	55.85	Zirconium	Zr	91.22

Table H.2 Ideal Gas Constant, R.

$1.987 \text{ cal/(mole} \times \text{K)}$

$8.314 \text{ kPa} \times \text{m}^3/(\text{kmole} \times \text{K)}$

$8.314 \text{ Pa} \times \text{m}^3/(\text{mole} \times \text{K)}$

$82.06 \text{ atm} \times \text{cm}^3/(\text{mole} \times \text{K)}$

$0.08206 \text{ atm} \times \text{L/(mole} \times \text{K)}$

Table H.3 Pressure Units

$$1 \text{ bar} = 10^5 \text{ Pa} = 100 \text{ kPa}$$
$$= 0.9869 \text{ atm}$$
$$= 750.06 \text{ mm Hg}$$
$$= 750.06 \text{ torr}$$
$$= 10^6 \text{ dyne/cm}^2$$
$$= 14.5038 \text{ psia}$$

Table H.4 Viscosity Units

$$1 \text{ cP} = 0.01 \text{ P}$$
$$= 0.01 \text{ (dyne} \times \text{s)/cm}^2$$
$$= 0.001 \text{ kg/(m} \times \text{s)}$$
$$= 0.001 \text{ Pa} \times \text{s}$$
$$= 10^4 \text{ } \mu\text{P}$$

Index

A

B

H

I, J, K

L

M

N

O

P

R

V, W

RETURN TO: **CHEMISTRY LIBRARY**

100 Hildebrand Hall • 510-642-3753

LOAN PERIOD 1	2	3
1-MONTH USE		
4	5	6

ALL BOOKS MAY BE RECALLED AFTER 7 DAYS.

Renewals may be requested by phone or, using GLADIS, type **inv** followed by your patron ID number.

~~DUE AS STAMPED BELOW.~~ NON-CIRCULATING UNTIL:		
NON-CIRCULATING UNTIL: FEB 17 2004		
MAY 22 2004		

FORM NO. DD 10
2M 4-03

UNIVERSITY OF CALIFORNIA, BERKELEY
Berkeley, California 94720–6000